Socio-Historical Studies of the Social and Human Sciences

This series is the first to focus on the historical development and current practices of the social and human sciences. Rather than simply privileging the internal analysis of ideas or external accounts of institutional structures, it publishes high quality studies that use the tools of the social sciences themselves to analyse the production, circulation and uses of knowledge in these disciplines. In doing so, it aims to establish Socio-Historical Studies of the Social and Human Sciences as a scholarly field in its own right, and to contribute to a more reflexive practice of these disciplines.

More information about this series at
http://www.palgrave.com/gp/series/15409

"This book is a pioneering one: based on original comparative research on the development of the social sciences in Europe and beyond, it shows the fruitfulness of an institutional approach shared without any dogmatism by all the contributors. Rich in fresh data and bold hypotheses, this work will be useful to all those who are interested in the social science of social sciences, an emerging and promising field."

—Jean-Louis Fabiani, *Central European University, Hungary*

"Comparativity in social sciences is akin to physical exercises: most scholars lament the lack of it – and do nothing about that. The authors of this book made a perfect job of producing a truly comparative history of social sciences, including both a wide range of national cases, from Argentina to Hungary, and an extensive spectrum of disciplines. This book discovers for social scientists how rich and diverse are the legacies of their intellectual enterprise."

—Mikhail Sokolov, *European University at St. Petersburg, Russia*

Christian Fleck · Matthias Duller
Victor Karády
Editors

Shaping Human Science Disciplines

Institutional Developments in Europe and Beyond

palgrave
macmillan

Editors
Christian Fleck
Department of Sociology
University of Graz
Graz, Austria

Victor Karády
Department of History
Central European University
Budapest, Hungary

Matthias Duller
Department of Sociology
University of Graz
Graz, Austria

Socio-Historical Studies of the Social and Human Sciences
ISBN 978-3-030-06515-7 ISBN 978-3-319-92780-0 (eBook)
https://doi.org/10.1007/978-3-319-92780-0

Printed on acid-free paper

This Palgrave Macmillan imprint is published by the registered company Springer Nature Switzerland AG
The registered company address is: Gewerbestrasse 11, 6330 Cham, Switzerland

Acknowledgements

This book is a result of the European research project INTERCO-SSH "International Cooperation in the Social sciences and Humanities", which was conducted by an international team of social scientists between 2013 and 2017. The project received funding from the European Union Seventh Framework Programme (FP7/2007-2013) under grant agreement no. 319974 (Interco-SSH).

The editors thank Thomas Klebel, Graz, for his ingenious handling and designing of the data the authors submitted to this volume.

Contents

Notes on Contributors

Fernanda Beigel is a principal researcher at CONICET and Head Professor at the National University of Cuyo (Mendoza-Argentina). Specialized in Sociology of Science and director of the Research Program on Academic Dependency in Latin America (PIDAAL). Her work is nurtured in the crossroad of Bourdieu's reflexivity and the Latin American tradition of Dependency Analysis. Recent publications: *The politics of Academic Autonomy in Latin America*. Ashgate: London, 2013; "Peripheral Scientists, between Ariel and Caliban. Institutional know-how and Circuits of Recognition in Argentina. The *career-best publications* of the researchers at CONICET", *Dados* (60:3), 2017; "Institutional expansion and scientific development in the periphery. The structural heterogeneity of Argentina's academic field" *Minerva*, 2018.

Mikael Börjesson is a professor in Sociology of Education at Uppsala University and is co-director of the research unit Sociology of Education and Culture (SEC) and director of the Swedish Centre for the Studies of the Internationalisation of Higher Education (SIHE). His main research domains are fields of education, transnational strategies and the internationalisation of higher education, elites and elite education, as well as applications of Geometric Data Analysis. He is currently directing the research project Swedish Higher Education. Financing, Organisation, Enrolment, Outcomes, 1950–2020 (SHEFOE), funded by the Swedish Research Council.

Donald Broady is a professor emeritus at the Department of Sociology, Uppsala University. Directing, with Mikael Börjesson and Marta Edling, the research unit Sociology of Education and Culture (SEC), see www.skeptron.uu.se/broady/sec/. Research on the sociology and history of cultural fields, elites, education, students' trajectories, transnational educational strategies, mark-up languages, internet applications.

Eric Brun is currently a postdoctoral researcher at the Curapp-ESS, a laboratory of the University of Picardie Jules Verne, France. He is the author of *Les Situationnistes, une avant-garde totale* (2014). A first axis of his research questions the political commitment of the intellectual professions by focusing on the artistic and political 'avant-gardes'. A second axis deals with the SHS Studies. A final line of his research concerns the sociology of youth.

Tobias Dalberg is a doctoral student at the Department of Education, Uppsala University, Sweden. He is currently finalizing the dissertation *Reaching the Pinnacle of Scholarship: Social, Educational and Professional Trajectories in the Humanities and Social Sciences in Sweden During the First Half of the 20th century*. He is the co-author on "Elite Education in Sweden—A Contradiction in Terms?" in Claire Maxwell and Peter Aggleton (Eds.) *Elite Education: International Perspectives* (Routledge, 2015) and "Higher Education Participation in the Nordic Countries 1985–2010: A Comparative Perspective" in the *European Sociological Review* (2017).

Matthias Duller is a postdoctoral researcher in the Department of Sociology at the University of Graz, a fellow at the Centre for Advanced Study in Sofia. His research focuses on the history and sociology of the social sciences during the Cold War in East and West. Further research interests are in historical sociology, sociological theory, and set-theoretic methods.

Christian Fleck is a professor at the Department for Sociology, University of Graz, Austria and Chief Research Fellow at the Poletayev Institute for Theoretical and Historical Studies, Higher School of Economics, Moskwa, Russia. Most recent publications: *Sociology in Austria*, Basingstoke: Palgrave Macmillan 2016; *Etablierung in der Fremde. Vertriebene Wissenschaftler in den USA nach 1933*, Frankfurt-New York: Campus 2015; *A Transatlantic History of the Social Sciences: Robber Barons, the Third Reich and the Invention of Empirical Social Research*, London: Bloomsbury Academic 2011.

Clarisse Fordant is a Ph.D. student at the Ecole des Hautes Etudes en Sciences Sociales (EHESS), France. She investigates the french scientific and political debates surrounding the measurement of integration and discrimination through the use of ethno-racial statistics between the years of 1995–2012.

Andrea Gallelli holds a Ph.D. in sociology from the University of Turin. His researches focus on social network analysis, cultural production and social exclusion, with a particular focus on the relational determinants of cultural products and activities. Currently based in Luxembourg he conducts consulting activity on survey design and data analysis for the non-profit and public sector.

Barbara Grüning is a postdoctoral research fellow at the University of Bologna (Italy). Her research fields range from the sociology of social sciences to the sociology of space and the sociology of memory.

Johan Heilbron is a historical sociologist, director of research at the Centre Européen de Sociologie et de Science Politique (CNRS, EHESS) in Paris and affiliated with the Erasmus University Rotterdam. He is currently a member of the Institute for Advanced Study (IAS) in Princeton. His research is in the fields of economic sociology, sociology of art and culture, and the sociology of knowledge and science. Book publications in the latter area include *The Rise of Social Theory* (1995, also in Dutch, French, and Portuguese), *The Rise of the Social Sciences and the Formation of Modernity* (co-edited, 2001), *Pour une histoire des sciences sociales* (co-edited, 2004), *French Sociology* (2015), and *The Social and Human Sciences in Global Power Relations* (co-edited 2018).

Victor Karády is emeritus research director of the French *CNRS* and distinguished research associate at the History Department of the Central European University in Budapest. He has made a dual career of historical sociologist at the Parisian *EHESS* and (since 1992) at the *CEU.* He served as principal investigator (associated with Peter Tibor Nagy) in two European research ventures: ELITES08 (on elite selection in six formerly socialist countries) and as the Hungarian partner of the INTERCO-SSH project (on the social sciences since 1945). His bibliography includes over 350 publications. Among the most recent ones see (with Adela Hincu, eds.), *Social Sciences in the Other Europe*, Budapest, CEU Press, 2018.

Marcus Morgan is a lecturer in Sociology at the University of Bristol, UK. His current research is in the areas of political sociology, social movements, and cultural sociology. He is the author of *Pragmatic Humanism: On the Nature and Value of Sociological Knowledge* (Routledge, 2016) and *Conflict in the Academy: A Study in the Sociology of Intellectuals* (Palgrave, 2015).

Peter Tibor Nagy is a doctor at the Hungarian Academy of Sciences, university professor and head of the Research Center for Sociology of Church and Religion at the John Wesley Theological College in Budapest (http://wesley.hu). He was co-leader of the European projects ELITES08 and (as a Hungarian partner) INTERCO-SSH (on the social sciences since 1945). His research foci include the history of educational provision in Central Europe, national schooling policies in a comparative perspective, secularization and the social standing of churches, historical sociology of the human sciences. He is author of 11 books in Hungarian and co-editor of 4 books in English, with altogether over 300 scholarly publications. See: http://nagypetertibor.uni.hu.

Marco Santoro is a professor of Sociology at the University of Bologna, Dipartimento delle Arti. He works on the history of the social sciences, cultural production, intellectuals, and on the mafia. He is a founding editor of "Sociologica. Italian Journal of Sociology". He has recently edited *The Anthem Companion to Everett Hughes* (2016, with Rick Helmes-Hayes).

Gisèle Sapiro is a professor of sociology at the École des Hautes Études en Sciences Sociales and Research director at the CNRS (Centre européen de sociologie et de science politique). The author of *La Guerre des écrivains, 1940–1953* (1999; Engl. *French Writers' War*, 2014), *La Responsabilité de l'écrivain* (2011) and *La Sociologie de la littérature* (2014; Spanish 2016; Japanese 2017), she has also (co)edited *Pour une histoire des sciences sociales* (2004), *Pierre Bourdieu, sociologue* (2004), *Translatio. Le marché de la traduction en France à l'heure de la mondialisation* (2008), *Les Contradictions de la globalisation éditoriale* (2009), *L'Espace intellectuel en Europe* (2009), *Traduire la littérature et les sciences humaines* (2012), *Sciences humaines en traduction* (2014), *Profession? Écrivain* (2017).

Rafael Y. Schögler is an assistant professor of translation studies at the University of Graz. In 2017 he was visiting researcher at the Centre for Translation Studies at UCL London and the Centre for Translation and Intercultural Studies at the University of Manchester. His research interests comprise sociology of translation, translation in the social sciences and humanities and sociology of SSH. Recent publications include "Les fonctions de la traduction en sciences humaines et sociales", in: *Parallèles* 29/2 (2017); "Translation in the Social Sciences and Humanities: Circulating and Canonizing Knowledge", in: Alif 38 (2018).

Gustavo Sorá is a tenured professor in the Anthropology Department at the National University of Córdoba and Researcher at CONICET (Argentina). His research focuses on the history and sociology of book publishing and translation. Book publications include *Editar desde la izquierda en América Latina. La agitada historia del Fondo de Cultura Económica y de Siglo XXI* (Siglo XXI: Buenos Aires, 2017), *Brasilianas. José Olympio e a gênese do mercado editorial brasileiro* (Edusp: São Paulo, 2010) and *Traducir el Brasil. Una antropología de la circulación internacional de ideas* (Libros del Zorzal: Buenos Aires, 2003).

Rob Timans is a sociologist and economist affiliated with the Erasmus Center for Economic Sociology (ECES) in Rotterdam the Netherlands. His Ph.D. thesis *Studying the Dutch Business Elite: Relational Concepts* *and Methods* won the Dutch Sociological Association's Prize for best dissertation of 2015–2016. His research interests are in economic sociology and the sociology of knowledge and science, in particular the sociology of research methods.

List of Figures

List of Tables

1

Introduction: Shaping Disciplines—Recent Institutional Developments in the Social Sciences and Humanities in Europe and Beyond

Christian Fleck, Matthias Duller and Victor Karády

The authors of this volume have collaborated for a period of four years within a European Union funded research project called International Cooperation in the Social Sciences and Humanities (INTERCO-SSH). Interco-SSH was dedicated to investigating particularities of the disciplines put together under the acronym SSH, and identifying past hindrances and future possibilities, to better the future collaborations beyond disciplinary fences and national borders. This volume reports on the results of one of the endeavors of our international collaboration; studying patterns of institutionalization across Europe and beyond. It analyzes the development of a sample of SSH disciplines in Argentina, France, Germany, Hungary, Italy, the Netherlands, Sweden, and the

C. Fleck (✉) · M. Duller
University of Graz, Graz, Austria
e-mail: christian.fleck@uni-graz.at

M. Duller
e-mail: Matthias.duller@uni-graz.at

V. Karády
Central European University, Budapest, Hungary

C. Fleck et al. (eds.), *Shaping Human Science Disciplines*, Socio-Historical Studies of the Social and Human Sciences, https://doi.org/10.1007/978-3-319-92780-0_1

United Kingdom. Two further volumes to be published in the same series will disseminate findings of other parts of Interco-SSH, one will be on internationalization and one on the transfer of paradigms, theories, key thinkers and methodologies across national fields of learning.

The primary focus of every chapter in this book concerns the institutional development of seven preselected disciplines from the social sciences and humanities in eight countries. They deviate from conventional routines, narrating the histories of the sciences, including the humanities and other 'softer' branches of scholarship. Most narratives of any scholarly past are presented as a succession of 'ideas,' research results and theories. Or, to say it in a more 'highfalutin' way, they try to offer explanations of how past scholars found the 'truth,' exemplified in a widely used history of sociology book's title: From Lore to Science (Barnes and Becker 1938/1961). Even if spokespeople for a so-called symmetry-perspective argue for that study of the causes of false propositions in the same way as one studies the causes for true ones, they follow paths ingeniously paved by intellectual historians. While not questioning the value of this approach, we have chosen to pursue a different one. 'Ideas' are certainly an essential part of what constitutes the field of scholarship. Nevertheless, this field is structured by other forces, most notably institutional, which also deserve to be taken seriously. This introduction outlines the main issues of the institutional perspective shared by the individual chapters of this book. In doing this we hope to make clear the meaning of institutions and why they are of crucial importance for a better understanding of the world of scholarship.

The bulk of the historiography of the social sciences and humanities (here, and throughout the book, SSH) has been written by proponents of the discipline under study, primarily for their disciplinary peers. Sociologists write histories of sociology for sociologists; anthropologists do the same for their tribe's fellows, and so on. The functions of these histories as disciplinary subfields range from identity-building, canonization of particular authors, to commentary on current debates on specific theoretical or empirical programs. In other words, the past fulfills services for the present, which was labeled 'presentism' by one of the leading exponents of the history of anthropology George W. Stocking (1965). Historicist versions of disciplinary histories, by which we mean scholarship interested in the historical genesis of the SSH as an object of investigation on its own right, have emerged comparatively later (see revealing autobiographical remarks by Blaug 1994). With the exception of the historiography of economic thought, and psychology/ psychiatry, this approach to most other SSH disciplines has not developed into specialized fields of scholarship of a larger size and have not been united into a specialization of the history of the SSH thus far. What 'presentist,' as well as 'historicist,' streams in the historiography of the SSH have in common is their predominant focus on authors and their scholarly output as the stuff out of which disciplines arise. Such narratives shimmer from the contents to the thinkers or vice versa. The standard version of histories of the SSH is thus modeled along the lines of intellectual history and very often, therefore slide into neighboring subfields like sociological/psychological theory, among others.

In contrast, the sociology of science, though it has been predominantly concerned with the natural sciences and latterly with technology, has at times taken the SSH into account. Robert K. Merton's detours into what he labeled 'sociological semantics' (Merton 1993 [1965]; Merton and Barber 2004) are promising examples. Recently, Charles Camic et al. (2011) proposed the transfer research modes that focus on scholars' daily practices in their work in the study of the SSH, continuing what two of the authors had called 'New Sociology of Ideas' some ten years earlier (Camic and Gross 2001). As revealed by the label, this approach differs from those already mentioned not so much with regard to the object of investigation as methodologically. Research practices—especially close attention to the environments in which scholars find themselves after their daily research, teaching, and writing—emphasize the sociological lens prevalent here for the *explanans*. The *explanandum* remains the ideational content and most often these studies focus on very particular (micro-) instances of SSH, i.e. individual researchers, concepts, mechanisms, practices, routines, etc. The importance of larger (macro-) contexts is generally admitted, but micro-contextualization—just as in the 'constructivist' sociology of scientific knowledge—is the explicitly favored perspective.

Our focus on the institutional analysis of the SSH does not deny the value of either perspective, but implies that they omit or sideline other aspects. Ours is thus a complimentary view that highlights structural shifts in the systems of higher education, as well as institutions of research and innovation (beyond the universities) within which the SSH make headway. As far as developments since 1945 are concerned, the structural conditions of the entire intellectual infrastructure of scholarly production, including the universities, underwent more profound change than ever before, by which we mean primarily its expansion and the immediate and indirect consequences of this growth. Science policy emerged only during and after WWII, and along with this emergence came debates about the best allocation of scarce resources. Until just before WWII, which was fought partly using science and scholarship resources, the world of learning was the privileged preserve of a tiny minority of upcoming generations. The 'republic of scholars' had been an enclave of sorts within society, communicating with ordinary people only in one direction, yet claiming to counsel the political class and guide the nation state spiritually. The quintessential locus of their reasoning was universities that, in most countries, only began to enjoy a level of autonomy from governmental interference from late nineteenth century onwards. Their intellectual practices, however, remained grounded in classical habits and areas until well into the twentieth century, often without much contact with new, extra-mural forms of knowledge production, notably within the emerging SSH. The expansion and transformation of the universities in order to make them respond to all kinds of societal demands from outside academe started, in most European countries, in the late 1950s and intensified in the 1960s. This led both to the multiplication of academic personnel to serve exploding numbers of students—initially because the universities became open to women—and to the decisive opening of academia to a set of new disciplines and branches of study. In times of quick expansion such as these, job opportunities for academics rocketed and it is safe to hypothesize that such conditions might affect the scholarly content cultivated by new entrants. If this assumption holds some truth, we need to know the institutional environment in which particular new approaches, methodologies, and research fields were proposed.

Innovation, it seems, is more likely to take place in an environment with an abundance of competitive positions than in situations of penury in the positional market. Outlining changing institutional conditions of intellectual pursuits thus offers a view of changing opportunity structures that, in some instances at least, can contribute to explaining changes in disciplines' intellectual landscapes. In this way, the institutionalization perspective can and does inform traditional representations of the source of scholarly options and individual creativity.

Probably the most productive aspect of the institutional perspective is its openness to systematic comparisons between developments in various disciplines and countries. Comparisons of this kind are significantly more difficult in ideas-centered approaches and absent in studies narrating an individual's performance. The expansion of the universities mentioned above gives rise to different responses if one analyses the chain of ideas or the birth and change of scientific paradigms. Although the post-war expansion affected all disciplines throughout Europe and beyond, it did so to different degrees. Indeed, the chapters that follow reveal remarkably different dynamics of disciplinary growth between countries as well as important interdisciplinary differences within countries. In addition, instances of institutional contractions and downturns can be observed, veritable breaks of continuity under authoritarian political regimes, as in the case of sovietized Hungary and the military dictatorship in Argentina. These are almost totally absent from narratives of individual disciplinary histories.

A perspective favoring the social structure instead of the expressions of the people observed does not detract from the utterances of those investigated, but claims that by considering the base that makes cloudy systems of ideas possible, we can better understand the ideas. Since there is no way to turn this claim on its head, one can thus argue that the institutionalization perspective is superior to its competitors.

In order to clarify the common perspective of this book's chapters, in the remainder of this introduction we will focus on two notions that are in need of further exposition: disciplines as the basic units on which our analyses rest; and the notion of institution, whose meaning has seen very diverse usage in different contexts.

Disciplines

A widely used classification calls specialized parts of science and scholarship 'disciplines,' defined, or at least marked, by specific topical foci, methodologies and intellectual approaches. Both 'natives' and observers see the overall field of science as consisting of an ensemble of disciplines. Some of these units are better-known and have a longer history than others. Mathematics, philosophy, and physics, for example, are longstanding while informatics or molecular genetics appeared only recently. Although it is hard to derive an exhaustive, general definition of what a discipline is, their functioning as building blocks of the larger 'house' called academia is generally accepted. They are, in the words of Rudolf Stichweh, 'the primary unit of internal differentiation of the modern system of science' (Stichweh 1992: 4).

The concept 'discipline' points immediately to at least three research areas. First, we need to explain their emergence, including new entities, second, we need to come to terms with the collaboration of scientists and scholars across the boundaries of disciplines, and what is debated under umbrella terms as inter-, multi- and trans-disciplinarity. Third, in a closer examination we see that the boundaries of any given discipline are anything but fixed and commonly agreed upon; disciplines can expand or contract with regard to the range of their explanatory claims. Since Thomas Gieryn (1999), debates about this problem are usually labeled 'boundary work,' since disciplinary frontiers are guarded and defended by 'boundary workers' and often redefined by those involved, even if in different ways to state borders.

Stichweh (1992) argues that it was only in the late eighteenth and early nineteenth centuries that the idea of scientific disciplines came to structure the field of scholarly pursuits, replacing a formerly hierarchical system with one based on functionally differentiated, horizontally coexisting units—each being concerned with different aspects of reality (ibid.: 7). The oldest disciplines in this sense were, then, formed from those scientific activities that were already well-established. Among nineteenth-century SSH these were philosophy, history, descriptive statistics, and early variants of geography, economics and political theory. Around the turn of the twentieth century, research into social and administrative problems was gradually imported into the universities. In some countries this formed the basis of new academic entities that were consolidated into special branches of study (with chairs, lectureships, study programs and occasionally university degrees) during the 1920s and 1930s. The post-1945 era also saw the internationalization and professionalization of SSH disciplines often following the American model (Wittrock 2001). If something becomes a model, it needs more detailed elaboration and more explicit reasoning. If this same entity is exported a higher degree of uniformity is desirable.

Heilbron (2004) argues that differentiation is only one of at least three mechanisms leading to the birth of disciplines. Economics might well be a case of differentiation, increasingly narrowing its scientific concerns from broad questions of social organization to market mechanisms. In German-speaking countries, business accounting thus split from *Volkswirtschaftslehre* (national economics) to form a new specialty initially called *Privatwirtschaftslehre* (private economics). The basis of sociology, on the other hand, is rather a case of a "specialty of generalities" in Comte's terms, synthesizing knowledge from scattered intellectual realms, conducive to the Durkheimian enterprise where sociology appears factually as demonstrated in the *Année sociologique* as the global umbrella of all kinds of scholarship related to matters social, including philosophy. A third mechanism is what Heilbron (2004) calls the "upgrading of practical activities to the status of a scientific discipline (as in the example of chemistry)" (ibid.: 35–36).

The processes of disciplinary genesis present a number of aspects and stages. They include the formulation of particular intellectual concerns, targets and perspectives, the formation of stable institutions at universities with a high level of public visibility, which secure the basic functions of the research, teaching, reproduction and canonization of specialists, as well as their professional organization in collectively self-promoting institutions. All of these aspects are characteristics of modern disciplines (Heilbron 2004: 30). Historically such processes in the SSH have been long-term, often not completed until the latter half of the twentieth century. Even if disciplinary differentiation has continued to change science systems in important ways, the post-World War II era can nevertheless be seen as a phase in which a relatively stable set of core disciplines, at least in the SSH, has occupied most of the scholarly terrain in Europe and beyond. In fact, it is one of the peculiarities of this era that particular scientific organizations (e.g. national funding bodies for basic research) have been globally imitated more effectively than ever before, a process label "isomorphism" by John Meyer and his collaborators (Drori et al. 2003). Such organizations usually strengthen established disciplinary differentiations but do not encourage new arrangements of the division and integration of the production of social knowledge. Vastly different national traditions notwithstanding (Lepenies 1988; Levine 1995), these traditions started to increasingly interact and recognize each other across national borders, contributing to international debate of how disciplines define themselves.

Without assuming that the disciplinary order of the post-war era is in any sense 'natural' in the SSH, i.e. one that corresponds to the differentiation of social realities themselves, the relative stability of the core disciplines provides a justification of sorts for international comparative research design. At the same time, one has to keep in mind that what hides behind a common disciplinary label can differ significantly between different countries. The rationale with which disciplines define themselves is anything but coherent. Abbott's book title *Chaos of Disciplines* (Abbott 2001) captures this insight well. While anthropology is held together via a common method (ethnography), political science follows the model of synthesizing knowledge of a common phenomenon (politics) from other disciplines. Economics is, today at least, unified by a theoretical assumption, famously put into one sentence by Lionel (later Lord) Robbins: 'Economics is a science which studies human behavior as a relationship between ends and scarce means which have alternative uses' (Robbins 1932: 15) but yet followed completely different cognitive paths before this paradigmatic shift. Sociology remains as vaguely defined as ever. In the first decades after WWII, however, several attempts to unify or streamline the discipline received significant attention (Celarent 2010; Calhoun and Van Antwerpen 2007; Pooley 2016; Steinmetz 2005).

This chaotic picture notwithstanding, it does appear to be established that disciplines, once stabilized, are broadly accepted categories that also form the basis for any inter-, trans-, or multidisciplinary endeavors. One of the factors responsible for this large-scale continuity are national and international institutions of research and science policy. The establishment of the American 'Social Science Research Council' in 1923 (Worcester 2001) or of the 'International Social Science Research Council' in 1952 (Platt 2002) are proof of the process of fossilizing disciplines by making instances of them members of such umbrella organizations. Organized cooperation between representatives of particular branches of science and scholarship has been practiced since the creation of the Royal Society and its counterparts elsewhere (the regional or national academies), but it is only recently that is has become a concern for policy pundits when science policy started to be a specialized, state-sponsored activity involving increasingly significant public funding. As long as scholars did not challenge the social or religious order, they could engage freely in the pursuit of their personal intellectual interests. We find early instances of systematic interventions from outside academia in Napoleonic France—where higher education formed part of the state bureaucracy—as well as in Wilhelminian Germany, where universities of the Humboldtian model were supposed to enjoy full intellectual autonomy but had to accept that the state decided who was allowed to occupy a chair. But worldwide science policy appeared worldwide in the decades following WWII as a basic public function destined to promote, frame and orientate the development of scholarly activities (cf. Drori et al. 2003, 2006).

One precondition for any kind of policy seems to be the clustering of those concerned in publicly recognizable social units. Politics is not concerned with individuals, but with larger assemblies of clients, at least if we follow the economic theory of democracy. Politicians execute policies, initiated only if a multitude of beneficiaries can be served, and they always take into account the anticipated impact on their electoral chances. With regard to science policy, 'discipline' functioned as the unit deserving of benefits. More recently, assemblages of disciplines occupy this place.

Very often a particular discipline was recognized as the provider of remedies to rising social problems. After the Sputnik Shock of 1957, for example, Western democracies invested in space sciences. When 20 years later unemployment rates did not recede, economics, psychology and sociology were funded to find cures. Each time disciplinary neighbors who did not get their share of public recognition and taxpayer's money reacted jealously, to say the least.

What might be seen from the outside as a convenient packaging technique often makes sense inside academia as well. The fact that universities were administered not as a plurality of disciplines, but in units called faculty, Fakultät, faculté, facoltà, facultad, faculteit, fakultet, or kar (to mention only those languages represented in this book) serves to uphold the outlined perspective.

What it at stake, then, is the autonomy of particular branches of scholarship *vis-à-vis* competitors. From the earliest days, some disciplines cooperated with neighboring "academic tribes" (Becher [1989] 1996; Becher and Trowler 2001; Müller 2014) more regularly and intensively than others. When universities first emerged in the late Middle Ages, they organized scholarship in Faculties, each bundling together a handful of disciplines inherited from Greek and Latin antiquity. Traditionally three 'professional' faculties became standard: Theology, Law, and Medicine, completed by Philosophy (*studium generale*) as an introduction to specialized studies. This organizational framework of learning was maintained well into the twentieth century. Collaboration was expected within these university subunits, which did not exclude competition for academic positions, funding or forms of canonization (prizes, distinctions, etc.). Philosophy and Law are the two Faculties within which most of the present-day social sciences and humanities started to become independent entities, a fundamental aspect of the process of institutionalization. As a corollary, one should keep in mind that, depending on the, sometimes contingent, location of founding fathers or major contributors to new disciplines, possibilities for successful independence were largely determined by power relations between established disciplines inside their Faculties, intellectual authorities and networks. In some universities, the Faculty of Law was a much more homogenous intellectual environment than the Faculty of Philosophy. In the Germanic academic pattern, dominant in continental Europe (outside France) since the nineteenth century, philosophy fizzled out into subunits focusing on languages (both living and dead), historical and geographical studies, as well as the early crystallizations of the natural sciences.

The American system of higher education, in which teaching and research are seen as two unrelated sides of one endeavor, molds the internal differentiation of the professoriate differently. Originally American liberal arts colleges placed less emphasis on departmental structure, only with the rise of research universities—and therefore disciplines in the present understanding—did they emerge (Jencks and Riesman [1968] 2002). The establishment of more-or-less streamlined curricula, one for each discipline, further supported the internal differentiation of science and scholarship.

A further aspect that needs recognition is the nature of the polity for which policies are designed. Europe and America functions differently with European states showing a higher degree of diversity, but science policies are even more differentiated than other aspects of government activities. For a long time after WWII, non-governmental—or 'private'—universities were big players only in the United Kingdom. In the rest of Europe such entities remained rare, at least in secular studies. Moreover, in federal countries, like Germany or Switzerland, the responsibility for all levels of education belongs to regional authorities. Traditionally, policies for higher education have been designed and applied to comparable institutional units, like universities, faculties, schools or departments within them. The post WWII expansionist policies founded new universities mostly for reasons unrelated to scientific considerations, as attested by the 'plate-glas' universities in the United Kingdom or the competition between German regions or sub-regions in the 1960s to get their own university as a vehicle of collective distinction. Incremental policies happened, then, by allocating funds to the subunits of each university on a largely egalitarian basis, given the lobbying power of local politicians and decision makers. It cannot come as a surprise that, after the 1989 German re-unification, German sociologists agreed upon, and enacted, the standard size of an ordinary department of their discipline (at minimum three full professorships) since they were sitting in the committees overhauling the universities in the former GDR.

With science and scholarship continuing to grow, science policy started to develop more aggregated units, such as that which now comprises the acronym SSH in European science policy. One could easily question the degree of kinship between the range of disciplines gathered under this umbrella. Try, for example, to identify commonalities between econometrics and archaeology, numismatics and linguistics, or mathematics and canon law. There are none, with one notable exception; in the European Research Area they all compete for the same pot of money destined for the SSH in a zero-sum-game.

The European Union's science policy reduced the universe of scholarship to three pillars: Life Sciences; Physical Sciences and Engineering; and Social Sciences and Humanities. Tensions between the traditional basic units of scholarship—'disciplines'—bundled together in these big containers (like squabbling relatives) is a built-in administrative device that facilitates the execution of science policies. It is nothing short of miraculous that the distribution of funds is much more even between containers than between smaller units. Thus the angry voices of those doing comparatively badly can be silenced, even before they start grumbling. No bundling forces can those located in the same container develop for common lobbying strategies in order to gain a voice at the tables where decisions are taken over their funding applications.[1] Such forced collaboration could, as another unintended consequence, result in the recomposition of inter-disciplinary relationships, or even the invention of new 'trans-disciplinary' identities.

At the same time, pleas to overcome specialism and narrow-minded expertise are routine in science policy language, and the funding of our Interco-SSH project by the European Union's General Directorate for Research and Innovation belonged precisely to proposals that might result in fragmentation being overcome. A goal of this relatively new discourse is the sidelining of disciplinary division. Sociologists of science, and science policy wonks, have been united in fighting 'Mode 1' research for some time. This is the type of research motivated by creating knowledge for its own sake, which is often executed within the

[1]See e.g., the list of stakeholders participating in the production of a document released in preparation of the post-2020 programme: *LAB-FAB-APP: Investing in the European future we want. Report of the independent High Level Group on maximizing the impact of EU Research & Innovation Programmes*, https://ec.europa.eu/research/evaluations/pdf/archive/other_reports_studies_and_documents/hlg_2017_report.pdf, Annex 2, p. 24.

previously fortified boundaries of traditional disciplines (Gibbons 1994; Nowotny et al. 2001).

Without mutual agreement, the traditional division of academic labor into disciplines has also been challenged 'from below.' Several attempts to create new fields of expertise—such as the line of 'studies' focusing on women, and later on gender, and cultural, and postcolonial problems, etc.—began within the fences of disciplines, but later joined forces under these new labels. Their common feature is the social movement-like appearance of calls for renewal. The most successful ones managed to gain a discipline-like status, at least in some universities in some countries. Other 'studies,' like those dedicated to peace or the future, failed to crystallize the same way or achieve similar public success. Similarly to when psychology and pedagogy split from philosophy to establish new disciplinary tracks within the historical organization of academia, the emergence of 'studies' is only the most recent example of such institutional transformation. The separation of management or business studies from economics—followed more recently by finance studies—striving to become entities of their own, shows a different trajectory. Compared to the sub-differentiation of psychology into about a dozen fields of expertise ranging from social and clinical psychology to brain research, all executed by scientists with a degree in psychology, management, finance and economics proper seem to follow a relatively unified, theoretically orientation. But, the recent increase of the number of professors of economics etc. holding a Ph.D. in mathematics could be interpreted as a weakening of its disciplinary identity since it has been long established as a field of study concentrating on human action.

Without further elaboration, we want to emphasize that a discipline is anything but a well-defined fact of nature. It can changes its boundaries, expand or contract, become imitated abroad or develop in a polygenetic mode in different places more-or-less simultaneously. Despite all of the conceptual fuzziness, the use and meaning of 'discipline' is relatively well established. In everyday conversation, scholars identify themselves by naming their disciplinary affiliation. From an observer's perspective, we see tense interaction and communication within networks of people who would identify themselves by mentioning the same disciplinary label. One could analyze the same constellation by highlighting that members of a discipline are all those who compete for symbolic capital from the same urn.

Such reasoning should be sufficient to legitimize the analyses of this book. We do not focus on individual scholars, or aggregates of variables related to those individuals, and we do not want to talk about university systems, scholarship, science, etc. on an abstract level. The intermediate level of disciplines should help us establish a collection of data that lends itself to comparative studies.

Institutionalization

In doing this we apply the language of institutionalization. Our perspective is to investigate the processes that contributed to the establishment, and further expansion, of new fields of science and scholarship via the emergence of institutional structures from common routines of the given branches of learning. The process of establishing a particular scholarly activity appears to have generally been the first step towards institutionalization. However, the term 'institution' lacks a clear meaning (Searle 2005). In economics it means something completely different than it does in sociology, for example. Philosophers, meanwhile, developed their own interpretation. In German philosophy-cum-sociology, the meaning has been elaborated in opposition to Anglophone sociologies. Where economists call everything that is not market-driven, self-interested rational behavior an institution, sociologists tend to reserve this term for relatively stable configurations that secure the mutual understanding of the behavior of others and us, and which thus shapes public conduct. Sociologists resort to a more diversified set of criteria when they talk of institutions. In most cases, the sociological terminology refers to particular instances of institutions instead of the overall concept; the family, market, educational systems, etc. are cases in point. The system of higher learning, the social organization of research and patterns of intellectual discourses are exemplary examples of institutional arrangements. This is what we propose to study in this book.

In analyzing the organization of scholarly disciplines we follow Edward A. Shils, who had the following to say about this process:

> By institutionalization of an intellectual activity I mean the relatively dense interaction of persons who perform that activity. (...) The high degree of institutionalization of an intellectual activity entails its teaching and investigation within a regulated, scheduled, and systematically administered organization. The organization regulates access through (...) for example study, teaching, investigation, publication, appointment, and so forth. It also entails the organized support of the activity from outside the particular institution and the reception or use of the results of the activity beyond the boundaries of the institution. (Shils 1970: 763)

Our analytic tools are outlined below, but first one has to refer to a specific feature of the overall process. By speaking about institutionalization, one could easily get the impression of an unidirectional process, which would be incorrect. Stemming from the same cultural environment as the more prominent concept of modernization, one regularly encounters a parallel understanding (Parsons and Platt 1973). Like the notorious 'take off' terminology of W.W. Rostow, several scholars using the institutionalization terminology hint at a single direction almost in form of a teleological process; development is seen as starting at different points on the time-line but after this continuous growth is the expected pattern.

Instead of this optimistic vision, we are aware of forces and developments acting against continuous processes of institutionalization, some even conducive to what could be called de-institutionalization. This might imply the disappearance or weakening of assets or resources necessary for further institutionalization in terms of professional jobs, funding, journals, curricula, public esteem or—in authoritarian regimes—freedom to pursue scholarly work, all things which have been around for a while. As far as the concept of institutionalization has been used to analyze the trajectory of particular SSH disciplines, several prominent authors have failed to discuss any mechanisms explaining what happened, very often 'telling the story' appeared to be enough (Bulmer 1984; Clark 1973; Drori et al. 2003; Oberschall 1972; Turner and Turner 1990).

Recognizing backlashes in the process of institutionalization is one thing, explaining them is quite another.[2] Why a data curve of relevant indicators turns down at a given time is often very difficult to determine. Still, the observation of such moves is a necessary first step, and the chapters of this book offer ample evidence that even the simple perception of such turning points has often escaped the attention of otherwise well informed insiders.

Sampling and Research Design

During the initial debates about the Interco-SSH project, we discussed two dimensions of comparison: the selection of countries; and the sampling of disciplines. The first was made before the second, because setting up an international collaborative research team is usually not an affair of scholarly legitimized sampling but arising from the heavy impact of previous joint collaborations. Happily, for us, our group consisted of representatives from countries with highly divergent intellectual and social histories, which we realized, taken together, could represent various aspects of larger developments. Larger European nations—such as France, Germany, Great Britain, and Italy—and smaller ones from the West, East, and North of the continent—such as the Netherlands, Hungary, and Sweden—display a reasonably diverse set of cases. In addition, we have been fortunate to also welcome partner scholars from Argentina, which has been for a very long time one of the overseas outposts of European learning, and which constitutes a considerable part of the Spanish-speaking academic world.

The choice of the disciplines upon which we focused was made following eclectic considerations. From all of the SSH discipline we sampled two classic humanities disciplines purposively—philosophy and national literature—then the core branches of modern social studies: economics; political sciences; and sociology. Anthropology and

[2]Compare Jacobs (2013), Turner (2014, 2015) with regard to developments in the USA. Both authors discuss downward turns but did not contribute much to conceptualize such events.

psychology were added, given their growing academic importance and scholarly market, as well as their impact on other types of social studies. Both have changed their affiliation to more aggregate units either over time or by country, to the extent that they produce divergences and ambiguities in the nature of the disciplines. What is anthropology in France is, in German-speaking countries, *Völkerkunde*, distinct from *Volkskunde*. During the last half century, these denominations underwent significant changes under various labels, such as ethnology, cultural anthropology, folklore, ethnography, or their local translations. The only rather permanent distinction here opposes studies of cultures outside and inside Europe. Our focus was laid on extra-European social anthropology, except in countries—like Hungary—where this had been, historically, a negligible intellectual pursuit as against national folklore. Almost everywhere psychology began as a spin-off from philosophy, but, over the following decade, meandered between the humanistic and the naturalistic pole. Today most of its subfields see themselves close to the natural sciences. The study of each national literature is an established field everywhere, but the subject itself lacks a homogenous transnational component.

Whereas philosophy and national literary studies are long established fields, the rest of selected disciplines are younger in origin. Economics might be regarded as the oldest, going back to the nineteenth century in most European countries, followed by anthropology and its surrogates, as well as psychology. Sociology and political science are the two most recently institutionalized disciplines.

In some reports on national developments we could not avoid taking other disciplines into account, our analyses were mostly concentrated on national samples of the above branches of the SSH.

Variables and Indicators

The crux of any comparative analysis is to fix the units and variables and thus permit comparisons across state or disciplinary boundaries. Our research strategy may be presented in three steps. First, expedient indicators for understanding and measuring institutionalization in general will be outlined. Second, the list of indicators used in the following chapters will be given in concrete terms. Third, a conclusion will be attempted with a short discussion about the limitations of our analyses.

To translate Shils' above quoted elaboration into measurable indicators, one could stipulate the following propositions: A discipline is more institutionalized if it: (1) can be studied in universities as a major subject; (2) has a specialized teaching staff; (3) there are opportunities for publication in specialized journals; (4) has financial, administrative, and logistic provision through established institutions; (5) has established and remunerated opportunities for the practice of the discipline; and (6) has a 'demand' for the results of its research. However, Shils' description is anything other then exhaustive and ignores several institutional patterns that might have a strong impact upon the development of a particular discipline. Further, one cannot make his explanations operational in order to interpret early stages of institutionalization. It cannot, for example, be used to explain the periods of founding figures like Émile Durkheim, Lester Ward or Stein Rokkan, and particular factors and vehicles of institutionalization—such as old-boy-networks, invisible colleges, teacher-pupil-relationships—are not portrayed at all.

As a starting point to analyze the different modes of institutionalization, we began with the organization unit level. In looking at this, we suppose that there are only few spaces where members of a discipline, or a discipline-in-the-making, can expect to meet regularly without individually arranged appointments; departments, professional associations or learned societies.

At what point did the discipline first appear in a university subunit, such as a department, institute, center, etc.? For how long did the name remain? How many similar units could be found at any given time? For a comparative analysis, it makes sense to restrict the search to institutions of higher education, classified by the OECD's International Standard Classification of Education (ISCED) as Tertiary-Type A. In practical terms, such a definition excludes teaching enterprises like *Fachhochschule*, colleges of Applied Science, and all others lacking the right to confer a doctoral degree or equivalent, that is performing training functions for would-be scholars. On the other hand, this definition has to accept the renaming and upgrading of institutions like the British Polytechnics, which became full-fledged universities in 1992. A related indicator would count the number of specialized research institutes outside the universities like Academies of Science institutes in communist regimes, the CNRS in France and its equivalents in Southern European countries, the Max Planck Institutes in Germany and special service agencies essential for the provision of basic data and information for several SSH disciplines, which would include Statistical Offices or the French *Institut national d'études démographiques*. Unfortunately, such research-oriented organizations are even more difficult to handle in a cross-country comparative analysis than universities. Their life span, governance, internal status hierarchies, etc. are much more complex than the historically developed diversities of the traditional higher education sector.

A second organizational framework that functions as a meeting space for members of a particular discipline is what has been invented in the USA as 'professional associations'—following its European antecedents of thematic regional, national or occasionally even transnational learned societies (since the nineteenth century). This type of institution is different from a trade union or popularization and advocacy agencies. Interestingly they diffused to other parts of the world and can be found in the latter half of the twentieth century in nearly every advanced economy. Since we focus on a comparison of nation states, the emergence of regional and international organizations as transnational gatherings lies beyond the scope of our project. Obviously, professional associations do not have the same function and operational structure in each country concerned. Some of them have remained inclusive, open to voluntary adherents, while others acted exclusively, like academies, with a limited elected, or invited, membership. (The German Sociological Association abandoned its policy to choose its members by invitation only in the late 1950s.) Specific features of the American scholarly professional associations are missing in most European countries, where it is usually the state that consecrates patent-like degrees.

The transfer of collectively accumulated professional knowledge to the next generation needs to be organized in a more stable fashion than what individuals can achieve alone. Only if there is some general agreement on both the content and *modus operandi* of this instruction can a discipline be considered established. Two further features should be considered here; admission procedures and preparation for a profession.

Currently one can differentiate between levels of education, like B.A., M.A., and Ph.D. or, as in the old European faculties, between graduation and the doctorate. For a discipline to assert its existence a curriculum is expected to be firmly organized at each of these levels. Closely related to this question is the problem of access to particular study programs. Some countries have historically practiced an open door policy at all levels, but nowadays there are several national higher educational systems with highly differentiated entrance exams (fixing criteria of preliminary scholarly achievement differently for different disciplinary tracks), or even general *numerus clausus*, or similar regulation. Third, the situation for particular disciplines at the university level differs according to its representation in the lower levels of schooling. If there is a separate subject, for philosophy for example, at the Upper Secondary Level (as has been the case in France and Germany since the nineteenth century) the situation for this discipline at the university or college level proves to be different from that of another SSH deprived of such antecedents. We can assume that the representation of particular fields of expertise at the Upper Secondary Level is a direct consequence of the public negotiating power of members of the given discipline. For example, Austrian professors of the Faculty of Law successfully banned political science as a separate study program up to 1971, and the discipline is still not taught in Austria at the high school level. If there is instruction at lower levels there are more jobs for graduates from the higher level, with teaching at the higher level also affected.

The institutionalization of a discipline has consequences for the collective profile of those involved in it. Besides the size and recruitment procedures of the professional staff concerned, their composition with regard to their competence and expertise are heavily related to the historical date and the conditions of the formal recognition of a discipline inside academia. The status of a discipline relative to others is usually decreed by the numbers of students or professors. Students could be counted either at entrance, as a stock, or at the exit doors as graduates. The most convincing indicator appears to be the number of graduates, since most generally this is the potential starting point of a scholarly career (especially a doctorate). Thus, comparing graduates makes more sense than any other measure of the student body. From a comparative perspective, we would be more successful if we concentrated primarily on the study of academic staff to the highest level. At least in continental Europe the full professor type university employees pass through similar rites of passage and their share in the teaching body seems to be equally comparable. (The structure of personnel in UK universities is different.) The way up the early stages of the academic ladder is transnationally comparable only with regard to the fact that there are some mandatory steps to be taken. But each European country established its own rules and habits of how one gets from a doctorate to professorship, which have evolved during the last seventy odd years. Despite the fact that even a simple headcount is not easy, all aspirations to explore *homo academicus* more comprehensively come up against a brick wall. Paraphrasing Bourdieu, one could say that those who classify do not want to be classified. The lack of data with regard to the social composition of the academic elite is astonishing in several countries and any attempts at an in-depth investigation of social changes in the composition of the professoriate are futile. Even trivial data for gender inequality have proved to be historically rare.

For their relative stabilization within often protean frontiers and further development of their content, disciplines need forums within which to exchange scholarly findings. Written communication plays a bigger role than verbal, or any other kind of, exchange. From a comparative perspective, one could either sample books, with all of their variants, or journals. From a practical perspective, journals are easy to identify, count, and analyze.

We are uncomfortably aware of the restriction of this small set of indicators and concede that others are worthy of consideration. Of particular importance are outreach of disciplines beyond the ivory tower and the different modes of funding academic work.

The interested reader will find further elaboration of indicators in the Handbook (Fleck et al. 2016) and short portraits of the development of individual disciplines in the countries under scrutiny on the INTERCO website.[3]

---

[3]http://interco-ssh.eu/short-histories-of-disciplines-in-the-world/.

In what follows, the reader will get comprehensive reports about the development of a sample of seven SSH disciplines in eight countries, all of them molded both by recent history and the European tradition of scholarship. Those interested in a single country will find the appropriate parts more easily than those who want to know what happened in and with a particular discipline. Our decision to ensemble the analyses along the dimension of the nation state was deliberate, because of our shared conviction that larger political, cultural and social conditions heavily influence the institutional shape that a particular discipline takes.

References

Abbott, Andrew. 2001. *Chaos of disciplines*. Chicago: University of Chicago Press.

Becher, Tony. 1996 [1989]. *Academic tribes and territories: Intellectual enquiry and the cultures of disciplines*. Milton Keynes: Open University Press.

Becher, Tony, and Paul R. Trowler. 2001. *Academic tribes and territories: Intellectual enquiry and the culture of disciplines*, 2nd ed. Buckingham: The Society for Research into Higher Education and Open University Press.

Becker, Howard, and Harry E. Barnes. 1938/1961. *Social thought from lore to science*. New York: Dover Publications.

Blaug, Mark. 1994. Not only an economist: Autobiographical reflections of a historian of economic thought. *The American Economist* 38 (2): 12–27.

Bulmer, Martin. 1984. *The Chicago school of sociology: Institutionalization, diversity, and the rise of sociological research*. Chicago: University of Chicago Press.

Calhoun, Craig, and Jonathan Van Antwerpen. 2007. Orthodoxy, heterodoxy, and hierarchy: "Mainstream" sociology and its challengers. In *Sociology in America: A history*, ed. C. Calhoun, 367–410. Chicago: University of Chicago Press.

Camic, Charles, and Neil Gross. 2001. The new sociology of ideas. In *The blackwell companion to sociology*, ed. Judith R. Blau, 236–249. Blackwell companions to sociology 5. Malden, MA: Blackwell.

Camic, Charles, Michèle Lamont, and Neil Gross (eds.). 2011. *Social knowledge in the making*. Chicago: University of Chicago Press.

Celarent, Barbara. 2010. Human behavior: An inventory of scientific findings, by Bernard Berelson and Gary A. Steiner. *American Journal of Sociology* 115: 1345–1350.

Clark, Terry N. 1973. *Prophets and patrons: The French university and the emergence of the social sciences*. Cambridge, MA: Harvard University Press ("privat").

Drori, Gili S., John W. Meyer, and Hokyu Hwang. 2006. *Globalization and organization: World society and organizational change*. Oxford: Oxford University Press.

Drori, Gili S., John W. Meyer, Francisco Ramirez, and Evan Schofer. 2003. *Science in the modern world polity: Institutionalization and globalization*. Stanford: Stanford University Press.

Fleck, Christian, Johan Heilbron, Victor Karády, Gisèle Sapiro, et al. 2016. Handbook of indicators of institutionalization of academic disciplines in SSH. *Serendipities: Journal for the Sociology and History of the Social Sciences* 1 (1): 1–30 (Material). http://serendipities.uni-graz.at/index.php/serendipities/article/view/20/17.

Gibbons, Michael (ed.). 1994. *The new production of knowledge: The dynamics of science and research in contemporary societies*. London: Sage.

Gieryn, Thomas F. 1999. *Cultural boundaries of science: Credibility on the line*. Chicago: University of Chicago Press.

Heilbron, Johan. 2004. Toward a historical sociology of disciplinary knowledge. In *The dialogical turn: New roles for sociology in the postdisciplinary age*, ed. Charles Camic and Hans Joas, 23–42. Lanham, MD: Rowman & Littlefield.

Jacobs, Jerry A. 2013. *In defense of disciplines: Interdisciplinarity and specialization in the research university*. Chicago: University of Chicago Press.

Jencks, Christopher, and David Riesman. 2002 [1968]. *The academic revolution*, with a new introduction by Christopher Jencks. New Brunswick, NJ: Transaction.

Lepenies, Wolf. 1988. *Between literature and science: The rise of sociology*. Cambridge: Cambridge University Press.

Levine, Donald N. 1995. *Visions of the sociological tradition*. Chicago: University of Chicago Press.

Merton, Robert K. 1993 [1965]. *On the shoulders of giants: A Shandean postscript*. Chicago: University of Chicago Press.

Merton, Robert K., and Elinor G. Barber. 2004. *The travels and adventures of serendipity: A study in sociological semantics and the sociology of science*. Princeton: Princeton University Press.

Müller, Albert (ed.). 2014. *Österreichische Zeitschrift für Geschichtswissenschaften*, vol. 25.3, *Die "Stämme" der Akademie – "Tribes" in Academe*. Innsbruck: Studien Verlag.

Nowotny, Helga, Peter Scott, and Michael Gibbons (eds.). 2001. *Re-thinking science: Knowledge and the public in an age of uncertainty*. Cambridge: Polity Press.

Oberschall, Anthony. 1972. *The establishment of empirical sociology: Studies in continuity, discontinuity and institutionalization.* New York and London: Harper and Row.

Parsons, Talcott, and Gerald M. Platt. 1973. *The American University.* Cambridge: Harvard University Press.

Platt, Jennifer. 2002. *Fifty Years of the International Social Science Research Council.* Paris: International Social Science Research Council.

Pooley, Jefferson. 2016. A "not particularly felicitous" phrase: A history of the "behavioral sciences" label. *Serendipities: Journal for the Sociology and History of the Social Sciences* 1 (1): 38–81.

Robbins, Lionel. 1932. *An essay on the nature & significance of economic science.* London: Macmillan.

Searle, John R. 2005. What is an institution? *Journal of Institutional Economics* 1 (1): 1–22.

Shils, Edward. 1970. Tradition, ecology, and institution in the history of sociology. *Daedalus* 99 (4): 760–825.

Steinmetz, George (ed.). 2005. *Politics, history, and culture, the politics of method in the human sciences: Positivism and its epistemological others.* Durham: Duke University Press.

Stichweh, Rudolf. 1992. The sociology of scientific disciplines: On the genesis and stability of the disciplinary structure of modern science. *Science in Context* 5 (1): 3–15.

Stocking Jr., George W. 1965. On the limits of 'presentism' and 'historicism' in the historiography of the behavioral sciences. *Journal of the History of the Behavioral Sciences* 1 (3): 211–218.

Turner, Stephen P. 2014. *American sociology: From pre-disciplinary to post-normal.* Houndmills, Basingstoke: Palgrave Macmillan.

Turner, Stephen P. 2015. Social scientists as experts and public intellectuals. In *International encyclopedia of the social & behavioral sciences*, 2nd ed., ed. J.D. Wright, 695–700. Oxford: Elsevier.

Turner, Stephen P., and Jonathan H. Turner. 1990. *The impossible science: An institutional analysis of American sociology.* Newbury Park: Sage.

Wittrock, Björn. 2001. Disciplines, history of, in the social sciences. In *International encyclopedia of the social & behavioral sciences*, 3721–3728. Elsevier.

Worcester, Kenton W. 2001. *Social Science Research Council, 1923–1998.* New York: Social Science Research Council.

2

The Rise of the Social Sciences and Humanities in France: Institutionalization, Professionalization, and Autonomization

Gisèle Sapiro, Eric Brun and Clarisse Fordant

To study the development of the Social Sciences and Humanities (SSH), three often related yet distinct processes must be differentiated: academic institutionalization; professionalization; and autonomization of a scientific field.

Institutionalization is understood here as the institutional development of disciplines within the academic system with the creation of curricula, faculty positions and diplomas. The notion of discipline

G. Sapiro (✉)
Centre Européen de Sociologie et de Science Politique (CESSP), Paris, France
e-mail: gisele.sapiro@ehess.fr

G. Sapiro
CNRS–École des hautes études en sciences sociales (EHESS), Paris, France

E. Brun
University of Picardie Jules Verne, Amiens, France

C. Fordant
École des hautes études en sciences sociales (EHESS), Paris, France

C. Fleck et al. (eds.), *Shaping Human Science Disciplines*, Socio-Historical Studies of the Social and Human Sciences, https://doi.org/10.1007/978-3-319-92780-0_2

is historically linked to the university rather than to science (Stichweh 1992; Boutier et al. 2006). The opposite of this process, de-institutionalization, can be observed in contexts of profound social transformation or increased political control—such as under the Latin American dictatorships—but also when the rise of new disciplines entails a reconfiguration of the division of academic labor.

Professionalization, or in Abbott's terms "professional development" (Abbott 1988), is characterized by the advent of professional organizations such as associations, which defend the interests of the profession but also play a regulatory role in defining a professional ethics. This process is not necessarily related to academic institutionalization, as the case of psychoanalysis shows, though by now most intellectual professions require academic training. An opposite process of de-professionalization can also be observed in authoritarian political regimes like the Nazi regime in Germany (Jarausch 1990), but such contexts may also favor state-driven professional development as in the USSR. De-professionalization can also result from inner struggles and divisions within a professional group, leading to alliances with the other disciplines.

Scientific development can be observed with the emergence of what Kuhn (1970) calls a "disciplinary matrix" (i.e. symbolic generalizations, a shared belief in the validity of some statements, shared values, shared examples of solutions to problems), and, to use Bourdieu's concept, the autonomization of a scientific field (Bourdieu 2001). A scientific field is defined by specific problems and tools, and a specific competition ruled by specific authorities such as scientific journals. For instance, psychology and sociology first developed as sciences before undergoing academic institutionalization. Autonomy is defined here both as characterizing a specific research area in the division of knowledge and as expressing a degree of independence from ideology, Religion, the State, or economic demand for applicable expertise. The latter refers to the opposition between fundamental and applied sciences (such as the sociology of voting or part of the sociology of organizations), though this opposition should be nuanced: research delivering expertise can still be relatively autonomous. In his study of the French academic field, Bourdieu (1984) distinguishes symbolic (or purely scientific) recognition granted by specific authorities in the field (scientific journals, committees for promotion, specific awards) from "temporal" success responding to external demand (from the State, from companies, from the elite, from the large public), the first being an indicator of autonomy, and the second of heteronomy. Loss of autonomy can happen in authoritarian regimes, but also in situations of accrued economic constraints. However, political or economic demand contributed to the development of the social sciences in many countries including France, where research is mostly State funded. The State can nevertheless grant a certain degree of autonomy to the academic field. This is the case in France, where scholars benefit from a special status as civil servants as their work is assessed by peers and not by their superiors.

These processes are embedded in the national and international division of scientific labor and in the power relations within the system of disciplines. This system of disciplines is also part of the larger social field, not only through funding and State recognition but also in terms of social recruitment. In this sense, its evolution is also linked to the changes in its recruitment and the patterns of reproduction of social hierarchies within the system. New areas of specialization emerge either as a result of processes of differentiation—what Abbott (2001) calls "fractionation" (borrowing his model from physics)—or by "hybridization," which means the importation of methods, problems, and conceptual tools from other disciplines or areas of specialization, as Ben-David and Collins (1966) suggested through their case study of experimental psychology in the nineteenth century.

Like professionalization, the nationalization of higher education and of what Bourdieu (1984) called the "academic field," encompassing both universities and research institutions, was part of a broader process affecting other fields such as literature, painting, and music throughout the nineteenth and twentieth centuries (Sapiro 2013). Alongside this process, a movement towards inter-nationalization—different from the eighteenth-century European republic of letters—began at the end of the nineteenth century and continued to develop during the interwar period. After the Second World War, the circulation of ideas, models, funding and scholars accelerated under the new American hegemony. Recent developments of the SSH in Europe are marked by the Europeanization of the academic fields (Heilbron 2013).

This chapter will analyze these embedded processes by taking a closer look at the case of French SSH.[1] The first part will concentrate on the historical phases of their institutionalization as academic disciplines, their professional development, and the autonomization of scientific fields in these domains, focusing primarily on seven disciplines: economics; political science; sociology; anthropology; philosophy; psychology; and literary studies. The second part will focus on morphological trends since the 1980s, providing quantitative data on the evolution of the seven disciplines studied here (faculty, researchers, students). Two major evolutions can be observed: feminization and internationalization.

Two Phases of Institutionalization

The processes described above occurred over the long term and in a non-linear and non-teleological manner, without necessarily overlapping, and in ways that varied for different disciplines. It is thus impossible to give a precise picture in one chapter. However, two moments can be identified as crucial in the growth of the SSH. The first started around the establishment of the Republican university in 1896 (90 years after the foundation of the State-run Napoleonic educational system with the university faculties), which favored the emergence and convergence of new disciplines and research institutions in the SSH through the introduction of the German scientific model in the academe (Digeon 1959), while fostering their nationalization in the context of accrued international competition. The nationalization of research was enforced in the interwar period by the model of internationalization promoted by the International Institute of Intellectual Cooperation of the League of the Nations. After an interruption, due to the German occupation and the reactionary policy of the Vichy regime, the second phase of growth of the SSH started in the 1950s–1960s under the new American hegemony.

[1]This chapter relies in part on the collective work done by the French team of the Interco-SSH project, which included: Lucile Dumont, Quentin Fondu, Constantin Brissaud, Jean-Michel Chahsiche, Corentin Bourdeaux, and Annabelle Bonnet. Madeline Bedecarré helped to translate and proofread it.

The First Phase of Academic Institutionalization of the SSH

The French SSH developed in large part thanks to the importation of the German model of research into universities (the Humboldtian unity of teaching and research), which fostered the introduction of a scientific model in what was then called the "sciences de l'esprit" (echoing the *Geisteswissenschaften*). This importation occurred in the phase of construction of the Republican university and the nationalization of French research marked by a large increase in the number of university personnel, which quadrupled between 1871 and 1911 (Karady 1983). The scientific paradigm began to gain recognition in the SSH in the 1870s, following France's defeat at the hands of Prussia. The humanities—philosophy, literary studies, history, and ancient languages—were characterized by a strong link between secondary and higher education, thereby focusing on the reproduction of the teaching personnel through the École Normale Supérieure (ENS—an elite school created during the Revolution) and the *agrégation du secondaire*, a national competitive qualification necessary to become a *professeur agrégé* in high school. The new scientific fields thus emerged outside of academia, through the foundation of scientific journals, learned societies, and specific research institutes and laboratories. These initiatives first developed in institutes for the natural sciences such as the Museum d'histoire naturelle (1793), which had a chair in anthropology since 1855, in elite schools (*grandes écoles*), which dominated (and continue to dominate) the French academic field, and in the newly founded École Pratique des Hautes Études (1868). The polarization between universities and research institutes, or research-oriented institutions that trained scholars, translated into a cleavage between professors and researchers that remains a structural property of the French academic field. Another property, that will not be developed here for lack of space, is the concentration of intellectual life in Paris and the ensuing dominant position of Parisian institutions in the academic field.

Economics held several chairs within elite schools like the Conservatoire des arts et des métiers (Jean-Baptiste Say's Chair, for example) since the first half of the nineteenth century. The Société d'économie politique, founded in 1842 and inspired by the free market, published a periodical, the *Journal des économistes*, which can be described as proto-scientific (Fourcade 2009). In 1877, economics became a compulsory subject in legal studies, entailing the creation of a chair of political economy in every faculty of law. In 1896, the possibility of specializing in economics was introduced within the law *agrégation du supérieur* (the exam that grants access to the position of university professor in legal studies). The *Revue d'économie politique* was launched in 1887. The prevailing approach in French economics was literary and juridical, focusing on the study of economic ideas and regulations. Despite the appearance of "economic calculus" among mathematicians and engineers, the mathematization of French economics occurred only after 1945 (Fourcade 2009).

During this time period, political science developed primarily outside of academia (where it gained autonomy from the teaching of public law only after the First World War), within the École Libre des Sciences Politiques (ELSP). Founded in 1871, with the intention of offering scholarly education to the new political and administrative elites, ELSP established political science as an applied science of administration rather than a research domain (Favre 1989). Scholarly pursuits remained rare.

The introduction of the scientific paradigm in the faculties of Letters (schools of humanities) was fostered by reorganization of the French universities in 1896 (Charle 1994). At the Nouvelle Sorbonne, the positivist model was promoted by partisans of modernization in three disciplines: history (Charles Seignobos and Charles-Victor Langlois); sociology (Émile Durkheim); and literary history (Gustave Lanson). This model was fiercely combatted by the advocates of the classical humanities, who denounced the "proletarianization" of these disciplines formerly reserved for the elite (Sapiro 2004). Until then, literary studies and history, like philosophy, had principally trained secondary teachers. The *agrégation* existed since 1821 for literary studies, 1828 for philosophy, and 1830 for history.

Illustrative of the nationalization of literature, the *Revue d'histoire littéraire de la France* and the Société d'histoire littéraire de la France (SHLF) were both founded in 1894. History of modern French literature was distinguished from that of "foreign literature," for which specific chairs had been created in 1830, before the establishment of specialized training in German and English languages in the 1840s and in Italian and Spanish at the turn of the twentieth century (Espagne 1993). "Comparative literature" developed by the end of the nineteenth century in France and England, as well as in Hungary, in order to study the reciprocal influence of national literatures. In France, the first chair of comparative literature was created in Lyon in 1896, but it was not until 1925 that a similar chair was established at the Sorbonne. In 1921, a journal was launched by Fernand Baldensperger and Paul Hazard: the *Revue de littérature comparée*, which is still the most important journal in this branch of "lettres modernes."

Alongside learned societies, many professional associations for teaching staff in secondary and higher education appeared, such as the Association des Professeurs de Lettres (APL), created in 1911, which included teachers of ancient languages and literatures (*lettres classiques*) and the new teachers of "French" trained in Modern language departments. The multiplication of such professional organizations was made possible by the 1901 law authorizing associations, which favored the professional development of many occupations under the Third Republic (Sapiro 2006).

Like psychology, sociology was taught as part of the philosophy curriculum. Philosophy had been incorporated into the Faculty of Letters under the First Empire as a way of controlling this discipline, which was suspected to have brought about the French Revolution (Ribard 2000). Most philosophers came from the literary department of the École Normale Supérieure, a state-run elite school first established during the French Revolution and re-founded by Napoleon (1808–1810) in order to train secondary school teachers. The creation of the *Revue philosophique* by Théodule Ribot in 1876 marked the autonomization of the philosophical field from the literary field in that it limited the discussion to peers. However, the rise of these new disciplines (psychology and sociology) challenged the hegemony of philosophy, which had to redefine its domains of inquiry given this new division of intellectual labor. The *Revue de métaphysique et de morale* was created in 1895 in order to reaffirm the supremacy of metaphysics in the face of these new sciences.

The field became roughly divided between metaphysicians, who allied with religion and literature, and philosophers of knowledge who engaged in a dialogue with the natural sciences. Careers were also torn between the figure of the professor, who trains secondary school teachers, and that of the philosopher who produces a system of thought: a paradigmatic example of the latter would be embodied by Sartre who resigned from the educational system to become a writer and independent thinker. While the first option requires passing the *agrégation*, the second option was favored by the rise of a specialized book market. Between 1875 and 1910, the annual number of philosophical books published increased almost threefold (from 50 to 140). This production was divided between textbooks, scholarly books for specialists, and popular science books for a broader lay audience (Fabiani 1988: 23–25). The number of published books in other SSH disciplines also multiplied (with publishers like Alcan).

Until the 1870s and 1880s, when Théodule Ribot imported the model of experimental psychology developed by Wilhelm Wundt in Leipzig, psychology was split between philosophy (derived from Cartesianism or English empiricism through Condillac) and "alienist" medicine (psychiatry), first organized around the Société médico-psychologique (1848) and its journal *Annales medico-psychologiques* (1843). The Laboratoire de psychologie physiologique at the Sorbonne was set up in 1889 and *L'Année psychologique* first appeared in 1894. At the same time, Ribot initiated the school of "pathological psychology," whose main representatives were Pierre Janet and Georges Dumas. The latter founded the Société française de psychologie in 1901 and the *Journal de psychologie normale et pathologique* in 1904, which became a platform for the collaboration between psychologists and researchers in the SSH (Carroy et al. 2006). Although the first French psychoanalytical society, the Société psychoanalytique de Paris (SPP), was founded in 1926, psychoanalysis was never recognized in France as an academic discipline, thus illustrating a case of professional development without state accredited training.

Like psychology, and following its model, sociology developed much more as a scientific field than as an academic discipline during this period. Its main academic agents were disciples of Durkheim, who launched the journal *L'Année sociologique* in 1898. Durkheim's competitors, René Worms, the founder of the first French learned society of sociology—the Société de sociologie de Paris—in 1895, who also launched the *Revue internationale de sociologie* in 1904, and Frédéric Le Play, who conducted several social inquiries, did not hold academic positions. Sociology was taught—combined with pedagogy—as a branch of philosophy in a few Faculty of Letters, and there were only four chairs explicitly dedicated to the discipline until the 1950s. Although Durkheim failed to be nominated for the first chair of "social philosophy" at the Collège de France, a prestigious institution dating back to the Old Regime that offered courses open to a large audience, Marcel Mauss obtained appointment there to the first chair of sociology (1931), having taught the "history of religion of uncivilized peoples" at the 5th Section of the École pratique des hautes études (dedicated to the secular study of religions). In 1924, the Durkheimians founded a learned society, the Institut français de sociologie, which was instrumental in perpetuating the discipline's scientific prestige (Karady 1976; Heilbron 2015).

The boundaries between ethnology and sociology were porous: *L'Année sociologique* offered a space for the dissemination of ethnological knowledge thanks to numerous reviews written by Henri Hubert, Robert Hertz, Émile Durkheim and Marcel Mauss, who also published certain founding texts of the discipline in France such as *The Elementary Forms of Religious Life* in 1913 or *The Gift* in 1925 and participated in the creation of the Institut d'ethnologie de l'Université de Paris in 1925. The institutionalization of anthropology was even more research-oriented than sociology. Chairs of anthropology had been created at the Museum d'histoire naturelle in 1855, and the Musée d'ethnographie du Trocadéro, inaugurated in 1882 (Conklin 2013). Most of the learned societies dedicated to anthropology published their own journals, for example, the *Bulletins et mémoires de la société d'anthropologie de Paris*, devoted to physical anthropology, created in 1859 by the learned society established the same year, or the *Journal de la société des américanistes* first published in 1895 following the foundation of this learned gathering.

The rise of the SSH did not result solely from the autonomous development of these scientific fields, but was fostered by political and governmental demand, which provided career opportunities in certain areas of expertise such as ethnology (another example being the case of demography discussed below). As in the United Kingdom, the French field of ethnology was closely tied to the Third Republic's colonial endeavors. For instance, the Institut d'ethnologie de l'Université de Paris, founded in 1925 by the Ministry of the Colonies, proposed a degree in colonial ethnology studies, intended to train civil servants and produce knowledge about the colonies in which they were stationed. Sites of investigation were limited to those within the borders of the colonial empire. In 1936, in Senegal, the Governor-General of French West Africa (AOF), Jules Brevié, founded the Institut Français d'Afrique noire (IFAN), whose mission was to promote "the scientific study of Black Africa and of French West Africa in particular, its countries, inhabitants, history, evolution, resources and productions"[2] (it was renamed the Institut fondamental d'Afrique noire in 1966, after de-colonization). The Office de la recherche scientifique d'outre-mer (ORSOM) was founded in 1942, under the authoritarian Vichy regime, and renamed ORSTOM in 1944. The IFAN and the ORSOM had to meet the local colonial Governors' demands for knowledge (Gaillard 1989).

The first university chair in ethnology was created in 1943 in Paris for Marcel Griaule. In 1944, an academic tenured position, the 'Maîtrise de conférences' in colonial ethnography was established in the geography department at the University of Lyon and held by André Leroi-Gourhan. Learned societies may no longer hold the driving role they did in the nineteenth century, but they continue to offer opportunities for publishing and intellectual exchange. Anthropological (but also archaeological) research has mostly taken place within public organizations (Campion-Vincent 1970).

The nationalization of French research reached its peak in the 1930s and 1940s. This period is also characterized by increasing direct state intervention in the organization of research under the Popular Front government and then the Vichy regime. In the first weeks of "The

[2]Arrêté 1945/E, 19 August 1936.

Phoney War" of 1939, the Centre national de la recherche scientifique (CNRS) was founded, which eventually became a prominent institutional agent in the French scientific landscape. In the decades following its inception the CNRS, came to play an original and eminent role in the rationalization and professionalization of research for all disciplines as an agency providing research means and as an organization regrouping research institutes, alongside French universities who, unlike their American counterparts, do not benefit from substantial private funding allocated to the development of research centers (Picard 1990). This nascent CNRS, which was first dedicated to the organization of physical sciences, only granted a marginal place to the SSH. The importance of the SSH grew after the war.

Parallel to this nationalization, the process of internationalization was given serious momentum in the interwar period by the International Institute of Intellectual Cooperation of the League of Nations and by the growing role of (mostly American) philanthropic foundations (Renoliet 1999; Fisher 1983; Fleck 2011; Tournès 2011). The intensification of international exchanges fostered the circulation of ideas, institutional models, and funds; a development that supported the academic institutionalization of sociology, political science, economics and psychology.

The Second Phase of Academic Institutionalization or the Golden Age of the SSH

After the Second World War, the SSH underwent a new phase of institutionalization, as well as one of professional and scientific development, due to various factors. The postwar movement of reconstruction mobilized the SSH. The demands of the welfare state's Fourth and Fifth Plans (for instance, studies on cultural consumption in order to foster the democratization of access to culture) alongside the economic demand of companies for the optimization of their organizational models were directed towards psychology, sociology, and economics. The American model replaced the German one. Through funding, philanthropic foundations encouraged the introduction of the scientific paradigm in the SSH and the development of empirical research modeled on the natural sciences all over Europe, in order—besides other targets—to combat Marxism (Pollak 1976). Moreover, during the 1950s and 1960s, the power relations between the natural sciences and the humanities became somewhat reversed; scientific disciplines gained prestige at the expense of the humanities in secondary education. Nevertheless, new career opportunities were offered in the SSH thanks to the development of the CNRS as well as the newly born 6th Section of École pratique des hautes études entirely dedicated to the SSH, the establishment of new state-run institutes and the full institutionalization of the SSH in universities with the creation of new bachelor degrees (*licence*).

In the postwar period, along with the CNRS whose mission was to support fundamental research, three other major state research institutes were founded, which were more oriented towards the production of expert knowledge for use by public administrations and policymaking purposes: the Institut national des études démographiques (INED); the Institut national de la statistique et des études économiques (INSEE); and the Institut national de la recherche agronomique (INRA).

The institutionalization of the science of populations illustrates the interconnectedness of scientific and political interests in a paradigmatic way. During the interwar period, demography held an important place in the political decision-making bodies receptive to pro-natalist and pro-family arguments, especially those supported by the figures at the borders of these two fields, like the parliamentarian and economist Adolphe Landry (Rosental 2003). This science of populations became institutionalized under the Vichy regime with the creation of the Fondation française pour l'étude des problèmes humains in 1941 by Alexis Carrel, a biologist, promoter of eugenics and winner of the 1912 Nobel prize in medicine. Although this research body ceased activity after the war, it provided institutional, scientific, and human resources for the foundation of INED in 1945. The political willpower to reverse the downward trend in birth rates that had characterized the French population since the beginning of the twentieth century indeed contributed to the renewal of the natalist policy put into place during the Vichy regime and inaugurated by the "*Code de la famille*," of which

Alfred Sauvy, statistician and the first director of the INED, was one of the main architects. The attention given to the French population went hand in hand with concerns over their most vital needs during times of food shortage. The organization of a nationally self-sufficient, rationalized and economically viable agricultural system was conducted under the auspices of the INRA founded in 1946 (Cranney 1996). In 1957, Sauvy also founded an Institut de démographie (IDUP) at the Sorbonne.

The INSEE was established in 1946 as a directorate-general of the Ministry of the National Economy in order to provide economic and social indicators. Its designation as an institute was intended to guarantee, in the public's eyes, and in those of possibly reluctant data providers, the statisticians' (at least partial) independence in their work (Touchelay 2000). Armed with its 18 regional-directorates spread throughout the entire territory (except the overseas departments), the INSEE handled administrative, social, macro- and micro-economic information emanating from, and/or requested by, the General Planning Commissioner as well as other social and institutional partners in order to conduct postwar reconstruction efforts under the aegis of Keynesian economics. Thus it offered an opportunity to experiment with survey methods imported from the United States and to initiate the first large-scale surveys. One example is the employment survey carried out at the beginning of 1950, which has been repeated every year since then.

Postwar sociology was also redefined by the importation of institutional models and methodological tools from the US. For example, Jean Stoetzel (1910–1987), inspired by the methods of American empirical sociology after spending time in the United States, set up the first poll institute in France in 1938: the Institut français d'opinion publique. As a consequence, sociology was split into two poles, an "intellectual" pole, under the influence of Jean-Paul Sartre and/or Marxism, and an "empirical" pole, related to the institutes for statistical studies and the CNRS, which created a laboratory, the Centre d'études sociologiques (Heilbron 2015).

Sociology, like other SSH disciplines such as anthropology, history, economics, general linguistics or later area studies, also developed in the

6th Section of the École pratique des hautes études for "*Sciences économiques et sociales*", created in 1947 with the support of the Rockefeller Foundation. First headed by Lucien Febvre, then, from 1956 on, by Fernand Braudel, this Section played a leading role in the development of scientific research in the SSH, fostering interdisciplinarity and internationalization through key figures such as Claude Levi-Strauss, Pierre Bourdieu and Roland Barthes, who would also be appointed later as professors at the Collège de France. In 1975, the 6th Section became an independent postgraduate school as École des hautes études en sciences sociales.

The 1950s and 1960s were also marked by the establishment of distinct bachelor degrees (*licence*) in several SSH disciplines, which thus became full-fledged university disciplines with specialized study tracks leading to a doctorate. In 1957, the law faculties became "*facultés de droit et de sciences économiques*", and the following year the "*facultés de lettres*" were renamed "*facultés de lettres et sciences humaines.*" In 1959, the latter amounted to fewer than 1000 tenured teaching personnel (*professeur*, *maître de conferences*, *maître-assistant* and *assistant*), a number that had almost quadrupled by 1968 (3908, of which 10% were full professors), and continued to grow until 1973–1974, when the numbers reached around 8000 (of which a quarter were full professors), but which stagnated for the next ten years (see Fig. 2.1). Within these faculties, the relative share of the "human sciences" (psychology and sociology) started to increase at the expense of philosophy, much like that of Modern history compared to Ancient history, or Modern literature compared to Ancient literature, indicating a growing interest in the study of contemporary societies (Soulié 2013; see Fig. 2.2).

Initiated by Raymond Aron (1905–1983), a *licence* in sociology was created in 1958, ten years after the *licence* in psychology (1947). In 1969, along with demography, sociology obtained a section of its own in the national body in charge of university professors' careers, the Comité consultatif des universités (now the Conseil national des universités), which guaranteed its institutional autonomy from philosophy. At the same time, the teaching and research staff in universities and the CNRS grew considerably (Heilbron 2015). The *licence ès sciences économiques* was established in 1959. In secondary education, a new

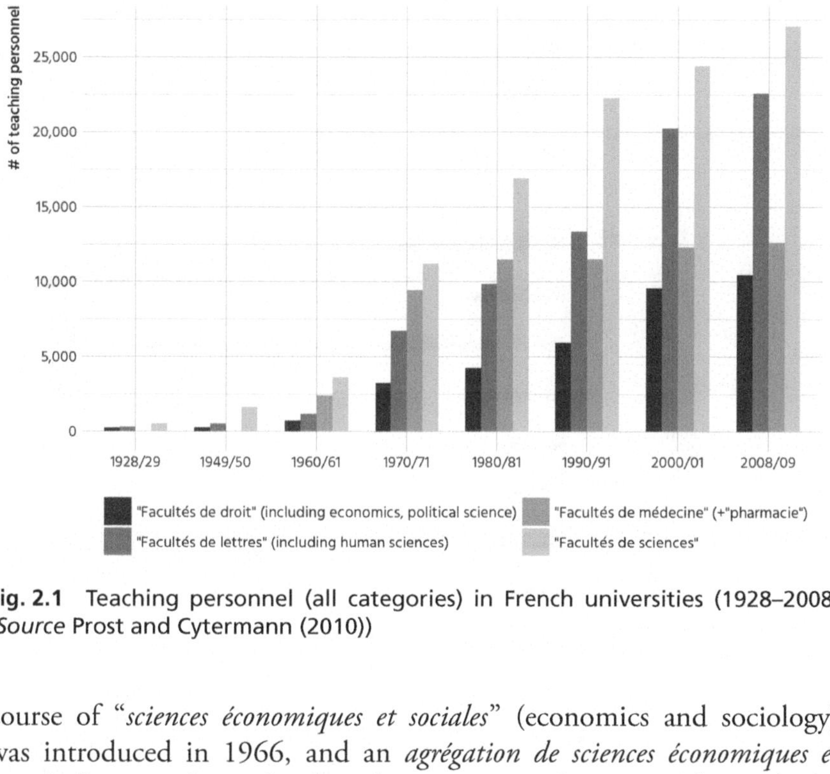

Fig. 2.1 Teaching personnel (all categories) in French universities (1928–2008) (*Source* Prost and Cytermann (2010))

course of "*sciences économiques et sociales*" (economics and sociology) was introduced in 1966, and an *agrégation de sciences économiques et sociales* for secondary school teachers was created in 1977, alongside the *agrégation de science économique* for university professors which, as previously mentioned, had been an area of specialization within the law *agrégation* since 1896. Undergraduate training in political science in Law faculties began in 1954. Two years later, in 1956, law faculties started to award a doctorate in political science and a graduate degree (DES or *Diplôme d'études supérieures*). In 1971, a separate *agrégation* of political science was set up, providing full professorships in universities.

The first *licence* in French language and literature was established in 1943 (*langue et littérature française*), but it did not give access to teaching in high schools. It was redefined as "*licence de lettres modernes*" in 1947 and the *agrégation de lettres modernes* was established in 1959, allowing professors of Modern literature to teach at the secondary level and in universities. The number of "licences" increased fivefold between

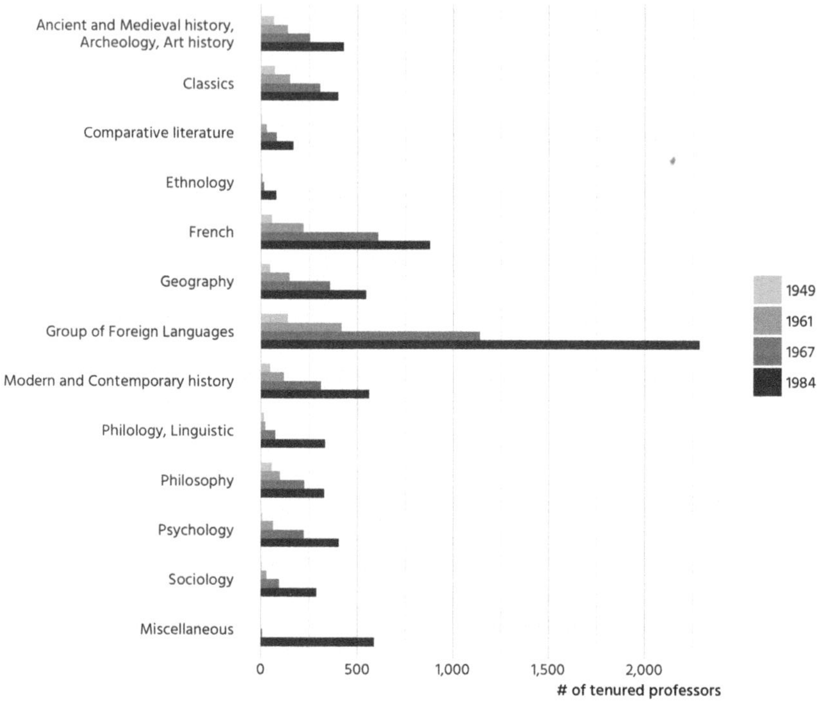

Fig. 2.2 Number of tenured Professors (senior and juniors) in the faculties of "Lettres et sciences humaines", per discipline (1949, 1961, 1967, 1984). We gathered under «languages» all foreign and regional languages (*Source* Soulié (2013))

1959 and 1965, and the number of diplomas conferred continued to grow until the mid-1970s. But, despite its increasing numerical superiority, modern literature remained inferior to classical literature in academic hierarchies for a long time. The monopole of the powerful Société des professeurs de français et de langues anciennes SPFLA (1946) over professional issues in secondary and higher education has been challenged by the AFEF (Association française des professeurs de français, first AFPF), established in 1967.

The professional development of the social sciences after 1945 can also be observed through the creation of associations, which mostly follow the American model. The Association française de science économique (AFSE) was born in 1950, following the example of the American Economic Association. Specialists of certain domains also gathered in associations to promote their research area, such as the Association française des économistes d'entreprise (AFED), established in 1969, but which remained quite small (around 150 members) in comparison to its American counterpart, the National Association of Business Economists.

Political science became more professionalized when the Association Française de Science Politique (AFSP) surfaced in 1949. French political scientists also proved active on the international stage through their early involvement in the founding of the International Political Science Association (IPSA) in 1949, in a context which saw the newly born UNESCO encourage the establishment of political science as an academic discipline in all countries in order to help pacify international relations.

Born in 1962, the Société Française de Sociologie (SFS) made several attempts at creating a vocational degree in sociology. However, in France the title of "sociologist" has been largely monopolized by scholars, unlike other countries where professional sociologists working outside academia (in companies or in consulting) are organized. The Société Française de Sociologie also failed to unify the discipline, despite its quick expansion throughout universities and within the CNRS, due to theoretical cleavages but also given the fact that, for a long time, sociologists came from different kindred disciplines. It was not until 2002 that an Association Française de Sociologie was founded, mostly composed of academics and Ph.D. students.

While the model of professional organization circulated from the United States to European countries, fostering the organization of research within national borders, the competition and changing power relations between countries were expressed in international associations. For example, in 1958 George Gurvitch (one of the two professors of sociology at the Sorbonne) founded the Association Internationale des Sociologues de Langue française (AISLF) in order to challenge the International Sociological Association, founded in 1948 with the support of UNESCO, which he considered to be over influenced by American sociology. The founding of IPSA was also marked by competition between French and British political scientists (Boncourt 2011).

The rise of relatively autonomous scientific fields within the SSH is best expressed by the mission and numbers of newly established journals. Between 1945 and 1975, many SSH journals were founded, most of them disciplinary, like *Revue économique* (1950), which is still one of the major journals in France, along with *Annales de l'INSEE* (1969) and *Revue française de science politique* (1951). At least six journals of sociology were launched (or relaunched) during this period: *Cahiers internationaux de sociologie* (1946); *L'Année sociologique* (1949); *Archives de sociologie des religions* (1956); *Sociologie du travail* (1959); *Revue française de sociologie* (1960); and *Archives européennes de sociologie* (1960). In 1975, Bourdieu started *Actes de la recherche en sciences sociales*, which promoted a unified conception of the social sciences, including political science, anthropology, and history (contrary to the presentism of mainstream social science in the United States), as well as literary studies and art history. During this period, the sociological field was divided into three competing schools led by Raymond Boudon (rational choice theory), Pierre Bourdieu (genetic structuralism, theory of social reproduction), and Alain Touraine (interactionist theory of action, social movements).

Within the very diverse editorial landscape of anthropology, where journals are often thematically organized around geographical areas, *L'Homme*, launched in 1961 by Claude Levi-Strauss and Émile Benveniste, has the largest disciplinary scope. An exile in the United States during the WWII, Lévi-Strauss taught at the École libre de hautes études and the New School for Social Research, where he discovered empirical Anglo-American anthropology and structural linguistics. In 1948, after he returned to France, he joined the staff of the 6th Section of the École pratique des hautes études, and in 1959 he was elected to the first chair of anthropology at the Collège de France. His journal *L'Homme* was modeled after *American Anthropologist* and the British journal *Man*, with the interdisciplinary aim of establishing links between anthropology and geography, linguistics and prehistory (Lévi-Strauss and Eribon 2001). *Ethnologie française*, the organ of the Société d'ethnologie française, was first published in 1971 after decolonization, which compelled anthropologists to redefine their discipline and objects of study.

In addition to journals, in the 1960s and 1970s the rise of the SSH was accompanied by a significant increase in book production. The creation of series of SSH books with trade publishers, edited by charismatic figures, which had an audience not only among the growing SSH student population but also outside of the academic field, was specific to France and Italy. Editorial interest in the SSH had already arisen in the first half of the twentieth century but developed mainly in specialized sectors around small scholarly publishing houses like Alcan, Vrin and Payot. Gallimard, the most prestigious French literary publishing house, began translating Freud and releasing Sartre and Merleau-Ponty in the 1930s and 1940s, and it started its first series, "Bibliothèque de philosophie," as early as 1950, with Sartre and Merleau-Ponty as editors. Two years later, Éric de Dampierre launched "Recherche en sciences humaines" at Plon. It was not until the middle of the 1960s, however, that this trend became significant with the creation of "Le champ freudien" by Jacques Lacan at Le Seuil in 1964, "Le sens commun" by Bourdieu at Minuit in 1966, and "Bibliothèque des sciences humaines" by Pierre Nora at Gallimard, also in 1966. This trend stimulated competition between publishers, especially between Gallimard, who entrusted Pierre Nora with another series devoted to history in 1971 ("Bibliothèque des Histoires"), and Le Seuil, who launched two new series, the first for poetics ("Poétique"), edited by Todorov and Gérard Genette in 1971, and the second for history ("L'Univers historique"), edited by Jacques Julliard and Michel Winock in 1973.

The academic institutionalization of the SSH thus partly overlapped with the rise of scientific fields in these disciplines, their professional development and the structuring of a specialized book market. Professionalization took different paths: secondary teaching for the humanities and to a certain extent for economics and sociology, applied sciences for economics, sociology and empirical psychology (through research and consulting), administration for political science (but also economics), and liberal professions for law and clinical psychology.

Morphology of the SSH in France Since the 1980s

After the second phase of its academic institutionalization and the 1969 University Reform, which phased out faculties and chairs and created disciplinary departments (the Sorbonne was divided in thirteen universities and many new universities were established in the provinces), the SSH became a very significant part of the French academic field. The rules that currently determine the organization of the profession of researcher were established in broad strokes in the 1980s. In 1984, CNRS researchers became civil servants, including two levels: *chargé de recherche* (junior) and *directeur de recherche* (senior). The same year, the status of university professors was redefined as "*enseignants-chercheurs*" (lecturer-researchers), reorganized in two levels instead of four: "*maître de conferences*" (junior) and "*professeur*" (senior). In theory, they divide their time between teaching and research, along with administrative tasks, which are increasing at the expense of research (Faure et al. 2006). Being civil servants, French academics are recruited through a competitive procedure, and tenure is almost automatically secured one year after being appointed as *maître de conferences* or *chargé de recherche*.

In 2000, the SSH accounted for one-third of all permanent professors and researchers in France (18.6% for the humanities, which was the second largest group of disciplines just behind the life sciences, and 13.2% for the social sciences, which came third (Barré et al. 2002: 12)). Most SSH scholars are located in universities and other higher education institutions such as the École des hautes études en sciences sociales and the École pratique des hautes études—in 2000, the SSH accounted for 37% of all permanent (tenured) professors (junior and senior altogether)—but they can also be found in research institutions (Barré et al. 2002). In 2007, the SSH academic sector was composed of 22,692 professors and 6032 researchers (junior and senior altogether); there were 36,628 Ph.D. students in the SSH in 2009, and 36,226 in 2015, amounting to almost half of the overall number of Ph.D. candidates in France (Weil and Cases 2012: 5). This section presents a morphology of the contemporary French academic field, tracing the principal evolutions since the 1980s and those that occurred in the years 2000 as a result of the Bologna process, which sought to harmonize and remodel the European academic landscape.[3]

[3]The data concerns comparable series from available official sources that are cited for each figure. For university professors, comparable series start in 1984, when the status of "enseignant-chercheur" was established.

In the survey that he conducted on the academic field in the 1960s, Bourdieu (1984) observed an opposition between a pole oriented towards reproduction (teaching, producing textbooks) and a pole oriented towards research (publishing research results in scientific journals and in the form of books). The first is mainly situated in universities while the second developed in research institutes like the CNRS, or at the École pratiques des hautes études and the École des hautes études en sciences sociales (EHESS). Although partly blurred, this opposition still structures the SSH landscape in France.

Currently there is a set of disciplines primarily based in the university system. These disciplines are composed by a high number of university professors, compared to the very small number of scholars working in research institutes like the CNRS. This is typical of disciplinary domains in which research is a minor activity compared to professional practice, like law or management. The different institutional anchorage corresponds, in certain cases, to an internal division within the discipline with respect to its orientation and approaches, as is the case for psychology: while "experimental psychology" and, since the 1980s, cognitive sciences, prevail at the CNRS, most university students in psychology study "clinical psychology."

With an emphasis on teaching, the humanities are also based mainly in universities. In 2000, professors in the humanities accounted for 22.2% of all permanent university professors (junior and senior altogether), thus forming the largest group of disciplines, followed by the social sciences with 15.1%, with medicine and life sciences joint third with 12% each (Barré et al. 2002: 12). The oldest and most important disciplines in the humanities—literary studies and philosophy—are traditionally organized around the *agrégation*, the national competition for the position of *professeur agrégé* in secondary schools; an informal (though not mandatory) requirement for access to higher education. This is also the case with history, although it developed more as a social science in France after the war, especially at the École des hautes études en sciences sociales, thanks to the *Annales* school.

The newer disciplines in the social sciences, such as economics or sociology, are split between research-oriented institutions, in which they initially developed, and universities, in which they were fully established as autonomous academic disciplines in the 1950s and gained a significant place thanks to the expanding undergraduate curriculum. This double anchorage explains the fact that, unlike the humanities which are more oriented towards academic careers in the universities, since the mid-1980s the CNRS and the EHESS have been regarded as the most prestigious places to work in sociology. Some of these disciplines now play a relatively minimal role within universities. This is the case for anthropology, which is often taught in sociology departments. This is also the case for political science, traditionally taught as part of the law curriculum in French universities and primarily attached to the Instituts d'études politiques, which developed following the model of Sciences Po, after it gained public status in 1945 (see below).

Though the disciplinary model is still dominant at the universities, it started to be challenged in the 1970s and 1980s by the rise of vocational training within universities (management, for example), and by the development of "studies" based on interdisciplinary approaches. Area studies emerged in the 1950s–1960s in the context of the Cold War (Popa 2015), theater, arts, and cinema studies developed after 1968 (theater studies were established even earlier at the Sorbonne (Fondu 2014)), and European studies appeared in the 1990s as master's programs. The institutionalization of gender studies is more recent in France, despite the creation by Hélène Cixous of a research center for women's studies at the University of Vincennes (Paris 8) in 1974, now known as the Center of Women's and Gender Studies, and of two other centers in 1985 at the University of Paris 7-Denis Diderot and Toulouse University, as well as specialized journals. Seven other research centers for "gender studies" were founded between 1999 and 2015, some of them with the support of the CNRS and other research institutions such INED. Gender studies started to develop in universities, especially in master's programs: some universities—Paris 8, Paris 7, Toulouse, Bordeaux and Lyon—offer their own master's degree, while in other places, such as the École des hautes études en sciences sociales, it is an area of specialization within another discipline (in this case sociology, but it is about to become autonomous). However, in what follows we will focus on the traditional disciplines, especially the seven studied in the Interco-SSH project.

Morphology of the Disciplines at the Universities

The current size of disciplines, measured by the number of university professors, reflects first and foremost their seniority at the university: law, the oldest, as well as literature and history, which developed at the end of the nineteenth century, oppose in this regard political science and ethnology, disciplines that were institutionalized within the academic world more recently. Thus, while the section entitled "private law" at the Conseil national des universités grouped together approximately 1800 professors in 2015, there were less than 200 in the section that includes prehistory, biological anthropology and ethnology. But seniority is a factor counterbalanced by that of the presumed "social utility" in a given configuration of relations between the academic field and the field of power, as evidenced by the importance of economics (which differentiated itself from law) whose headcount reaches 1800 professors, outweighing philosophy (400), a much older discipline (see Fig. 2.3). Still, university economists constitute the dominated pole in the field of economists in France compared to "engineer-economists" trained in elite schools (École Polytechnique and/or École Nationale de la Statistique et de l'Administration Économique, or ENSAE) and belonging to the "grands corps" of civil service, that is to say the higher administration (Lebaron 1997, 2000).

In keeping with the observed trends in the years 1950–1960 (Soulié 2013), the relative share of literary disciplines decreased between 1984 and 2013. For example, while the number of professors in all SSH sections combined increases, the number of professors in the French language and literature section has remained the same, at about 1000 professors, since the end of the 1990s, that is to say a decrease in their relative share by 4.3 points between 1984 and 2013. This decrease also concerns Anglo-American languages and literature (−3.4 points), classical languages (−2.6 points) and philosophy (−2.2 points). In the case of classical languages and literature, the total number of professors has decreased continuously since 1984.

On the other hand, the number of university professors in interdisciplinary studies and/or applied fields, like management (whose numbers have multiplied by 5 between 1984 and 2013), education

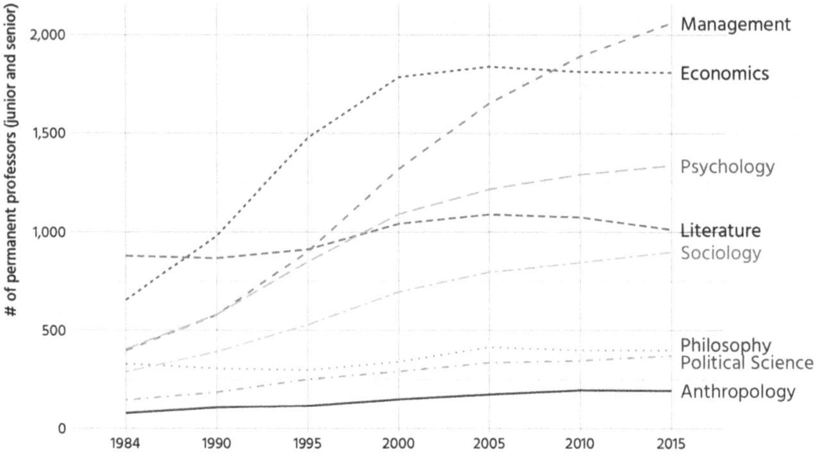

Fig. 2.3 Number of permanent (junior and senior) professors in French Universities in the seven SSH disciplines, compared with management (1984–2015) (literature = French language and literature; sociology = sociology, demography; anthropology = ethnology, prehistory, biological anthropology) (*Sources* For Years 1984–1990, Prost and Cytermann (2010); for years 1995–2015, Gesup, DGRH A1-1, French Ministry of National Education, Higher Education and Research)

sciences, or even communication studies have increased. Psychology and sociology sections have also experienced a relatively significant growth since the 1980s (the number of professors has multiplied almost threefold), while the number of professors in economics, which experienced major growth until the beginning of the 2000s rising from 650 professors in 1984 to 1834 in 2002, has since stabilized or even slightly decreased.

The differences in numbers of professors according to the disciplinary sections within universities do not merely reflect the power relations between disciplines, however. The disciplines' positions relative to each other in the broader scientific field and their scientific or vocational appeal at a given moment must be equally taken into account. For instance, while management has become attractive from a vocational standpoint, ethnology lost the scientific aura it had at the end of the 1950s following the high recognition achieved by Levi-Strauss' structural anthropology, which offered ethnology a theory that freed it from the colonial context in which it was embedded in France. The same holds for linguistics, which had served as a model to structuralism.

Ethnology as a discipline found refuge at the EHESS where it rose from 9 to 17% of chairs between 1975 and 2016, which places it second behind history (which went from representing 30–41% of chairs during the same period).[4] This growth is all the more significant given that the number of professors at the EHESS went from 149 to 240 between 1975 and 2015, and considering that other disciplines were welcomed there during the contemporary period, namely law, philosophy (in particular analytical philosophy, still marginally present in France) and cognitive psychology (which replaced social psychiatry), whereas linguistics henceforth has very few representatives. Sociology comes third at the EHESS with 12% of chairs in 2015, just like in 1975,[5] while it experienced significant growth in the university system. However, it remains "temporally dominated" (in Bourdieu's sense) by economics, on the one hand, and political science on the other.[6] These developments contradict Foucault's predictions (1966) which, in light of the success of structural linguistics and anthropology, heralded the death of man in the episteme and a return to language as a non-transparent signifier in the postmodern era.

It should also be noted that a small number of professors in a discipline can also be a sign of elitism, as is the case for philosophy which, despite being dethroned from its status of the "queen-discipline" (Fabiani 1988) by the development of the SSH since the 1960s, has retained its prestige. The share of full professors within the body of professors is noticeably larger here than in the other disciplines, and they are heavily concentrated in Parisian universities, which are the most renowned.

[4]Data kindly provided by Pierre-Cyrille Hautcoeur, president of the EHESS, who conducted these tallies. The count that we made based on the school's website in 2016 confirms the 2015 data.

[5]This proportion increased at the end of the 1980s and then decreased (Backouche and Topalov 2008).

[6]As evidenced by the recruitment in sociology, which is more open to degree holders in political science than political science is to sociologists, and the fact that the calls for tenders that finance research increasingly promote socioeconomic analyses.

The Number of Students and Degree Holders in France Since the End of the 1990s

The differential growth in the number of professors across disciplines corresponds to the evolutions in the number of students enrolled in each of them. Although the overall number of students in the SSH grew 60% between 1970 and 2001 (Weil and Cases 2012: 6), this evolution is not homogeneous among the disciplines. A marked increase can be observed in the relative share of the number of diplomas recorded in the areas of management science, information and communication sciences, education sciences and political sciences in the SSH between 1996–2015. For management sciences and the information and communication sciences, this increase mainly concerns undergraduate "vocational" degrees (which accounted for 47% of students majoring in management science in 2014–2015). That is to say that the varying progression in numbers of university professors is less an expression of the relative scientific prestige of disciplines, than of the rise of vocational tracks and degrees (Gingras and Gemme 2006). While in science, technology, and health, close to half of all master's programs combine research and vocational orientation, in 2013 this was only the case for 15% of master's programs in the SSH. However, vocational master's degrees multiplied in the SSH, attracting many students, cultural management being a typical example (Dubois 2013).

Conversely, the majority of the more classic SSH disciplines—like history, economics, psychology, French literature, and philosophy—now see their role diminish in proportion to the amount of degrees issued in the SSH tracks, as well as in proportion to the number of enrolled students. Two types of disciplines can nevertheless be distinguished here. In the case of economics, psychology, French literature, or even, albeit to a lesser extent, sociology, we are dealing with tracks that, despite being proportionately in decline, still attract a significant number of students. In the 1980s and 1990s, during the so-called second "massification" of higher education, these disciplines served as entry tracks (Chenu 2002). A large proportion of the degrees issued in these disciplines are to undergraduates. Conversely, disciplines such as anthropology and political science are characterized by small numbers of undergraduate students and by the fact that they issue a large proportion of their degrees to master's students. This explains that the numbers of university professors are relatively large in the first case and limited in the second. Thus, during the academic year 2014–2015, undergraduates accounted for 70% of the overall number of students enrolled in history, French literature, psychology, and sociology tracks, but only 14% of students in political science and 36% of students in ethnology, prehistory or biological anthropology. Anthropology and prehistory, like classical languages and literatures or even religious studies, now constitute essentially specialization tracks for graduate students: 37% of the students in anthropology and prehistory are enrolled in master's programs and 26% in doctoral programs (see Fig. 2.4). Graduate training in the social sciences is also offered by the École des hautes études

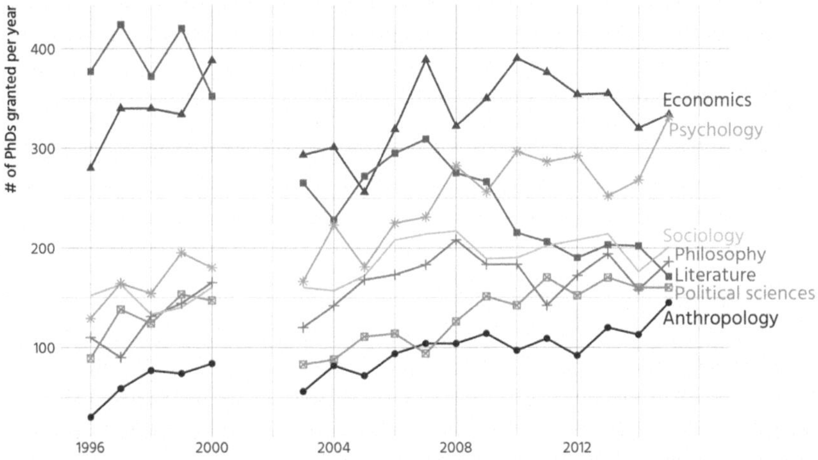

Fig. 2.4 Number of Ph.D.s granted per year and per discipline, 1996–2015 (literature = French language and literature; sociology = sociology, demography; anthropology = ethnology, prehistory, archeology). The 'École des hautes études en sciences sociales' (EHESS) is not taken into account. As a result, the number of Ph.D.s granted in sociology and anthropology/ethnology is underrepresented. For the record, from 2006 to 2015, EHESS awarded an average of 40 Ph.D.s per year in each of these disciplines (*Source* French Ministry of National Education, Higher Education and Research, Système d'information sur le suivi de l'étudiant (SISE). *Scope* French universities and the 'Institut d'études politiques')

en sciences sociales: from 2006 to 2016, 4152 students earned master's degrees, and 2120 earned Ph.D.s, more than half of them in history (22%), sociology (17%) and anthropology (16%).

It was not until the end of the 1990s that an undergraduate curriculum in political science independent of the law curriculum was introduced in some universities, like the University of Paris 8. This limited supply of undergraduate programs is related to the history of the institutionalization of the discipline in France. After the nationalization in 1945 of the École libre de sciences politique and the creation of the Institut d'études politiques (IEP) in Paris (also called Sciences Po Paris), many other IEPs have opened: in Strasbourg in 1945, Bordeaux, Toulouse, Grenoble and Lyon in 1948, Aix-en-Provence in 1956, Lille and Rennes in 1991 and St. Germain-en-Laye in 2013. Today, along with the IEP in Paris, the IEPs still attract more than 60% of all students in political science in France.

Finally, the position held by different disciplines in the SSH within the space of higher education is also related to students' social properties. Compared to the other SSH disciplines, political science and philosophy have a larger proportion of students having earned their *Baccalauréat* one or several years in advance as well as a larger proportion of students enrolled in an institution of higher education based in Paris, which points to elitist recruitment given the high concentration of the most prestigious institutions in Paris. On the contrary, among students in sociology, psychology or management, there is a larger proportion of students who earned their *Baccalauréat* one or several years late and/or have a technical or vocational *Baccalauréat* (*vs* general), an indicator of disadvantaged social origins. Large disciplines, based on the number of students (management, psychology), can be distinguished from "dominant" disciplines in terms of academic and social recruitment (political science, philosophy, or classical languages and literatures) (Renisio 2015). These dominant disciplines also attract more men, although other principles of division may also be adduced (and in particular the gendered opposition between mathematical disciplines and literary ones) as detailed below.

Research Institutions

Although they are primarily located in universities and in other institutions of higher education like the EHESS and the EPHE, the SSH are also present within research institutions. In the case of the humanities, this largely concerns the CNRS (where they accounted for 11.5% of researchers in 2000) and the IRD (Institut de Recherche pour le Développement, formerly Orstom, where their share is more limited, with 4.6% of the personnel). As for the social sciences, they are divided between several institutions: the CNRS (7.3% of all researchers); the IRD (22.2%); INRA (8.5%); and INED (100%) (Barré et al. 2002: 12). Indistinguishable from sociology in the university system, demography is closely associated with the Institut national des études démographiques (INED), which hosted 54 permanent researchers in 2014. As for economists, they are very present within the Institut national de la statistique et des études économiques (INSEE), whose headcount (all categories of personnel included) skyrocketed after 1960, going from 2500 to 5600 in 1976 (Fourcade 2009). Currently, INSEE employs 5770 people, 1515 of whom belong to category A, which groups together "administrators" and "attachés-statisticiens."[7]

At the top of the hierarchy of research institutes that recruit personnel in France stands the Centre national de la recherche scientifique (CNRS). Periodically evaluated by interdisciplinary sections (but where disciplines retain their identity), these researchers are affiliated with research centers (laboratories). Following the project to draw the CNRS and public universities closer together, these laboratories—more than 300 in the SSH at the beginning of the 2000s—have become, for the most part, mixed units under the jurisdiction of two authorities that integrate, in addition to researchers, "engineers" and technicians from the CNRS proper, and a large share of university professors, to whom the CNRS offers fellowships for one to two years. As a result of a voluntary policy implemented by the CNRS administration, the number of research centers dwindled to 194 in 2010 (through merging or removal).

[7]Source: 2015 INSEE activity report.

Until the reforms linked to the Bologna Process, the CNRS was a major actor in structuring French research in SSH via its cross-disciplinary organization, the assessment of laboratories, the evaluation of most of the scientific journals (for ranking, funding and giving the label of the CNRS), as well as the funding of a certain number of research programs (in addition to the recurring credits to labs) and of academic events (conferences, exhibits, and so on). It was largely deprived of this function with the creation of two new agencies in 2007: the Agence nationale pour l'évaluation de la recherche de l'enseignement supérieure (AERES), which was replaced by the Haut Conseil de l'évaluation de la recherche et de l'enseignement supérieur (HCERES) in 2013; and the Agence nationale pour la recherche (ANR) in 2005, which finances projects that meet open, thematic or international calls. Yet, the share of SSH in funding remains rather low.[8]

Likewise, within the CNRS, although the department of Human and Social sciences (renamed the Institut des sciences humaines et sociales in 2008) is one of the largest, the position of the SSH has been regularly challenged since the postwar years (particularly at the beginning of the 1970s following a report by the OCDE (Martin 2002)). It is also, unsurprisingly, the least endowed financially.[9] Currently the number of researchers in the SSH at the CNRS is decreasing (from a peak of 2200 in 2002 to 1648 in 2014). Their share in the scholarly staff of the CNRS decreased from 19% in 1999 to 16% in 2010 and 15% in 2014. These researchers are concentrated in the Paris region: this was the case for two-thirds of them in 2015 (with Lyon and Aix-Marseille following far behind).[10]

---

[8]In the first five years, SSH projects were financed up to 175 M€ all categories combined (close to half of them having responded to the open call) (Weil and Cases 2012, 33). In 2015, the 35 projects selected in the frame of the "challenge" "Innovative, integral and adaptive societies" (where most of the projects proposed by researchers in the SSH were submitted), represented only 5.3% of all projects accepted in 2015 (665) and, in terms of financing, 3.4% of funding for the generic call for projects, that is to say about 10 million euros out of 300 million (outside co-financing). The success rate of projects submitted by sociologists and political scientists was markedly lower than that of projects submitted by economists. Data presented by the head of the ANR's SSH department François Héran, in the report of the Conseil scientifique de l'INSHS, "Analyse des résultats des projets SHS suite aux AAP "défis sociaux" de l'ANR pour 2015-2016 et propositions de réformes," February 1st 2016 [online].

[9]*Le CNRS en chiffre*, CNRS, 2000, p. 6.

[10]See the INSH's website: http://www.cnrs.fr/inshs/presentation/carte_chercheurs.htm.

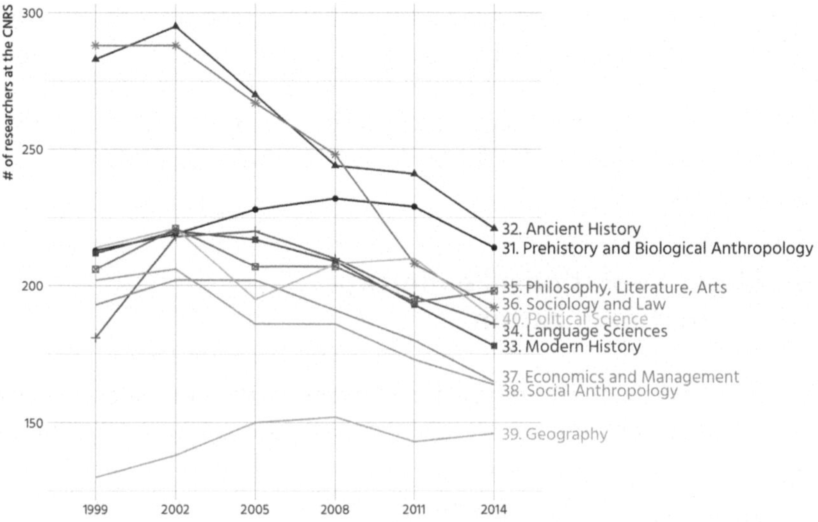

Fig. 2.5 Number of researchers at the CNRS per SSH sections (1999–2014) (*Source* CNRS annual employee reports)

CNRS researchers belong to multidisciplinary sections. If one observes the evolutions of scholarly staff by "section", it appears that sections that together host the largest clusters have experienced the biggest decrease since the 2000s (see Fig. 2.5). Such is the case of the 36th section of the CNRS, for example, which groups together sociology with legal studies, and which, until the beginning of the 2000s, formed one of the most important SSH sections numerically speaking (such was also the case with the "ancient and medieval worlds" section). This decrease results from a proactive policy implemented by the CNRS in order to promote new disciplines such as cognitive sciences.

Can we draw the conclusion that sociology, an academic discipline that developed post-1945 in the 6th section of the EPHE and at the CNRS, has become primarily a discipline for instruction in universities, resembling the first model evoked earlier? Indeed, in the university system, the figures grew significantly, going from less than 300 tenured professors in 1984 to almost 900 in 2015. It has been proven that sociology at universities played for many students the unofficial role of preparing them for admission into schools of 'social work' (Convert 2010). Nevertheless, sociology must be counted among the SSH disciplines that enrolled the most doctoral students in 2014–2015 (almost 2000), more even than economics or psychology (1920 and 1800 respectively). One can see in this a sign of this discipline's prestige in the scientific field.

Two Major Evolutions: Feminization and Internationalization

One of the major morphological evolutions of the contemporary academic field is the process of feminization, more pronounced in the SSH than in the natural sciences. The overall share of women among the teaching personnel in higher education in France was of 6% in 1946, 18.7% in 1968, 29.3% in 1981, and 27% in 1992. In numbers, it means a multiplication by 50 (from 182 in 1946 to about 10,000 in 1991). Focusing on the faculties of "*lettres*," the share of women in 1946 was similar to the average (6%). It was in 1934 that a woman was appointed to a chair in a faculty of "*lettres*" for the first time, but the feminization was interrupted by the Vichy regime, and until 1947 there were no women in the Paris faculty. At that time, half of the students in the Paris faculty of "*lettres*" were female. In 1967, women accounted for 7.5% of *professeurs* and *maîtres de conferences* in this faculty, but 21.6% of the *maîtres-assistants*. In 1981, 16.4% of the *professeurs* and 37.9% of the *maîtres de conférences* in the faculty of *lettres* were female, a percentage that rose to 23.7 and 41.4% respectively in 1992 (Charle 1999).

The process of feminization accelerated in the 1990s, affecting all SSH disciplines, although they are unequally feminized. Between 1999 and 2013, the share of women went from 25 to 42% among all professors (junior and senior) in sociology, 20–30% among those in political science, 25–34% among those in economics, and 46–56% among those in psychology. Within the department of SSH at the CNRS, which is the most feminized, the percentage of women went from 41.7% in 2001 to 46.5% in 2014, compared to 30.6 and 33% respectively for all researchers at the CNRS. This evolution is therefore faster

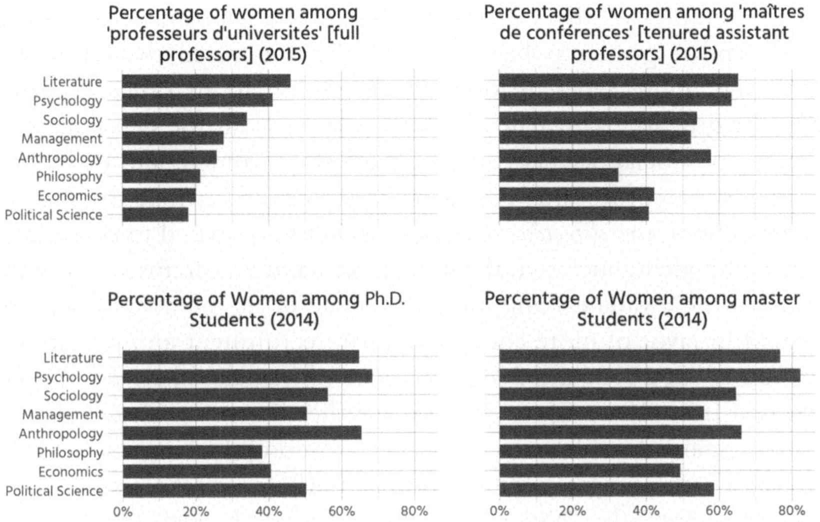

Fig. 2.6 Percentage of women among full professors, tenured assistant professors, Ph.D. students, and master students in French universities. (literature = French language and literature; sociology = sociology, demography; anthropology = ethnology, prehistory, biological anthropology). (*Sources* For professors, GESUP, DGRH A1-1, French Ministry of National Education, Higher Education and Research; for students: DGESIP/DGRI, French Ministry of National Education, Higher Education and Research, SISE)

than for the CNRS where, since 1991, the share of women has barely increased from 30.1%.[11]

However, one finds that the higher one moves up the hierarchy of professional (or educational) careers, the smaller the share of women. In other words, the share of women is lower at the senior level than at the junior level in all institutions (university, CNRS, EHESS), as it is lower among the doctoral candidates than among the students in master's programs (see Fig. 2.6). Moreover, males tend to become full professors earlier in their career, though the tendency is for the gap to diminish.

[11]Annual employee report of the CNRS, 2001 and 2014.

The examples of the CNRS and the EHESS illustrate the impact of an explicit gender policy.[12] The CNRS pioneered this evolution, but it had more repercussions in the SSH than in other domains. Consequently, in 2014, half of the researchers from the department of Human and Social Sciences at the CNRS were women, and their share reached 39.6% of research directors. At the EHESS, where the share of women among the *directeurs d'études* is smaller compared to other institutions of higher education, the more recent and less dramatic evolution is due to the awareness and action of some of its members (men and women) in favor of more equality both in recruitment and promotion: between 2009, 2012, and 2015, the share of (female) *directrices d'études* increased from 19 to 25 to 30%, while the share of *maîtresses de conférences* remained stable at around 40%.[13] Among the "*directeurs d'étude cumulants*" (meaning a joint position at EHESS for scholars holding a position in any other higher education or research institution), which is even more selective given the smaller number of positions offered, an evolution can also be noted, from 3 women (5%) in 2009 to 5 (9%) in 2010, and then to 9 (16%) in 2012, the number stagnating to 8 between 2013 and 2015 (15–16%).

Beyond the gendered divide between the natural sciences and the SSH, the share of women is also unequally distributed across disciplines within the SSH. One can compare the most "masculine" disciplines: philosophy, economics and political science, where in 2013 women made up between 27 and 34% of all university professors (junior and senior altogether), to the most "feminine" ones: psychology and French literature, where women represent more than 50% of university professors (junior and senior). In management, sociology/demography and ethnology the ratio is around 40%.

This distribution also corresponds to the ratio of women among students enrolled in Ph.D. programs between 2007 and 2013: their share is

---

[12]It is not a matter of affirmative action, but the growing awareness of discriminatory mechanisms against women.

[13]The analysis of the age pyramid confirms this analysis, the distribution between men and women being more equal among the youngest. See the EHESS's annual employee reports from 2010, 2014, and 2015.

the highest in psychology (69%) and literary studies (68%), followed by anthropology (62%), sociology (54%), political science (45%), economics (41%) and philosophy (39%). A similar opposition can be observed at the largest graduate school in the social sciences in France, the EHESS (see above) where between 2006 and 2015, 59% of master's students enrolled were women, whereas only 54.6% of those who earned a master's degree were women. Their share was slightly smaller—55%—among the enrolled Ph.D. students during the same period,[14] and the rate fell to 51.3% of those who obtained their Ph.D. These figures illustrate once again the narrowing of the percentage of women towards the top of the academic hierarchy. The most feminized disciplines are literary studies (women account for 70% of second year master's students) and anthropology (two-thirds in M2 and in Ph.D.), followed by sociology (60% in M2 and in Ph.D.). The most masculine disciplines are political studies, economics and philosophy where between one-third and 40% of students are women. The share of female Ph.D. students in economics has increased, however, from one-third in 2006 to half in 2015–2016. History is in the middle with half of female students enrolled in M2 and Ph.D. programs.[15]

This division reflects gendered stereotypes, which oppose female literary critics and male scientists since the 1960s, when scientific tracks gained a dominant position. This gendered division also tends to exclude women from disciplines that have the most direct access to power—namely the disciplines which originated in law schools (economics, political science) but also philosophy—and reserves for them disciplines such as psychology, which as early as 1940 seems to have been more welcoming for women (Jeanpierre 2016: 147). In other words, far from having completely *disappeared* from the French academic field, the gendered boundaries and hierarchies have *shifted*.

The other noteworthy demographic evolution is the internationalization of students and professors. From 2007 until 2013, the share of

[14]Given that not all masters students who enrolled completed their master's degree, and that not all of them chose to continue at the doctoral level, while some students coming from elsewhere in France or abroad enrolled directly in Ph.D. programs.

[15]Data kindly provided by Audrey Gauthier (EHESS).

foreigners from educational systems outside of France among students enrolled in master's programs went from 30 to 34% in economics, 16–20% in sociology, 9–13% in anthropology–archeology–prehistory, 5–7% in psychology, 9–17% in French literature, and 14–17% in philosophy, but remained stable, around 13% for the master's programs in political science. As for students enrolled in doctoral programs, a degree program in which the share of foreigners is generally high (39% of all doctoral students in France in 2014/15, twice as many as in master's programs), one finds the same increase in each of these disciplines. Thus, the share of students from foreign educational systems went from 52 to 53% for economics, 35–38% for sociology, 28–31% for anthropology, 21–26% for psychology, 35–37% for political science, 40–44% for French literature and 36–40% for philosophy.

If we look at the EHESS, which is one of the most international graduate schools in France, foreign students represent 39% of those enrolled in master's programs and 52% in the Doctoral School between 2006 and 2015. The percentage is a little smaller, but still significant, for those who obtained their degree: 37% of master's laureates and 46.6% of the Ph.D.s. The EHESS seems to be particularly attractive to female foreign students (they represent around 60% of the foreign students enrolling at all levels—6% more than the share of female among French students—58% of those who got a master's degree, but 52% of the Ph.D.s). Literary studies (where only one out of ten students is a foreigner in the second year of the master's program), cognitive science and economics are the least internationalized within the EHESS. For the two latter, the share of foreign students raises from one quarter enrolling in the second year master's to one-third in Ph.D. In other disciplines (history, sociology, anthropology, philosophy), it rises from one-third in the second year master's to half in Ph.D. In 2014/15, a third of the foreign students came from the European Union, America was the second best represented area with 30.3% of the international students (20% from Latin America), Asia accounted for about 20% and Africa 10% with Oceania the least represented at 0.5%. The high share of European countries in this distribution reflects the effect of the regionalization of research with the construction of a European research area (Heilbron 2013), but the distribution also results from specialization (half of the research centers at the EHESS are areal) and longstanding contacts and partnerships with certain countries like Italy or Germany as well as with areas like Latin America or Eastern Europe. The ten best represented countries were Italy (224), Brazil (129), Colombia (80), Turkey (63), Germany (62), China (56), the USA (54), Chili (50), Romania (42), Mexico (42), these countries accounting for close to a quarter of the students. This rather low concentration rate attests to the diversity of geographic origins.[16]

Among researchers at the CNRS, all disciplines combined, the percentage of foreigners reached 10% for the first time in 1999 and grew to 16.3% in 2015, 12.5% of which are natives of European countries (almost all from the European Union), and close to 4% are originally from a non-European country (1.8% from the Americas, 1% from the African continent, 0.8% from Asia, 0.3% from the Middle East and 0.1% from Oceania). At the EHESS, the share of foreign nationals among the professors (independently of where they completed their Ph.D.) went from 17% in 2013–2014 to 19% in 2015–2016.[17]

The share of foreigners among university professors recruited also tended to increase over the course of the second half of the 2000s, according to the data on annual recruitment of *maîtres de conferences* (assistant professors) at universities. Since the end of the 2000s, more than 15% of the *maîtres de conferences* hired each year—all disciplines combined—were foreign nationals, versus approximately 10% in the first half of the 2000s. It should be noted that the vast majority of them (about two or three) earned their Ph.D.s in France. Consider also that the recruitment of foreigners to *maîtres de conférences* jobs primarily concerns the disciplines in the natural sciences. That said, within the SSH, economics sets itself apart with an increased ratio of foreigners recruited (around 17% for the period 2001–2014). The group formed by economics and management is distinct in its recruitment of foreigners for the *maître de conférences* jobs, given that a considerable part (about 30%) of these foreigners come from former colonies in North

[16]Data kindly provided by Audrey Gauthier (EHESS).

[17]Data kindly provided by the human resources department of the EHESS.

Africa (whereas, all disciplines combined, the great majority of foreign *maîtres de conférences* come from countries in the European Union, in particular Italy).

In conclusion, it appears that all seven disciplines achieved full academic institutionalization in France during the twentieth century, with the creation of distinct undergraduate and graduate degrees. Philosophy was the earliest at the end of the nineteenth century, political science was the last to have a full undergraduate training at the turn of the twenty-first century. Sociology, anthropology, psychology and economics achieved full institutionalization as academic disciplines by the end of the 1950s. As we have seen, relatively autonomous scientific fields developed in these areas much earlier, starting in the mid-nineteenth century, although a significant part of their representatives were also responding to external demands, especially in the case of anthropology within the colonial context and economics during and after the Second World War. The rise of disciplines like demography and political science was even more dependent on the state's demands and it took them longer to achieve scientific autonomy. Apart from psychoanalysis, which has not achieved academic recognition as a discipline (except within Paris 8), the professionalization in these disciplines was largely state-controlled because of the quasi-monopoly of the academy over the titles and dominant positions of scholars in most disciplines, compared to non-academics (secondary school teachers, for example, among which those holding national titles like «*agrégé*» or «*certifié*» are also dominant). In economics and in demography, however, the dominant researchers come from elite schools (Polytechnique, ENSAE). Political science and economics gained autonomy from law, a field dominated by professionals. In the field of political science, though Sciences Po and the other IEP are "temporally" dominant in Bourdieu's sense (meaning they hold high economic and political capital but less scientific capital), university professors have become "symbolically" (that is to say scientifically) dominant (Bourdieu 1984). Despite their symbolically dominant position in the academic field (with research institutes), universities occupy a temporally dominated position within the field of power, where elite schools (such as Polytechnique, Ponts et chaussées, Mines, École normale supérieure, and Sciences Po) still provide access to the highest positions (Bourdieu 1988), whereas the business schools play a growing role in the economic field. University degrees have much less value on the job market than education in elite schools, and while a master's degree significantly increases the chances of getting a job, a Ph.D. has almost no added value as it is not recognized in collective labor agreements, contrary to many other European countries. Attempts to change this situation began in 2010, and a decree issued in 2013 appointed a commission to examine the conditions for this, however these attempts were dismissed by the neoliberal labor law passed in 2016. Meanwhile, the neoliberal tendency to reduce the number of civil servants has impacted the academic job market, which is less and less able to absorb the growing population of Ph.D. holders.

Moreover, the academic disciplinary model in the SSH is now challenged by the reconfiguration of the division of knowledge on the one hand, with the rise of "studies" and, on the other, by the injunction to professionalize academic training (the two being linked in some cases, for example European studies). The spread of this trend towards professionalization (i.e. vocational orientation), which has gained prominence since the 1980s in the context of the second "massification" of higher education (Faure and Soulié 2006), and can be illustrated by the meteoric ascent of Management studies, has accelerated with the introduction of "New Public Management" in universities and the Bologna process. The restructuring of the French academic field into large units (the Pôle de recherche et d'enseignement supérieur or PRES, which evolved into Communautés d'Universités or COMUES), principally to improve the position of French academia in international rankings, destabilizes the whole system, favoring the advent of a second tier academic bureaucracy and the imposition of a management model while simultaneously concentrating the resources in a few so-called "poles of excellence." In this new academic landscape, the SSH are even more dominated by the natural sciences and the growing cognitive sciences, and some disciplines in the humanities, like classical literature, are threatened with disappearance or marginalization. These evolutions reflect the decline of cultural capital which had achieved a relative autonomy in France thanks to the development of the universities and the meritocratic system (Bourdieu 1979), a decline that can be observed in the reconfiguration of the field of power (Denord et al. 2011).

Furthermore, although scientific knowledge still informs public policy, it has been progressively replaced by, or combined with, management, as is the case in the United States (Medvetz 2012). Take for example French prime minister Manuel Valls' statement in the aftermath of the terrorist attack at the Bataclan on 13 November 2015: "I am fed up with those who permanently look for excuses and cultural or sociological explanations to what happened."[18] If the Ministry of Education and Research's response was to commission a report from the CNRS about research on radicalization, Valls' quote indicates that the anti-social science rhetoric, which is typical of the far right, has spread to the center-left, even in a country like France where the social sciences still inform public policy and political reflection.

References

Abbott, Andrew. 1988. *The system of professions: An essay on the division of expert labor*. Chicago: The University of Chicago Press.

Abbott, Andrew. 2001. *Chaos of disciplines*. Chicago and London: The University of Chicago Press.

Backouche, Isabelle, and Christian Topalov (eds.). 2008. *Vingt ans d'élection à l'École des hautes études en sciences sociales (1986–2005). Synthèse des résultats d'enquête*. Paris: Ecole des hautes études en sciences sociales (EHESS).

Barré, Rémi, Michèle Crance, and Anne Sigogneau. 2002. *La Recherche scientifique française: les enseignants-chercheurs et les chercheurs des EPST. Situation démographique le 31.12.2000*. Paris: Observatoire des Sciences et des Techniques (OST).

Ben-David, Joseph, and Randall Collins. 1966. Social factors in the origins of a new science: The case of psychology. *American Sociological Review* 31 (4): 451–465.

Boncourt, Thibaud. 2011. L'Internationalisation de la science politique: une comparaison franco-britannique (1945–2010). PhD thesis, Institut d'études politiques de Bordeaux, Bordeaux.

[18]At the French Senate, on 26 November 2015. Sonya Faure, Cécile Daumas, Anastasia Vécrin, «Culture de l'excuse: Des sociologues répondent à Valls», *Libération*, 12 janvier 2016.

Bourdieu, Pierre. 1979. *La Distinction. Critique sociale du jugement de goût.* Paris: Minuit. English transl. by Richard Nice. 1984. *Distinction: A social critique of the judgement of taste.* Cambridge: Harvard University Press.
Bourdieu, Pierre. 1984. *Homo Academicus.* Paris: Minuit. English transl. by Peter Collier. 1988. Stanford: Stanford UP.
Bourdieu, Pierre. 1988. *La Noblesse d'État: Grandes écoles et esprit de corps.* Paris: Minuit. English transl. by Lauretta C. Clough. 1996. *State nobility: Elite schools in the field of power.* Stanford: Stanford UP.
Bourdieu, Pierre. 2001. *Science de la science et réflexivité.* Paris: Raisons d'agir. English transl. by Richard Nice. 2004. *Science of science and reflexivity.* Chicago: The University of Chicago Press.
Boutier, Jean, Jean-Claude Passeron, and Jacques Revel (eds.). 2006. *Qu'est-ce qu'une discipline?* Paris: Editions de l'EHESS.
Campion-Vincent, Véronique. 1970. Organisation de l'ethnologie en France. *L'Homme* 10 (3): 106–124.
Carroy, Jacqueline, Annick Ohayon, and Régin Plas. 2006. *Histoire de la psychologie en France XIXe–XXe siècles.* Paris: La Découverte.
Charle, Christophe. 1994. *La République des universitaires, 1870–1940.* Paris: Seuil.
Charle, Christophe. 1999. Les femmes dans l'enseignement supérieur. Dynamiques et freins d'une présence 1946–1992. In *Avenirs et avant-gardes en France XIX-XX siècles, hommage à Madeleine Rebérioux*, ed. Vincent Duclert, Rémi Fabre, and Patrick Fridenson. Paris: La Découverte.
Chenu, Alain. 2002. Une institution sans intention. La sociologie en France depuis l'après-guerre. *Actes de la recherche en sciences sociales* 140–142 (1–2): 46–61.
Conklin, Alice. 2013. *In the museum of man: Race, anthropology, and empire in France, 1850–1950.* Ithaca, NY: Cornell University Press.
Convert, Bernard. 2010. Espace de l'enseignement supérieur et stratégies étudiantes. *Actes de la recherche en sciences sociales* 183: 14–31.
Cranney, Jean. 1996. *Inra, 50 ans d'un organisme de recherche.* Paris: Editions Quae.
Denord, François, Paul Lagneau-Ymonet, and Sylvain Thine. 2011. Le champ du pouvoir en France. *Actes de la recherche en sciences sociales* 190 (5): 24–57.
Digeon, Claude. 1959. *La Crise allemande de la pensée française (1870–1914).* Paris: PUF.
Dubois, Vincent. 2013. *La Culture comme vocation.* Paris: Raisons d'agir.
Espagne, Michel. 1993. *Le Paradigme de l'étranger. Les chaires de littérature étrangère au XIXe siècle.* Paris: Cerf.

Fabiani, Jean-Louis. 1988. *Les Philosophes de la République*. Paris: Minuit.
Faure, Sylvia, and Charles Soulié. 2006. La recherche universitaire à l'épreuve de la massification scolaire. *Actes de la recherche en sciences sociales* 164: 61–74.
Faure, Sylvia, Charles Soulié, and Mathias Millet. 2006. Rationalisation, bureaucratisation et mise en crise de l'ethos académique. *Regards sociologiques* 31: 107–140.
Favre, Pierre. 1989. *Naissances de la science politique en France, 1870–1914*. Paris: Fayard.
Fisher, Donald. 1983. The role of philanthropic foundations in the reproduction and production of hegemony: Rockefeller foundations and the social sciences. *Sociology* 17 (2): 206–233.
Fleck, Christian. 2011. *A transatlantic history of the social sciences: Robber Barons, the Third Reich and the invention of empirical social resarch*. London: Bloomsbury.
Fondu, Quentin. 2014. L''invention' de la discipline universitaire des études théâtrales en France (1948–1981). Master thesis, EHESS, Paris.
Foucault, Michel. 1966. *Les Mots et les choses*. Paris: Gallimard.
Fourcade, Marion. 2009. *Economists and societies: Discipline and profession in the United States, Britain and France, 1890s to 1990s*. Princeton and Oxford: Princeton University Press.
Gaillard, Gérald. 1989. Chronique de la recherche ethnologique dans son rapport au centre national de la recherche scientifique, 1925–1980. *Cahiers pour l'histoire du CNRS* 3: 85–129.
Gingras, Yves, and Brigitte Gemme. 2006. L'emprise du champ scientifique sur le champ universitaire et ses effets. *Actes de la recherche en sciences sociales* 64: 51–60.
Heilbron, Johan. 2013. The social sciences as an emerging global field. *Current Sociology* 62 (5): 685–703.
Heilbron, Johan. 2015. *French sociology*. Ithaca and London: Cornell University Press.
Jarausch, Konrád H. 1990. *The unfree professions: German lawyers, teachers and engineers, 1900–1950*. Oxford: Oxford University Press.
Jeanpierre, Laurent. 2016. L'aventure des sciences de l'homme. In *La Vie intellectuelle en France, 2, de 1914 à nos jours*, ed. Christophe Charle, 139–166. Paris: Seuil.
Karady, Victor. 1976. Durkheim, les sciences sociales et l'Université: bilan d'un semi-échec. *Revue française de sociologie* 17 (2): 267–311.

Karady, Victor. 1983. Les professeurs de la République: le marché scolaire, les réformes universitaires et les transformation de la fonction professorale à la fin du 19ème siècle. *Actes de la recherche en sciences sociales* 47–48: 90–112.
Kuhn, Thomas S. 1970. *The structure of scientific revolutions*. Chicago: University of Chicago Press.
Lebaron, Frédéric. 1997. La dénégation du pouvoir. Le champ des économistes français au milieu des années 1990. *Actes de la recherche en sciences sociales* 119: 3–26.
Lebaron, Frédéric. 2000. *La Croyance économique. Les économistes entre science et politique*. Paris: Seuil.
Lévi-Strauss, Claude, and Didier Eribon. 2001. *De près et de loin*. Suivi de *Deux ans après*. Paris: Odile Jacob.
Martin, Olivier. 2002. Introduction. *La revue pour l'histoire du CNRS* 7: 1–4.
Medvetz, Tom. 2012. *Think tanks in America*. Chicago: University of Chicago Press.
Picard, Jean-François. 1990. *La République des savants : la recherche française et le CNRS*. Paris: Flammarion.
Pollak, Michael. 1976. La planification des sciences sociales. *Actes de la recherche en sciences sociales* 2: 105–121.
Popa, Ioana. 2015. Aires culturelles et recompositions (inter)disciplinaires. La 6e section de l'EPHE et les études sur l'espace russe, soviétique et est-européen. *Actes de la recherche en sciences sociales* 210: 60–81.
Prost, Antoine, and Jean-Richard Cytermann. 2010. Une histoire en chiffre de l'enseignement supérieur. *Le Mouvement social* 233: 31–46.
Renoliet, Jean-Jacques. 1999. *L'UNESCO oubliée. La Société des nations et la coopérations intellectuelles (1919–1946)*. Paris: Publications de la Sorbonne.
Renisio, Yann. 2015. L'origine sociale des disciplines. *Actes de la recherche en sciences sociales* 210: 10–27.
Ribard, Dinah. 2000. Philosophe ou écrivain? Problèmes de délimitation entre histoire littéraire et histoire de la philosophie en France, 1650–1850. *Annales. Histoire, Sciences Sociales* 55 (2): 355–388.
Rosental, Paul-André. 2003. *L'Intelligence démographique*. Paris: Odile Jacob.
Sapiro, Gisèle. 2004. Défense et illustration de 'l'honnête homme'. Les hommes de lettres contre la sociologie. *Actes de la recherche en sciences sociales* 153: 11–27.
Sapiro, Gisèle. 2006. Les professions intellectuelles, entre l'Etat, l'entrepreneuriat et l'industrie. *Le Mouvement social* 214: 3–24.

Sapiro, Gisèle. 2013. Le champ est-il national? La théorie de la différenciation sociale au prisme de l'histoire globale. *Actes de la recherche en sciences sociales* 200: 70–85.

Soulié, Charles. 2013. Des humanités à 'l'économie de la connaissance'? Les transformations du corps enseignant en lettres et sciences humaines en France (1949–2010). In *Faut-il brûler les Humanités et les Sciences humaines et sociales?* ed. Marc Conesa, Pierre-Yves Lacour, Frédéric Rousseau, and Jean-François Thomas. Paris: Michel Houdiard Editeur.

Stichweh, Rudolf. 1992. The sociology of scientific disciplines: On the genesis and stability of the disciplinary structure of modern science. *Science in Context* 5 (1): 3–15.

Touchelay, Béatrice. 2000. L'I.N.S.E.E. et le C.N.P.F. de 1946 à 1961: l'histoire d'une alliance modernisatrice. *Le Mouvement Social* 191: 25–47.

Tournès, Ludovic. 2011. *Sciences de l'homme et politique. Les Fondations philanthropiques américaines en France au XXe siècle*. Paris: Editions des Classiques Garnier.

Weil, François et Chantal Cases. 2012. *La Structuration de la recherche en sciences humaines et sociales. Rapport du groupe Structuration de l'Alliance Athena*. Online.

3

Germany: After the Mandarins

Matthias Duller, Christian Fleck and Rafael Y. Schögler

Struggling with the Burden of a Nazi Past

The end of the Third Reich has defined all aspects of Germany's public life for all decades since this momentous event. According to M. Rainer Lepsius, the three post-Nazi societies reacted very differently.

We want to thank the following individuals for either providing expertise in preparing the report or commenting an earlier version of this paper: Mitchell Ash, Gerhard Göhler, Harald Hagemann, Manfred Heinemann, Andreas Hess, Nicole Holzhauser, Dirk Kaesler, Dirk Kemper, Marianne Kneuer, Heinz D. Kurz, Wolfgang L. Reiter, Reinhold Sackmann, Katharina Scherke, Irmingard Staeuble, Richard Sturn, and John Torpey. Of course, responsibility for any errors solely lies with the authors of this chapter.

M. Duller (✉) · C. Fleck · R. Y. Schögler
University of Graz, Graz, Austria
e-mail: matthias.duller@uni-graz.at

C. Fleck
e-mail: christian.fleck@uni-graz.at

R. Y. Schögler
e-mail: rafael.schoegler@uni-graz.at

C. Fleck et al. (eds.), *Shaping Human Science Disciplines*, Socio-Historical Studies of the Social and Human Sciences, https://doi.org/10.1007/978-3-319-92780-0_3

The Austrians "externalized" Nazism by attributing it to the invading Wehrmacht, while the two Germanys generated opposing responses. The Eastern Communist regime blamed capitalism for using fascism as the last measure to avoid its own end and, in doing so, "universalized" the past; the West Germans sought the causes of the preceding catastrophe by exclusively processing their own history to understand the social forces that brought Hitler to power. Lepsius denominated this "internalizing" (Lepsius 1993: 232–233). However, even within this habit of taking an internal perspective, different explanations for the recent past have emerged. Initially, the debate focused almost exclusively on the demonic personality of Hitler and the topic of collective guilt. Later the weaknesses of the political system and the political parties of the Weimar Republic entered center stage, followed by a more intensive debate and soul-searching concerning the Holocaust, the role of anti-Semitism and the lack of civil resilience vis-à-vis the 'hangmen'; after the Eichmann and the first Auschwitz trial, attention turned towards the role of the exterminators and their organizations. Next came a heated controversy about the historians' narratives. The *Historikerstreit* of the mid 1980s was followed by a switch towards ordinary soldiers and their role in the mass-killings, accompanied by a broader coverage of the Wehrmacht as part of the extermination machinery. Feminist historians made the role of women, as perpetrators and victims, the subject of debate. Only after the military had been dealt with did German industry turn into the locus of attention of historical research when their role in connection with forced labor and exploitation caught attention. Post-trauma and the role of handing over guilt and sorrow from one generation to the next was another move expanding the scope of coverage of the Nazi past. At a relatively late phase of *Vergangenheitsbewältigung*, were the universities and—even later—the state bureaucracies investigated in more detail concerning their involvement and active role in stabilizing the terroristic machinery.

Exponents of the social sciences and humanities initiated or commented on nearly every turn of attention. The fervent devotion to coming to terms with the collective past was particular strong within the well-educated strata of West Germany. This was not, however, accompanied by a similar fervent investigation of the past of the academic disciplines and professions themselves, which only began much later.

Re-construction and Re-education

After the unconditional surrender of the Third Reich on 7 May 1945, Germany's academic system was confronted with severe problems. The end of a regime always leaves behind both the social structures and people as agents of continuity connecting the two sides of the rift between yesterday's dictatorship and today's democratic structures. Democracies cannot rid themselves of large parts of their citizens. Nor is renewing institutions an easy task, particularly if the survival of these very institutions requires the expertise that is, quite literally, embodied in members of the economic, administrative and intellectual elite of the old regime. In the German case, the transition period was granted its own legal status. From May 1945 until the establishment of the two German states—the Federal Republic of Germany and the German Democratic Republic—legal powers and authority were in the hands of the Occupation Forces. Almost all traditional institutions were scrutinized and their future turned into a matter of debate inside the planning staff of the Allied Forces. The fate of those surviving was similarly at stake: who should be allowed to continue working in their occupation? How? For which measures should supporters of the past regime be punished? For how long? And who should be allowed to educate the next generation? The latter was a question that once again became pertinent four decades later when the communist governed part of Germany collapsed. Certain detailed plans had already been developed before the end of armed fighting. However, improvisation was dominant in the early post-war years. It might be argued that Germany's reconstruction was the first instance of purposeful nation building, which turned, relatively quickly, into a success story. The defeated regime did not retain any legitimacy. Only certain sub-public spaces, so to speak, offered space for a kind of nostalgia, which is best caught by the (verbatim translation of) the phrase "it wasn't all bad under Hitler". Pollsters turned this phrase into a questionnaire item to regularly measure how far people had distanced themselves from the past. People did, and they could in part because the post-war society offered them sufficient opportunities to forget the past and turn their attention to the future.

It is well known that exiled scholars and foreign experts blamed the educated strata of the German population for their servility and conformism, with some expanding these accusations towards the primordial institution in which these moods were formed and elaborated, German academia (Gerhardt 1993; Hartshorne 1937; Hartshorne and Tent 1998; Ringer 1969; Weinreich [1946] 1999). Neither the organizational structures, nor the power relations in decision-making procedures, were subject to an overhaul in West-German universities after the end of the Third Reich; one of the surprising facets of the early post-war years. The only interventions of visible institutional consequence were the dismissals of the most ardent Nazis inside the professoriate and the appointment of exiled scholars to these positions, or into newly established ones.

Substituting key figures was the chosen, and successful, option for re-education. Despite later criticism, its success should be emphasized. Within one-and-a-half academic generations, the habitus of the former German Mandarins (Ringer 1969; Elias 1996) was entirely eradicated and a new one emerged. This cultural change was accompanied and supported by a decline of professors' social status, which decreased from extreme elitism to upper middle-class ordinariness. According to Hesse (2013) and Sohn (2016), during the twentieth century the relative earnings of German professors declined continuously. From before World War I until the end of the 1950s, the average income of a German professor was higher than the remuneration a deputy under-secretary of state, the highest obtainable rank in state bureaucracy. From the 1960s onwards, this relationship reversed with top-ranked civil servants earning more than professors (Hesse 2013: 123–125). Whereas the number of positions at the highest levels of bureaucracy remained almost the same, the professoriate expanded with a very high growth rate. Sohn shows that, at the beginning of the twentieth century, the salary of an average professor qualified him to belong to the top 1% of income; some even reached the top 0.5%. From the Great Depression onwards, the professors lost this lead position, but remained in the top percentile. From the start of the 1960s, the average professor's income moved him (the number of female professors started to increase only from this point) out of the top 1%; a professor seldom managed to remain in "the illustrious group of the top 0.5% by means of their professorial earnings alone" (Sohn 2016: 91). An increase from less than 2000 professors in 1950 to more than 40,000 professors some 65 years later makes it highly improbable this group could maintain their elite position in the income hierarchy.[1] No elitist attitude of an "aristocracy of the mind" (Thomas Mann propagated such as a remedy for Germany in 1945) could be preserved during such a dramatic expansion of the professoriate.

The Development of the University System

Fundamental changes to the political-administrative systems also affected universities. The Allies rezoned Germany, not only into occupation zones, but later into new *Länder*. The responsibility for all levels of the education system remained with these regional governments, but the power relations between the *Länder* moved from domination by Prussia to a more balanced system in what then became the Federal Republic. The elimination of Prussia from the map had a direct effect on the universities. After 1945, the re-established *Kultusministerkonferenz*—an inter-governmental coordination body for educational affairs—did not have to resolve quarrels between a dominator and the rest, but functioned well as a coordination forum.

Two of the ten Prussian universities were located outside of the new borders of both parts of Germany: Königsberg and Breslau.[2]

[1]In 1950, the working population of West-Germany amounted to about 23 million, it grew up to 1989 to about 30 million (before unification) and by the year 2000 has grown further to more than 38 million. In other words, the working population grew by the factor of 1.6, whereas the number of professors multiplied by the factor of 20.

[2]Three more *Reichsuniversitäten* were outside of the German-speaking territories after 1945: The old German Charles University in Prague, the University of Strasbourg, which became German speaking after the defeat of France, and the short living university in Posen, later: Poznan.

Together with 14 universities from the other 9 former Länder, the remaining 8 Prussian universities—Berlin, Bonn, Göttingen, Greifswald, Halle, Kiel, Marburg, and Münster—plus the 2 special cases of Cologne and Frankfurt (funded partly by municipality and foundations) were reorganized according to the new zoning: Berlin, Greifswald, Halle, Jena, Leipzig and Rostock were in the Eastern Zone. The remaining Universities belonged to the bi-zone of the American and British Occupation Forces. While no new universities were established in the bi-zone in the early years of the Federal Republic (FRG), in the French zone two were re-opened or newly created: Mainz and Saarbrücken. The only new founding took place in the Western sectors of Berlin when, in 1948, the Free University opened its doors as a counterweight to the now Humboldt University, which was located in the city's eastern part under Soviet control.

After the collapse of Hitler's regime, Old Prussia could not find a way to continue their interest in establishing the late founded Friedrich-Wilhelms University in Berlin as the leading institution on German-speaking territory; power and prestige was distributed more evenly between the universities. Since no center remained, neither in politics nor culture, universities' levels of attraction changed. Historically small-town universities like Freiburg, Göttingen, Heidelberg, Tübingen or Marburg, as well as relatively young universities in the larger cities of Frankfurt (founded 1914), Cologne and Hamburg (both founded in 1919), profited from this process. Only Munich was a larger city with a distinguished university established in the early nineteenth century. The traditional academic merry-go-round—changing institution for better salaries or work-conditions—started in post-war Germany, and only from the 1960s onwards.

The expansion of the higher education (HE) system began in the 1960s when new universities opened their doors. The boom continued until the middle of the following decade. After reunification, a number of new universities were established in the new *Länder*—the former East Germany. Around 2015, there were about 80 universities with promotion rights (advanced graduate degree of Ph.D.), where the traditional range of disciplines is represented. There are also about 14 Technical Universities, which now also run non-technical study degrees and departments. Not included in these numbers are smaller private, or

Table 3.1 Increase in number of professors 1953–2015, percentage of women

Discipline	Number of professors	Increase until 2015 by the factor of	Percentage of women	
	1953		1953	2015
Economics	66	6	0	14
Philosophy	63	5	0	19
German literature	49	10	0	42
Business economics	38	49	0	22
Psychology	19	37	0	39
Sociology	17	18	0	38
Political sciences	16	19	0	29
Anthropology	5	7	0	35

Sources Statistisches Bundesamt (1953); Data for 2015 provided to the authors by the German Federal Statistical Office; for anthropology 1953: Haller (2012)

semi-private, universities.[3] There are about seven times more (so called full) universities in Germany today than there were at the end of the Nazi regime. During roughly the same period, the number of the two highest ranked professors at universities increased by a factor of 21, the overall research and teaching staff by a factor of 61 and the student population at universities by a factor of 27.

Considering only those disciplines that interest us here, the picture differs slightly. For the year 1953 we identified a total of 273 male professors, and not a single woman, in 8 disciplines (including business economics). Six decades later, the number of professors climbed to 4415. This amounts to a pool of professors 16 times larger, although female applicants remain severely underrepresented in some disciplines. Women make up less than 20% of professors in philosophy and even less, 14% in economics. In German Literature, women make up over 42% of the professoriate (see Table 3.1).

One more characteristic of academic personnel needs to be mentioned to understand the development of the West-German academic scene over time. De-Nazification started under the management of military

[3]Excluded are also Universities for Applied Science and their counterparts.

governments, which perceived relatively quickly that the initial plan to scrutinize past personnel by forcing everyone to fill out a questionnaire did not work (Ernst von Salomon made this obligatory questionnaire the subject of a novel: *Der Fragebogen*). Thus, they handed the responsibility of deciding who could continue working in their profession to local examination boards, which had strong incentives to be very generous to most of those under scrutiny. As a consequence, more individuals were rated followers, *Mitläufer*, than were legitimized according to the spirit of the regulations. A side effect of this generosity, however, was the swifter and more thorough re-integration of sidestepped individuals into the new society; not a small success, if looked at from a comparative perspective on regime changes and their consequences (comp. Lübbe 1983).

A policy of offering compensation jobs to those who had lost theirs was even written into the *Grundgesetz*, the simulacrum of a constitution, and a following implementation law in 1951. The beneficiaries were civil servants, including professors, who were dismissed during Nazi rule, refugees from the former eastern parts of the Reich, and those having left the Soviet zone. People affected by the initial de-Nazification measures were also taken into account. It is hard to believe that such a mishmash of whitewashing former Nazis with compensation for victims of dictatorship could have found legitimization. Other unintended consequences, however, are of more imminent interest for universities. Along with former Nazi party members, these measures brought older folk of questionable reputation back into tenured positions, it offered refugee scholars an opportunity to return, but also narrowed the market for the next generation over a period of several years.[4]

By the middle of the 1950s, the West-German university system had reobtained stability. Some of the most prominent followers and ideologues of the Nazi era remained dismissed; others were given positions in lower ranked places or in research institutes only slightly related to neighboring universities, like the *Sozialforschungsstelle Dortmund*.

[4]In the early 1950s about one-tenth of the personnel belonged to this composite group of refugees and dismissed professors. Unfortunately, there are no detailed statistics for the number of returnees for the disciplines under consideration here. For economics, see Hesse (2010); for sociology, see Fleck (2011), Chapter 7.

Two decades after the break-down of all German state institutions, its universities were restored. However, at much the same time, discontent with them and the larger education system started to rise. Germans joined the move to catch up with competitors in the aftermath of the Sputnik shock. A memorandum authored by later prominent social scientists was published by the Socialist German Students Organization SDS under the programmatic title "University in Democracy" (Nitsch et al. [1961] 1965) and, in the same year, the Frankfurt Institute for Social Research published an empirical study on the political consciousness of students (Habermas et al. 1961). Several authors expressed their concern about what had been named by Georg Picht the "German educational catastrophe" (*Bildungskatastrophe*) years before the students started with their protest against the professors' authoritarian attitudes and behavior. In reaction to the "educational emergency" (*Bildungsnotstand*) declared by the same author in a series of articles published in a Protestant weekly in 1964 the following year a then young Ralf Dahrendorf published a small brochure "Education as a civil right" (*Bildung als Bürgerrecht*) mimicking the linguistic conventions of constitutional rights for this proposal.

From an institutionalist perspective, the student movement did not contribute much to the changes which took place in the years after 1968, but from a cultural perspective "1968" is a watershed; some even called it a cultural revolution. In the following decades, the habitus of the professoriate became more informal and permissive. Several of the movement's spokespersons became professors at a very young age because the expansion of the HE system needed personnel. Thus, the professoriate moved politically to the left for the first time since the 1848 revolution. So-called "reform universities", in particular in Bielefeld, Konstanz and Bremen, acted as avant-garde. During the 1980s, however, these sentiments were soothed again. Budget restrictions affected the higher education system, but it still produced more aspirants for professorships than available positions. West Germany's professoriate acted almost without any dissent during the next big challenge: the unification of the two parts of the country. This was accompanied by a complete overhaul of the universities in the East of Germany, which enabled Westerners to get positions otherwise unimaginable for this cohort. For the rest of the twentieth century, the reconstruction of the Eastern part of a now unified Germany garnered all available attention.

The last consequential change in German academia occurred around the end of the twenty-first century's first decade. At this time the federal government intervened by establishing a so called "Excellence Initiative" with the clear intention of ranking universities according to their research performance (fiercely criticized by others, e.g. Münch 2007). To this end, a handful of universities (not departments or meso-organizational units like schools or faculties) received high amounts of research funding. Within a few years, an unplanned but foreseeable consequence was the overproduction of young researchers who could not find appropriate positions after leaving their specialized graduate schools. Within 30 something years, roughly the span of one academic generation, Germany's higher education system overproduced talent once again with the migration of the surplus, especially to neighboring German-speaking countries, causing some harsh xenophobic reactions there. Competition in the job market seldom goes hand-in-hand with a "foreigners welcome" attitude.

Seven Disciplines in Comparison

Let us now turn to a more detailed description of the development of the seven disciplines that form our sample from the social sciences and humanities (SSH).[5] We will proceed similarly in each case, providing data for the beginning of the period under consideration and trajectories over time indicating the main developments with regard to the scholarly profile of each discipline and highlighting particularities that seem characteristic for the academic world of Germany. We have restricted our presentations mostly to West-Germany, but will add a summarizing narrative about the SSH disciplines in the Eastern part

[5] If not otherwise indicated data have been collected for the present purpose inside the INTERCO-SSH project, http://www.interco-ssh.eu/.

under communist rule. Statistical data for the 1940s are not available, so in most cases we start with data from the first gathering of statistics by the post-war German statistical office.

Philosophy

In Germany philosophy traditionally held a highly prestigious position. Germans regularly designate themselves using the phrase "the country of poets and thinkers". After 1945, philosophers can be split into those extinct, those returning and those who tried to make sense of the recent past. Germany's most famous philosopher of the twentieth century, Martin Heidegger, was banned from his professorship and subsequently forced into retirement. This administrative exclusion did not affect his influence in Germany or abroad. Several prominent philosophers returned from exile but none of them enjoyed the same prestige as Heidegger. Nevertheless, Helmuth Plessner, Karl Löwith and Theodor W. Adorno became prominent and influential figures in post-war West-Germany. Karl Jaspers, who was dismissed by the Nazis in 1937 and not allowed to publish, survived the Nazi years in inner emigration, a term he didn't invent, but which does fit his case better than others. He left Germany in 1948 for neighboring Basel, but remained a public intellectual in his home country. None of the influential philosophical schools of the early twentieth century, Neo-Kantianism, Phenomenology, and Logical Positivism, regained standing in the early post-Nazi years. The rise to prominence of the Frankfurt School of Critical Theory only began in the 1960s. The same can be said for Hermeneutics, embodied by Hans-Georg Gadamer, and analytic philosophy established by Wolfgang Stegmüller in Munich.

The high prestige of philosophy is mirrored by the 63 professorships already existing in the early 1950s. It is also highlighted by the strong position of philosophy in several of the traditional university curricula. For a very long time, every candidate for a doctorate at the Philosophical Faculty had to take an exam in philosophy, which resulted in the representation of philosophy in all of these Faculties. One also needs to take into account that at this time in some cases philosophy also included pedagogy, psychology or sociology as part of the denomination of a chair. Max Horkheimer and Adorno both held professorships of philosophy and sociology in Frankfurt from 1949 onwards. Plessner and Jürgen Habermas held consecutive professorships in philosophy and sociology and Plessner was president of the new founded German Philosophical Association (from 1950) and the German Sociological Association (1955–1959). Habermas never served in any professional organization.

Philosophy experienced the lowest growth rate in the sample of disciplines under consideration here, which should not be interpreted as the by-product of an elitist strategy of keeping the numbers of guild members low. It might be more accurate to say that the West German philosophers were not able to persuade the guardians of the taxpayer's treasure to expand the professoriate further than was strictly necessary. In East Germany, the situation was similar. Classical philosophy was substituted by branches of Marxist–Leninist doctrines, but more technical parts of philosophy continued to be taught at the universities.[6]

The undramatic history of German philosophy over the last seven decades can also be seen in the moderate expansion of the journals. Some of the traditional journals, e.g. *Archiv für Geschichte der Philosophie* and *Kant Studien*, were published without much interruption, even during the Nazi years. Other titles started only after WWII: *Zeitschrift für philosophische Forschung* in 1946, *Philosophische Rundschau* and *Deutsche Zeitschrift für Philosophie* both date to 1953, the later was edited in East Germany initially by Ernst Bloch and Wolfgang Harich but came under party control very quickly.

Economics

With 66 professors in 1953, economics was, back then, the largest social science discipline within our sample. However, this number is disputable considering the boundaries of economics have always been

[6]There is no comprehensive history of German philosophy after 1945, see for shorter overviews Wolf-Gazo (1980) and Schnädelbach (1994).

contested and statisticians changed the classification procedures more than once. Subdivisions of the field, internationally established at present, did not have an equivalent in German universities in the 1950s for several reasons. Firstly, German economics saw itself as part of the larger area of *Staatswissenschaften* which, besides economics, encompassed fields known elsewhere as political theory, economic and social history, constitutional law and sociology. The *Verein für Socialpolitik*, founded in 1873 and silenced during the Nazi period, but re-established in 1948, functioned for a very long time as the meeting point for this diverse group of scholars. At its annual meetings all leading social scientists of the time participated and its proceedings, *Schriften*, run to 187 volumes from 1873 to 1939.[7]

Nevertheless, in the 1950s economics was well established in the German universities and was represented at all universities with a separate curriculum and a first certificate somewhere below doctorate level: the *Diplom Volkswirt*.

The level of institutionalization is similarly reflected by the landscape of professional journals available at the time. Several of these were founded back in the nineteenth or early twentieth centuries and continued to be published during the Nazi regime.[8] Therefore, they did not need to be re-established after 1945; only the editors changed in some cases.

- *Zeitschrift für die gesamte Staatswisssenschaft*, founded 1844, since 1986 *Journal of Institutional and Theoretical Economics*, English only.
- *Jahrbücher für Nationalökonomie und Statistik*, founded 1862, today with an English subtitle and bilingual contributions.
- *Jahrbuch für Gesetzgebung, Verwaltung und Volkswirtschaft im Deutschen Reiche*, founded 1871, since 2016 *Schmollers Jahrbuch–Journal of Contextual Economics*, English only.

---

[7]Continued from 1949 onwards, with the subtitle *Neue Folge*, published by Duncker & Humblot.

[8]The exception is the famous *Archiv für Sozialwissenschaft und Sozialpolitik*, lately edited by Joseph A. Schumpeter and Emil Lederer, which stopped publishing immediately after Hitler became Reichskanzler in 1933. Plans of the publisher to reopen it after the war did not materialize.

- *FinanzArchiv*, founded 1884, since 2000 English only.
- *Schmalenbachs Zeitschrift für betriebswirtschaftliche Forschung*, founded 1906, since 2000 *Schmalenbach Business Review*, 4 issues a year in English (complemented by 8 issues remaining in German).
- *Weltwirtschaftliches Archiv*, founded 1913, since 2003 *Review of World Economics*, English only.

These titles remain the leading economics (or management) journals in Germany, with the switch to English indicating that it lasted long until German economics became globalized.[9]

Given the strong tradition of the "historical school", it is not surprising that, in the 1950s and even 1960s, this particular tradition dominated economics. The "neoclassic synthesis", coined by Paul A. Samuelson and embodied in his textbook, did not immediately find traction in Germany's economics. Erich Schneider's three-volume textbook, originally published between 1947 and 1952, functioned as the ambassador of a moderate Keynesianism, which conquered the field of economic theory much earlier than economic policy (Hagemann 2017).

With regard to economic policy, one school played an idiosyncratic role in West Germany. So-called ordo-liberals from the University of Freiburg tried to establish a distinct perspective located between the historicists and the neoclassical school. Members of this milieu gained political power in Germany, in particular when its disciple Ludwig Erhard became minister for economic affairs in the first FRG government, remaining in this position for 14 years. "Social market economy" and "Germany's economic miracle" became catchphrases associated with Erhard's politics.

As in many other countries, academic economists obtained political influence both by changing sides and turning into professional politicians, but also as members of advisory boards. The Board of

[9]There is one originally Austrian journal *Zeitschrift für Nationalökonomie*, now: *Journal of Economics* which was founded in 1929 and one postwar outlet from Switzerland *Kyklos*, founded in 1947.

Academic Advisors to the Federal Ministry for Economic Affairs (*Wissenschaftlicher Beirat beim Bundesministerium für Wirtschaft*) claims to be older than the government it advises. In 1943, Erwin von Beckerath assembled an informal group of fellow economists to discuss the country's economic future after the end of the war.[10] 17 well-known economists established this Board a year before the *Bundesrepublik* was officially founded. Today the Board consists of around 40 members, all professors. There are personal overlaps between this board and the extramural institutes for economic research. The German Council of Economic Experts (*Sachverständigenrat*) is the second advisory board, created in 1963. It consists of only five members, two of them recommended by the trade unions and industrialists' organization, but all appointed by the government for tenure of five years. Unlike its American counterpart, the German Council is independent of the government. In the vernacular, it is referred to as the "Five Sages" because it publishes an annual report, which receives much public attention. The third institutionalized form of economic expertise in Germany, business cycle prognosis, is in the hands of a group of specialized extramural research institutes, who compete every five years for this well-paid contract.

During the 70 years after the establishment of the *Bundesrepublik*, economics expanded less dramatically than other fields, which is related to its established nature at the beginning of the period and the differentiation within the field of economics, most notably the rise of business economics from the 1960s onwards and finance around the turn to the twenty-first century. Up to 2010, the number of economics professors had increased only by a factor of six. What is remarkable is the increase of business economics, both with regard to faculty and students. In 2015 not only were there 49 times as many professors in this field as there were in 1953, there were also nearly a quarter of a million students, number one out of all study programs, compared to 90,000 students in all other branches of economics put together.

---

[10]It is worth mentioning that German economists could manage to secure some niches even during the Nazi regime, see Janssen (2012).

Sociology

After the end of the Nazi system, the discipline and the profession of sociology had to be reconstructed fundamentally. Founded in 1909, the German Sociological Association (*Deutsche Gesellschaft für Soziologie*), which had been pared to the bone in 1934, was re-established by its last surviving pre-Nazi board member Leopold von Wiese, who had not become compromised during the dictatorship. One of his former (American) students, Howard P. Becker (not to be confused with Howard S. Becker), served as university liaison officer for Hessen and had been instrumental in the quick reconstruction of sociology. As early as September 1946, the first post-war *Soziologentag* (the eighth in total) was held in Frankfurt and von Wiese was elected as president of the Association (Borggräfe and Schnitzler 2014).

The American occupation forces were very interested in sociology. Whereas the Soviets in the Eastern part feared sociology as a bourgeois alternative to their own Marxism–Leninism and banned sociology at the first opportunity, following the practices in the Soviet Union, the Americans thought that it could help re-educate the Germans. Several (later) well-known American and British sociologists worked for shorter or longer periods in Germany, either within the occupation forces' administration or in academia as consultants or visiting professors.

The return of exiled sociologists brought new centers of sociological research into existence. The so-called Frankfurt School under Horkheimer and later Adorno established itself as one of three competing/collaborating schools. In Cologne, René König, another émigré, took over the leadership and created a type of sociological research that avoided any philosophical underpinnings. Sociology as an empirical undertaking became the motto of this group, which produced a remarkable number of followers, taking over professorships in other universities from the late 1950s onwards. The Cologne (or König) School championed the American type of empirical social research, setting aside the fact that its founder spent his exile in Zurich and only familiarized himself with US sociology in the 1950s. The third "school" was organized by Helmut Schelsky, a more philosophically oriented scholar, who brought together former Nazis and young researchers.

In 1952, 17 professors represented sociology, the majority of whom taught not only sociology, but also other disciplines like *Staatswissenschaften*, philosophy, etc. In 2015, the number of sociology professors had reached 299, an increase of the same (relative) size as the average for all professors at German universities.

The establishment of proper academic sociology programmes at several universities from the 1960s onwards brought members of the post-war generation into professorships. After they started as professors these people—who at the time were relatively young (in their early 40s)—became very quickly challenged by representatives of a generation socialized in the student and youth movements. From the late 1950s to the late 1960s, German sociology underwent several disintegrative controversies, the most famous being the so-called Positivism Dispute.

From the 1970s onwards, sociology flourished at several newly established universities, in particular at Bielefeld where a large Faculty of Sociology became one of the new focus points of sociological research. Extra mural research units beyond the domain of "*Demoskopie*" (public opinion research) and market research were created. Some of them still exist; several Max Planck Institutes, the *Wissenschaftszentrum Berlin*, and infrastructural units like GESIS, for example. Sociology took pleasure in becoming the recipient of one private initiative when Jan Philipp Reemtsma used his inheritance to found the *Hamburger Institut für Sozialforschung*, which played a crucial role in problematizing the role of the Wehrmacht during WWII. This research institute and its journal *Mittelweg 36* occupies a remarkable place in public intellectual debate beyond academic disciplinary boundaries since its founding in 1984. The German Sociological Association is currently structured in 35 sections and has about 2400 members (80% of them holding a Ph.D.). The oldest and most prestigious journal is the *Kölner Zeitschrift für Soziologie und Sozialpsychologie,* re-founded by von Wiese in 1948 and based on its precursor the *Kölner Vierteljahreshefte für Soziologie*, which was published for twelve years from 1921 to 1934. The journals *Soziale Welt*, created in 1950, and *Zeitschrift für Soziologie*, established in 1972, are other journals broadly covering sociology. All of these journals publish nearly exclusively in German.

Over the last three decades, German sociology stretches from more public commentary oriented authors like Ulrich Beck, through strictly inwards academic discourse oriented post-Parsonian system theory, championed by Niklas Luhmann and with an ironically broad recognition in disciplines far from sociology such as literature or law, to a broad spectrum of more or less mainstream sociology, divided into specialized branches ranging from education to urban sociology (the professional association runs 35 different sections). International resonance is earned more frequently by specialized, empirically oriented, sociologists than by theoreticians. Habermas and Beck are two contemporary lighthouses in the latter section, with broad international audiences and a large number of translations. Habermas, for instance, has been translated into more than 30 languages.

Psychology

The Nazi dictatorship transformed psychology massively; large numbers of scholars were expelled or persecuted for racist or political reasons. While psychoanalysis was almost extinguished, with most of its largely Jewish exponents killed or expelled, and Gestalt psychology obliterated, a kind of applied psychology flourished. The number of jobs for psychologists in the Wehrmacht rose from 69 in 1935 to 170 in 1938. The air force employed some additional 150 psychologists in 1942. In addition to military psychology, the Nazis employed psychologists in welfare programs dealing, among other things, with the rising numbers of orphans. The increase resulted in the introduction of a diploma degree program on the university level with standardized curricula in 1941 (Geuter 1984).

After the war, psychologists were only marginally concerned by denazification because those contributing heavily towards Nazism, like Erich Jaensch, were no longer alive. In accordance with the conservative intellectual reconstruction programme of the late 1940s and early 50s, many professors participated in the search for remedies to the moral aberrations of the Nazi era in the philosophical traditions of the pre-war years, which resulted in the predominance of characterology, holistic psychology and

Gestalt psychology. This philosophical re-traditionalizing, however, did not survive its aged proponents (Ash 2008) with the younger generation preferring a more "scientistic" understanding of psychology, very often following behaviourism and experimental designs. The size of academic psychology around 1950 was remarkably small: 19 chairs in 1953 and 56 a decade later. In 2015, the number had increased to 698; the second largest increase in our sample of disciplines.

Many psychologists educated in the 1950s had spent study visits in the USA and developed a new style of research by rejecting philosophical psychology. Experimentation and test construction, both of which had a previous place in German psychology to a small degree, were championed by the younger generation.

Germany's psychologists are professionally organized in the German Society for Psychology (*Deutsche Gesellschaft für Psychologie*), following the Society for Experimental Psychology, founded in 1904, which intentionally changed its title in 1929 to become more inclusive. This organization runs its own journal *Psychologische Rundschau* (founded in 1949) which still publishes in German only.

Only very few psychology journals have been published for longer. The *Zeitschrift für Psychologie* was founded in 1890 by Hermann Ebbinghaus, initially with an additional "*und Physiologie der Sinnesorgane*". It has appeared ever since, with a short interruption in the early post-war years. From 1954 on it was edited in East Berlin and survived the German unification without further interruption; since 2007 it has published in English. *Psychologische Forschung* was founded by Gestalt psychologists in 1922. Since 1974, it has been named *Psychological Research* and publishes in English only. *Zeitschrift für experimentelle Psychologie* (f. 1953) became *Experimental Psychology* in 2002. *Psyche* started in 1947 and is still the most prominent German psychoanalytic journal.

Though some young psychologists and many of the students were politically active, the discipline's intellectual development was not disturbed by Marxist or other versions of psychology (a Marxist, so-called Critical Psychology got some resonance in West Berlin under its founder Klaus Holzkamp). Instead, the 1970s were marked by another wave of extensive professionalization, this time prevailingly in the clinical and therapeutic sectors that were developed only marginally until then. This opened up a large array of new professional fields for psychologists from industry to business (marketing), setting psychology further apart from the other SSH disciplines.

The unworldliness of the discipline was once again demonstrated during the unification process, which did not see an exchange of the academic personnel in the former GDR so typical of the rest of the SSH disciplines. East German psychologists had, in fact, adopted the same methodological tools as their Western counterparts, which made their adaption to the newly established standards easier.[11]

Political Science

Politics as a subject of academic research and teaching reaches back to the establishment of universities in the middle ages. It gained prominence in the first half of the nineteenth century, when many of its proponents were actively involved in the liberal movements leading to the revolution of 1848. During the autocratic restauration following the defeat of the revolution, scholars of law, history and the economy successfully claimed authority over the academic study of politics.

In interwar Germany, the extramural *Deutsche Hochschule für Politik* (German Academy for Politics) in Berlin became the leading institution for a new attempt to establish political science as the science of democracy, pursuing the dual function of raising a democratic awareness among political elites and the public at large and developing political science as a scholarly discipline. Many of its prominent faculty members emigrated after 1933, some of them becaming crucial figures in the (re-) establishment of political science in Germany after 1945.

The US-American occupation powers supported the introduction of political science as part of their re-education policies, which led to the first chairs in political science by the late 1940s. Hessen pioneered institutionalizing political science at German universities assigning chairs

[11] The history of German psychology is comparably well researched, see from Ash and Geuter (1985) to Lück and Guski-Leinwand (2014).

to the universities in Marburg, Darmstadt and Frankfurt. In West-Berlin, the *Deutsche Hochschule für Politik* was re-established, firstly as an extra-mural institution and later integrated into the Free University (FU, founded in 1948). By 1960, half of all chairs in political science were assembled in West-Berlin where the discipline was able to differentiate early on (Bleek 2001: 271). With more than 30 professors by 1989, the FU continued to house the largest department in political science, even after it had lost more than half of its professors following the 1990 reunification. Scholars from the FU Berlin were highly influential. Ernst Fraenkel, who returned from the USA in 1951, is considered as one of the "founding fathers" of the discipline and Otto Stammer became an important pioneer in empirical political sociology. In Heidelberg, Alfred Weber convinced his former student Carl Joachim Friedrich to spend his summer terms in Germany while continuing his position at Harvard during the winter terms from 1954 until 1966. As is the case in sociology, in political science at least one returning émigré received a professorship at each university; they acted as kind of spearhead for the Westernization of West-Germany (see Winkler 2014).

Chairs in political science increased by a factor of 19 between 1953 and 2015. The standard endowment of political science departments at universities consists of three chairs; political theory, political institutions, and international relations.

Apart from studies on contemporary political institutions in Germany (parties, elections, etc.), prominent topics of research have always been contemporary history of German politics, particularly studies on the Weimar Republic, its breakdown and the subsequent Nazi dictatorship. Habilitation theses by Bracher (1955) and Sontheimer (1962), two of the leading scholars of the then younger generation, are exemplary within this research focus. West Berlin-based political scientists dealt with the Soviet occupation zone and the emerging German Democratic Republic, a field that remained prominent under the label *Deutschlandforschung* (research on Germany) until the fall of the Berlin Wall, and even beyond within studies on the reunification process. Only gradually did international politics gain importance with C.J. Friedrich as the most prominent proponent. This later developed into peace studies, development studies, and area studies, the latter being of continuing importance. The field of normative and empirical studies of democracy also flourished during the institutional expansion of political science in West-Germany.

Influential extramural research is being done at several institutes of the Max Planck Society, even if disciplinary boundaries are porous, especially with sociology, (e.g. the Max Planck Institute for the Study of Societies in Cologne, the MPI for Research on Collective Goods in Bonn, the MPI for Social Law and Social Policy in Munich, or the MPI for Comparative Public Law and International Law in Heidelberg), and at the *Wissenschaftszentrum Berlin*, a leading research institution in various fields of policy analysis.

The German Association for Political Science (*Deutsche Vereinigung für Politische Wissenschaft*, DVPW) was founded in 1951. It remains the main professional association for political scientists in Germany with more than 1700 members. As the result of various disputes, the German Society for Political Science (*Deutsche Gesellschaft für Politikwissenschaft*, DGfP) was founded in 1983 as a counter-association, enjoying a generally more conservative reputation.

The oldest journal is the *Zeitschrift für Politik*, created in 1907. The *Politische Vierteljahresschrift*, first issued in 1960, is another general-purpose journal and the official organ of the DVPW. Additionally, many sub-disciplinary sections of the professional associations publish specialized journals, e.g. the *Zeitschrift für Internationale Beziehungen*. German political science journals still publish in German (see Bleek 2001, 2012).

Anthropology

Anthropology, in the sense of Anglo-Saxon social and cultural or French anthropology, does not have an exact counterpart in Germany. The subject most closely corresponding to it has been called *Völkerkunde* or *Ethnologie*, which is distinct from *Volkskunde* (folklore). Traditionally, *Volkskunde,* the study of the people (singular), studied German folk culture and *Völkerkunde,* the study of peoples (plural), visited far away cultures and peoples, mostly extra-Europeans. The two disciplines, though showing some intellectual family resemblance, have had little personal or institutional overlap. During the twentieth century, *Anthropologie* denoted research in medicine and biology with roots in physical or biological anthropology, and palaeoanthropology. In addition, we find a very German specialty in this area: philosophical anthropology. The latter's influence on ethnology has been rather limited, while its key authors Max Scheler, Helmuth Plessner, and Arnold Gehlen did have some followers in sociology.

Since the end of Nazism, German anthropologists lacked any influence on the discipline internationally. Global histories of the discipline, even if written by a German scholar like Werner Petermann's (2004), refer to historical precursors from Germany of the eighteenth to the early twentieth century, to Franz Boas's German origins or to the influence of Dilthey on Clifford Geertz, but do not include German developments in the discipline more recent than the 1920s (Haller 2012: 14).

In 1869,[12]Adolf Bastian became the first university docent and in the same year he founded the *Berlin Society for Anthropology, Ethnology and Prehistory* with Rudolf Virchow. An autodidact as an ethnologist, Bastian undertook extensive journeys as ship's doctor and produced a massive opus of more than 80 partially multivolume books and hundreds of articles. Bastian's obsession with empirical detail and material collection with little or no theoretical pretensions became typical for a second thrust of German ethnologists, who found their home in an uncommonly high number of ethnological museums and collections in many small and mid-sized towns.

The first full professorship was established in 1920 in Leipzig. By 1945, this number had grown to five.

After 1945, three theoretical traditions dominated German ethnology: *Kulturmorphologie*, *Kulturhistorie* and a German variant of what was called *Ethnosoziologie*. Centered in Vienna before the Nazi regime, *Kulturmorphologie* became dominated by Catholic clerics who sought to prove the historical primacy of monotheism as the original form of human religion. As members of the missionary order *Societas Verbi*

[12]It is worth noting that this was before Tylor in Oxford (1884) and Boas at Clark University (1888).

Divini (SVD), they carried on an older tradition called diffusionism—the idea that cultural similarities are the product of cultural contact between neighboring peoples or migration. After WWII, its exponents re-established themselves in Frankfurt at the Frobenius-Institut named after its founder. A journal associated with this group, *Anthropos*, has been published since 1906.

Kulturhistorie emphasized studying cultural change as a process and reaches far back into human history.

The third tradition, *Ethnosoziologie*, was introduced by the Austrian born Richard Thurnwald who sought to incorporate sociological concepts into ethnological studies. During the Nazi regime, Thurnwald's student Wilhelm Emil Mühlmann was an outspoken proponent of a racist ethno-sociology. He later assumed a more appropriate terminology and became a towering figure in German ethnology (Haller 2012: 58).

With the old generation still in place after 1945, the younger generation did little to challenge their teachers, but rather sought their intellectual satisfaction in overseas expeditions that became increasingly available during the 1950s through money from the German Research Fund (DFG). In the context of a modernizing discipline in the line of Western anthropology, German historicist thought appeared hopelessly outdated. It wasn't until the 1960s that there were any serious attempts to catch up theoretically with international trends. Nevertheless, the discipline remained in a rather stagnant mode, both intellectually and institutionally. The number of academic personnel hardly rose, even in times of great educational expansion.

German anthropology produces at least two journals of some standing. As early as 1869 the *Zeitschrift für Ethnologie* started to publish overseas research; the journal still exists. There is also *Sociologus*, founded 1925 by Thurnwald, which still appears in print. It now carries the subtitle *Journal for Social Anthropology* and publishes bilingually in German and English. The field of physical anthropology organized in its own association and in 1949, with *Homo. Zeitschrift für die vergleichende Forschung am Menschen*, created a follow-up journal to the discredited *Zeitschrift für Rassenkunde*. It ceased to exist in 1990, re-appearing after the German unification, which brought with it the unification of East and West biological anthropologists under a new umbrella—the Society for Anthropology—and a new journal *Anthropologischer Anzeiger: Journal of Biological and Clinical Anthropology.*

German Language and Literature (*Germanistik*)

Examining German language and literature goes back far longer than most other disciplines under consideration here. The discipline's predecessors can be found as early as the seventeenth century and it acquired contour between 1820 and 1880 when chairs, journals, and graduates were institutionalized (Rompeltien 1994). The most prominent figures referred to, concerning the early institutionalization of the discipline, are usually Jacob and Wilhelm Grimm, and Karl Lachmann. Lachmann especially is said to have made a considerable contribution to the academic institutionalization of the discipline when he got a chair in philology and Altdeutsche (old German) literature in 1816 in Berlin, where another key figure Friedrich Heinrich von der Hagen had taken the first position of this kind in 1810.

Academic journals were also created very early, some of which, like the *Zeitschrift für deutsches Altertum und deutsche Literatur* (1841–) or *Archiv für das Studium der neueren Sprachen und Literaturen* (1846–) still exist today and—unsurprisingly—publish in German.

The academic and professional association *Deutscher Germanistenverband* (association of German Studies scholars) was founded in 1912 in Frankfurt. Under Nazi-rule it became incorporated into the *Nationalsozialistischer Lehrerbund* (National Socialist Federation of Teachers). In 1952, the academic wing was reinstated as the *Vereinigung Deutscher Hochschulgermanisten* (German association of academic German Studies scholars) and has existed ever-since. This national association is complemented by German Studies associations in other countries. In 1951, the *Internationale Vereinigung für Germanische Sprach- und Literaturwissenschaft* (IVG, International Association for Germanic Studies) was founded on the recommendation of the UNESCO (Batts 2000).

After WWII, the number of professors and students in this discipline outnumbers all other humanities disciplines and made *Germanistik* a dominating field inside academia. The interpretation of "works in themselves" (*werkimmanente Interpretation*) dominated methodologically during the 1950s, led by important figures such as Wolfgang Kayser and Emil Staiger. This approach was ahistorical and contextual factors were not perceived as important.

In the 1960s, the old generation of professors was concerned about inquiries into their Nazi past and tried to cover this by reorienting the discipline towards more philological considerations. This was also the time when a younger generation began to question the historical role of their discipline and professoriate. A peak of that movement was reached in 1966 and 1968 at the "*Germanistentag*", the annual conference of the association of German Studies where, in 1966, personal confrontations took place and in 1968 protesters called off lectures and requested that the political past be dealt with as well as the methodological reorientation of the discipline carried on (Hermand 1994: 155). In the 1970s, curricula were reformed and became more structured. The reforms that took place between 1969 and 1975 led to a stronger focus on contemporary literature instead of classical texts.

Germanistik provides a rather remarkable case of a career trajectory in academia: In 1992, it was revealed that a leading member of its professoriate, Hans Schwerte, was a former leading SS officer who used a false identity from 1945 onwards.[13]

Due to the relatively high number of personnel in the early years after WWII, the expansion of Germanistik has been modest. It expanded to keep pace with the expansion of secondary education and because the number of university sites grew.

SSH in the German Democratic Republic

When the GDR fell apart in 1989–1990 neither university professors nor students had contributed to its demise. Unlike the situation in Poland, Czechoslovakia or Hungary, East German academics were in

[13]Jäger (2007). An even more prominent case, Romance studies scholar Hans Robert Jauß who played a crucial role in interdisciplinary research, made headlines in 2016 when his former university commissioned a historian to find out the truth about his Nazi past, see Westemeier (2016).

almost complete agreement with the communist regime until its end, which is also true of SSH scholars. This was the result of a meticulous and successful engineering of a communist intelligentsia during the early phase of the GDR (Connelly 1997). Key to this development was one of the most radical personal and institutional discontinuities in recent European history of higher education: 83% of the teaching personnel at East Germany's six universities lost their jobs after the end of WWII (Jessen 1999: 261). This made it possible for the authorities to install a new intelligentsia that had relatively few ties with the pre-war conservative elites. Its social constitution, to a large degree drawn from those with a workers' or peasants' background, and the sincere efforts at political indoctrination led to the most loyal intellectual elite of the Eastern bloc. As a result, East German SSH scholars underwent another wave of unprecedented discontinuity after 1990, when most of its academic institutions were colonized by the abundant academic cadres from West Germany.

In the early years of the division of Germany, real sympathizers with the new regime were extremely rare among the professoriate. Instead of winning over the old professors, emphasis was placed on the students. Most of them received state subsidies and were integrated in the party structures. The first scholarship program, established in the late 1940s, emphasized the promotion of the SSH (58% of scholarships were awarded for it). The political reliability of the candidates was the major selection criterion, sometimes overtrumping academic standards (Jessen 1999: 56). The political pressure was intensified with the university reform of 1951 when a new career track for young academics was introduced, the so-called *Aspirantur*. Different to the traditional assistantships, *Aspiranten* were centrally selected by the Socialist Unity Party (SED) and not by the professors. Again, the SSH were most concerned comprising 60% of these positions at the peak of the program in 1954 (Jessen 1999: 59).

The success of political infiltration of the SSH is best depicted by the growing rates in party membership among the university personnel. At the Philosophical Faculties, the share of SED members among professors rose from 38% in 1954 to 67% in 1965. This number had already been up to 63% among higher assistants (*Oberassistenten*) in

1954, rising to 77% in 1965. At the ideologically even more important Faculties of social sciences,[14] 74% of professors were SED members in 1954, rising to 85% in 1965. Among docents and higher assistants, this number was well above 90% at times during the same years (see Table 3.2). The share of former members of the Nazi Party among professors in the philosophical (11%) and social science faculties (17%) was significantly lower than in the mathematical and natural science faculties (31%) and most notably at the medical faculties, where almost half had been Nazis (46%; all dates for 1954). In addition, it is notable to mention the concern for promoting women among students and academic staff.

Finally, while the SED was exceptionally attentive to creating a new intelligentsia loyal to the party-state, repression and, in particular, the open border with West Germany allowed all 'bourgeois' and 'reactionary' academics to be 'got rid of' before the building of the Berlin Wall in 1961. Those who remained enjoyed sizable salaries up to 60(!) times the average salary during the 1950s, very unlike the meagre financial rewards for professors in neighboring Poland or Czechoslovakia (Connelly 1997: 331, 337). This set of measures proved highly effective. During the 1953 workers' uprising, intellectuals were either passive or sided with the regime. When the first generation of fully GDR-schooled professors took office in the 1960s, the loyalty of academics was not a point of contention. Institutionally, the 20-year project of engineering a loyal communist university elite was completed with the 1968 law on the unified education system, after which no major reforms were introduced until the collapse in 1989–1990.

Political indoctrination notwithstanding, East German SSH did produce scholarship in certain areas that was recognized abroad. Typically, research was pursued at the various institutes of the Academy of Science, where even officially disgraced 'revisionists' could sometimes continue their research. Universities, on the other hand, were largely confined to teaching.

[14]In 1954, they comprise economic sciences, law and the 'sciences of society' (Gesellschaftswissenschaften); in 1962, they were labeled as Marxism–Leninism, journalism, economics, and law.

Table 3.2 University teachers in the GDR 1954–1965: SED membership, working class background, women (percentages)

		Philosophical faculties				Faculties for social sciences, incl. Marxism–Leninism			
		Professors	Docents	Higher assistants	Assistants	Professors	Docents	Higher assistants	Assistants
SED members	1954	37.6	47.8	63.2	42.7	73.8	87	90.8	79.7
	1962	63.4	76.4	66.1	65.1	81.1	95.9	94	81.2
	1965	66.5	65.5	77.1	61.1	85	89.8	87.9	76.9
Working class back-ground	1954	12.8	15.2	38.6	37.9	26.2	50.7	60.9	56.5
	1962	21.1	38.5	33.9	42.7	48.6	58.6	59.6	57.6
	1965	30.3	37.9	45.7	47.5	50.9	58.1	63	58.2
Women	1954	5.5	21.7	10.5	23.3	4.8	8.7	10.3	15.5
	1962	7.7	6.9	16.4	32.1	5.4	6.1	7	27
	1965	10.3	4.8	19.3	35.5	2.4	4.5	14.5	23.6

Source Jessen (1999: 463–491)

Among the prominent SSH scholars, one should first mention Marxist revisionist philosopher Ernst Bloch, who openly embraced the SED state by returning from the USA in 1949. Bloch came into conflict with the party during the latter's anti-revisionist campaign in the aftermath of the Hungarian uprising in late 1956. After the Berlin Wall had been erected in 1961, Bloch stayed in the West. Even later than Bloch, literary scholar Hans Mayer left Leipzig for the West in 1963. Despite an army of orthodox Marxists–Leninists, GDR philosophers were recognized in the fields of logic and the history of philosophy.[15] The other disciplines from our sample did not produce many researchers or results that received attention beyond the Communist party.

A Comparative View on the Academic Personnel

Plotting the size of the highest ranked status groups[16] in German universities, similar to full professors elsewhere, for each of our SSH disciplines over 30 years according to data from the Federal Statistic Office of Germany, we see trends and differences. While every care has been taken in the compilation of these data, it should be added that some fluctuations probably remain the product of changing classification routines that can only be interpreted by digging deeper into the particularities of each discipline—something beyond the scope of this chapter. The effects of the German unification process as an outstanding critical juncture in recent decades, however, may serve as an example for statistical particularities. Unsurprisingly, most disciplines experienced a jump in academic personnel due to the growth of the country. For the professoriate, in Fig. 3.1, this jump is visible in the years after 1991 when, for the first time, institutions of the former GDR were included in the statistical reports.

As a general trend, all disciplines display a moderate growth over the period (factor 1.9) with the steepest increase taking place in three

[15] For an institutional history of philosophy in the GDR see Maffeis (2007).

[16] In official nomenclature: C3, C4 and W3, W2 respectively.

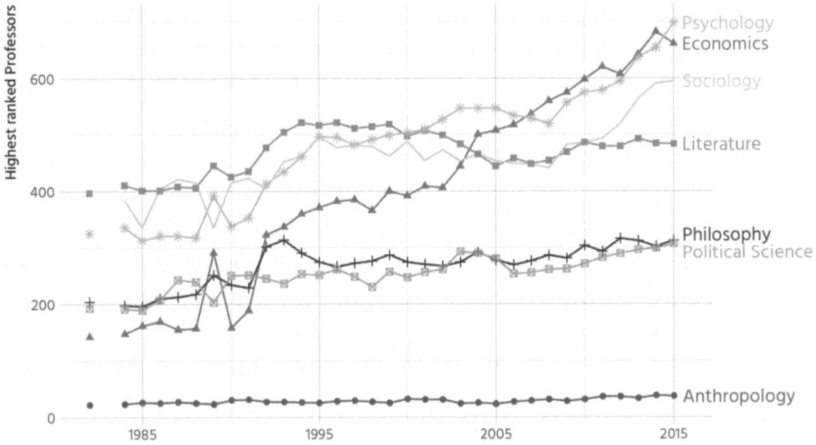

Fig. 3.1 Highest ranked Professors for disciplines, 1982–2015 (*Source* Federal Statistic Office of Germany: Scientific and artistic personnel)

disciplines after 2006: Economics, psychology and sociology, with only economics and psychology showing a remarkable growth over the whole period of 30+ years. Comparing specific disciplinary developments, the marginality of German anthropology, as well as the large size of economics (narrowly defined[17]) and psychology, become obvious, while the relative weight of all other disciplines appears as contested and volatile. Economics and psychology show the greatest and most steady increase in professors, in absolute and relative terms (factors 4.6 and 2.1, respectively). The increase of professors in the disciplines of literature, philosophy, and political science is very moderate (between 1.2 and 1.6).

Another dimension of interdisciplinary comparison concerns the feminization of the SSH. Here, we want to point out two general trends. First, while all disciplines show a steady and significant increase in their share of women over the last three decades, the degree of feminization differs strongly between disciplines and status levels. Anthropology is the

[17]Economics in this figure comprises *Volkswirtschaftslehre* (political economy), *Finanzwissenschaft* (public economics), *Ökonometrie* (econometrics) and *Internationale Wirtschaft* (international economics). *Betriebswirtschaftslehre* (business economics) is not included here.

most feminized discipline in our sample. The share of women among its full-time personnel grew from roughly a quarter in the early 1980s, passed the 50% mark in 2004, and reached its highest number in 2014 with 61%. Psychology and literature show very similar figures with a female majority of 60 and 58%, respectively, among full-time personnel in 2015. At the lower end are philosophy and economics, where the share of women has been below 10% in the early 1980s and is now slightly below one-third (2015). Sociologists are rather gender-balanced today (2015: 47% women); more than political scientists (41%).

The second finding is that all disciplines display striking *internal* gender inequalities. This is illustrated by the much higher share of women among the lower ranks of research associates and assistants than among professors in all disciplines. The larger the gap between the upper and lower lines, shown in Fig. 3.2, the more gender predicts the position of any member in the disciplinary staff within the academic hierarchy. Irrespective of the overall degree of feminization, perfect internal gender equality would be reached if the three lines overlapped. In such a case, women and men in lower ranks of the disciplinary hierarchy would have statistically equal chances of reaching a professorship. Looking at these hierarchically graduated feminization patterns over a period of more than three decades, it appears that inequality has hardly decreased. Further, inequality is also continuously visible in highly feminized disciplines.

In psychology, a clear majority of full-time personnel are women. However, this percentage has been largely the result of the many women taking positions as research and teaching associates. In the latter categories, almost two-thirds (65%) are women, while the percentage of female professors only reached 39% by 2015. Viewed in process, the gap between the feminization among the lowest academic ranks and the one among professors has been growing from the 1980s until the first decade of the twenty-first century and appears to be slowly declining over the last few years. Internal gendered power imbalances are an ongoing concern in the psychological academic profession despite its generally high degree of feminization. Almost all other disciplines display similar patterns with slightly rising or constant levels of inequality over most of the period under consideration and timid (possibly elusive) signs of reversing this trend since the 2010s.

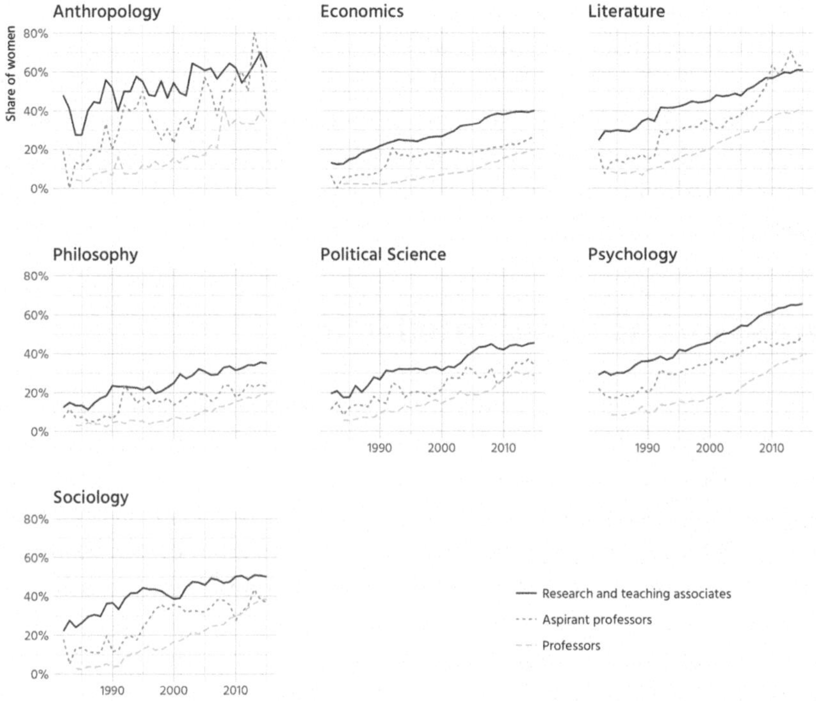

Fig. 3.2 Percentage of women among professors, aspirant professors, and research or teaching associates, seven disciplines, 1982–2015 (The categories contain, according to the German classification system, for 'professors' C4/W3 and C3/W2; for 'aspirant professors' those with more or less stable university positions (C2/W1, docents and assistants); 'research and teaching associates' contain what is labeled 'wissenschaftliche Mitarbeiter'. We are aware of the danger of underreporting in the middle group of aspirants because an unknown proportion of so called Privatdozenten earn their living outside the HE system; there are no indications that one sex is represented there stronger than another one.) (*Source* Federal Statistic Office of Germany: Scientific and artistic personnel (own calculations))

The only discipline where we observe more robust signs of power convergence between the genders is sociology (comp. Lutter and Schröder 2016). With well over a third (38.5%, 2015) of professors being women, sociology is among the most feminized in our sample, along with literature and psychology. But these numbers are accompanied by lesser percentages of females among the lower academic ranks. This means, while fewer women enter the academic profession in sociology, the relative chances of being promoted to professor for those who do are better than those of their female colleagues in psychology and anthropology. Put differently, in the sociological academic profession gender is less associated with any researcher's hierarchical position than in cognate disciplines.

Professors as Public Intellectuals

Many academics not only followed their professorial obligations but used the public sphere to spread their message or to participate in public debates. The years of Nazi dictatorship, when only those standing with the governing party could raise their voice, did not stop this behavior. After WWII, professors sometimes acted as educators of the nation, as concerned citizens, or just enjoyed performing publicly. To offer this story in a condensed form, we used the name index of ten books about the history of (mostly West) Germany since 1945.[18] Some of these titles will be influential in how Germany's past will become its written history. Future generations might be able to consult other sources, but it is safe to assume that the 'big' books written by professional historians whose names will be remembered in the future will continue to play a role.

We found over 50 different names when the analysis was restricted to those who were alive at least for some time after the war and narrowed the sample further to 35 by selecting only those whose names appeared in more than five of the ten books.[19]

The top five most frequently named representatives of the SSH-field comprise Carlo Schmid (named 70 times), Jürgen Habermas (48), Ralf Dahrendorf (45), Rudi Dutschke (42) and Helmut Schelsky (34). The first three are without doubt scholars and public intellectuals who

[18]Bender (2007), Brechenmacher (2010), Conze (2009), Glaeßner (2006), Görtemaker (1999), Herbert (2014), Kielmansegg (2000), Wehler (2008), Winkler (2014, 2015).

[19]Any name, which intuitively seems to be missing, is with the highest probability amongst those that were mentioned less frequently in this sample.

had a reasonable influence in West Germany after 1945. All three did not start their public career until the mid 1950s. The leader of the student's movement Dutschke was seriously wounded by an assassination attempt; after his recovery he published a book and gave interviews, but it is difficult to call him a scholar. Schmid, on the other hand, was an active politician for a very long period of his life, not highly successful, but might have received attention more for his actions as such than for his publications in theory of the state or his translations of Baudelaire.[20] Dahrendorf took an active role in politics for some time, but he could have been quoted for his role as an analytic interpreter of the historical development of Germany. This is also the case for the historians on the list of the most quoted SSH scholars, Hans Mommsen and Heinrich A. Winkler, but also the historical sociologist M. Rainer Lepsius. They might have received their share of attention (each of them has been quoted 11 times) from historians because of their opinions and interpretations of Germany's development and nothing else.

Sociologist Schelsky (34) surpasses these historians. He delivered catchy phrases—such as the "sceptical generation" and "levelled middle class society"—to interpret the current state of society. Furthermore, he did not refrain from intervening in public debates from the perspective of a concerned citizen rather than offering detached analytics of social developments.

It remains puzzling that sociologists are so well represented in the list. Three members of the Frankfurt School: Theodor W. Adorno (18), Herbert Marcuse (16) and Max Horkheimer (14) are present, as is Ulrich Beck (14). Political science is represented by Karl D. Bracher (25), Richard Löwenthal (22) Ernst Fraenkel (14), Eugen Kogon (11) and Wolfgang Abendroth (6). Each of them played a decisive role in the process of re-education and they all, more or less, belonged to the political left. Philosophers are less prominent than someone familiar with German intellectual history would expect: Karl Jaspers (27), Ernst Bloch (8), Martin Heidegger (6), and Karl R. Popper (5) did

---

[20]We excluded from further consideration two more politicians with a background in academia, Ludwig Erhard and Karl Schiller, because they became known for their political doings and not for any of their scholarly contributions.

not get much space in German historians' narratives. Only two men made it on the list who lived in the Eastern part of Germany: Robert Havemann (16) and Wolfgang Harich (8), who held academic positions for a while but became well-known primarily for their role in opposing the Communist party in the East. Both, however, claimed to remain communists.

Most surprising is the very weak representation of other SSH disciplines. From economics, only Alfred Müller-Armack (21), and Joseph A. Schumpeter (8) are mentioned in more than half of the selected books. Müller-Armack is seen as one of the founding fathers of Germany's social market economy (and three of his compatriots in this camp Walter Eucken, Wilhelm Röpke and Alexander Rüstow got some, though less prominent, coverage). Schumpeter did not play a role in post-war Germany, so his mention in half of the books indicates more a recognition of him as a reference author than as a public intellectual (which is true also for the giant John M. Keynes, whose name is mentioned 15 times). It is nevertheless surprising that other well-known economists like Friedrich A. Hayek, who taught for some years in Germany, did not receive much attention, to say nothing of the silenced others from this field. Psychology is in an even weaker position than the disciplines discussed so far. Jaspers is seen as a philosopher despite his psychiatric expertise and psychoanalyst Alexander Mitscherlich is mentioned seven times, with his co-author and wife Margarete named only twice.

Overall it seems appropriate to conclude from these data that only some disciplines functioned as seed beds for scholars as public intellectuals, but the echo scholars gained beyond the ivory tower is modest.

Concluding Remarks

Looking at a sample of Germany's SSH disciplines over a long period shows that the relative weight changed markedly. Philosophy and German literature lost some of their initial standing and the comparably new disciplines of sociology and political science conquered much territory. Economists lost their dominating position as the interpreters of economic affairs due to the rise of management, and more recently, finance studies. By reason of its internal shift away from the 'real' economy into an abstract world of modeling and, more recently, experimentation, university-based economists no longer interact as intensely with politicians or shape public debates in a sustainable manner. Psychology took a different trajectory by, on the one hand, occupying space in education and, on the other, channeling its graduates into professional careers of counseling, handing over the public and the interdisciplinary spaces to other disciplines.

From an international comparative point of view, Germany does not take lead in the percentage of 25–64 year olds with a tertiary degree. With less than 28%, Germany is ranked 29th of 44 OECD member states: In 2015, only 15% of Germans held a bachelor's equivalent, 11% a master's degree, and 1% a doctorate (OECD 2016: 37 and 41). In spite of this moderate development of expanding the base for a knowledge society, there is a continuing decline of the German Mandarins: a professoriate educating the nation and leading its destiny. One could argue that the Mandarins have been erased forever by their involvement in the 12 years of totalitarian dictatorship with its extermination and expulsion policies, and the 40 years of communist domination of the smaller, eastern part of the nation. Their vacant space has been occupied by professionally educated experts, more concerned with their status inside their reference groups than within their society.

References

Ash, Mitchell G. 2008. Psychologie. In *Kulturwissenschaften und Nationalsozialismus*, hrsg. Jürgen Elvert und Jürgen Nielsen-Sikora, 823–862. Stuttgart: Steiner.

Ash, Mitchell G., and Ulfried Geuter (eds.). 1985. *Geschichte der deutschen Psychologie im 20. Jahrhundert. Ein Überblick*. Opladen: Westdeutscher Verlag.

Batts, Michael S. 2000. *Fünfzig Jahre IVG. Die Geschichte der Internationalen Vereinigung für Germanische Sprach- und Literaturwissenschaft 1951–2000*. Wien: Ed. Praesens.

Bender, Peter. 2007. *Deutschlands Wiederkehr. Eine ungeteilte Nachkriegsgeschichte 1945–1990*. Stuttgart: Klett-Cotta.

Bleek, Wilhelm. 2001. *Geschichte der Politikwissenschaft in Deutschland.* München: C.H. Beck.

Bleek, Wilhelm (ed.). 2012. *Schulen der deutschen Politikwissenschaft.* Wiesbaden: VS Verlag für Sozialwissenschaften.

Borggräfe, Henning, und Sonja Schnitzler. 2014. Die Deutsche Gesellschaft für Soziologie und der Nationalsozialismus. Verbandsinterne Transformationen nach 1933 und 1945. In *Soziologie und Nationalsozialismus. Positionen, Debatten, Perspektiven*, hrsg. Michaela Christ und Maja Suderland, 445–479. Berlin: Suhrkamp.

Bracher, Karl D. 1955. *Die Auflösung der Weimarer Republik. Eine Studie zum Problem des Machtverfalls in der Demokratie.* Stuttgart: Ring-Verlag.

Brechenmacher, Thomas. 2010. *Die Bonner Republik. Politisches System und innere Entwicklung der Bundesrepublik.* Berlin, Brandenburg: Be.bra-Verl.

Connelly, John. 1997. Ulbricht and the intellectuals. *Contemporary European History* 6 (3): 329–359.

Conze, Eckart. 2009. *Die Suche nach Sicherheit. Eine Geschichte der Bundesrepublik Deutschland von 1949 bis in die Gegenwart.* München: Siedler.

Elias, Norbert. 1996. *The Germans. Power struggles and the development of habitus in the nineteenth and twentieth centuries.* Cambridge: Polity Press.

Fleck, Christian. 2011. *A transatlantic history of the social sciences. Robber barons, the Third Reich and the invention of empirical social research.* London: Bloomsbury Academic.

Gerhardt, Uta (ed.). 1993. *Talcott Parsons on national socialism.* New York: Aldine de Gruyter.

Geuter, Ulfried. 1984. *Die Professionalisierung der deutschen Psychologie im Nationalsozialismus.* Frankfurt am Main: Suhrkamp.

Glaeßner, Gert-Joachim. 2006. *Politik in Deutschland.* Wiesbaden: VS, Verl. für Sozialwiss.

Görtemaker, Manfred. 1999. *Geschichte der Bundesrepublik Deutschland. Von der Gründung bis zur Gegenwart.* München: C.H. Beck.

Habermas, Jürgen, Ludwig von Friedeburg, Christoph Oehler, and Friedrich Weltz. 1961. *Student und Politik. Eine soziologische Untersuchung zum politischen Bewußtsein Frankfurter Studenten.* Luchterhand: Neuwied am Rhein.

Hagemann, Harald. 2017. Ordoliberalism, the Social Market Economy, and Keynesianism: Germany after 1945. In *Liberalism and the Welfare State. Economists and Arguments for the Welfare State*, edited by Roger Backhouse, Bradley W. Bateman, Tamotsu Nishizawa, and Dieter Plehwe, 57–74. New York: Oxford University Press.

Haller, Dieter. 2012. *Die Suche nach dem Fremden. Geschichte der Ethnologie in der Bundesrepublik 1945–1990.* Frankfurt am Main: Campus Verlag.

Hartshorne, Edward Y., Jr. 1937. *The German universities and national socialism.* London: George Allen & Unwin.

Hartshorne, Edward Y., und James F. Tent. 1998. *Academic proconsul Harvard sociologist Edward Y. Hartshorne and the reopening of German universities: His personal account.* Trier: Wissenschaftlicher Verlag.

Herbert, Ulrich. 2014. *Geschichte Deutschlands im 20. Jahrhundert.* München: C.H. Beck.

Hermand, Jost. 1994. *Geschichte der Germanistik.* Reinbek bei Hamburg: Rowohlt.

Hesse, Jan-Otmar. 2010. *Wirtschaft als Wissenschaft. Die Volkswirtschaftslehre in der frühen Bundesrepublik.* Frankfurt am Main: Campus Verlag.

Hesse, Jan-Otmar. 2013. German university professors' salaries in the 20th century. A relative income approach. *Jahrbuch für Wirtschaftsgeschichte* 54 (1): 111–127.

Jäger, Ludwig. 2007. Schneider, Hans. In *Neue Deutsche Biographie*, hrsg. Historische Kommission der Bayerische Akademie der Wissenschaften, 296–298.

Janssen, Hauke. 2012. *Nationalökonomie und Nationalsozialismus. Die deutsche Volkswirtschaftslehre in den dreißiger Jahren des 20. Jahrhunderts.* Marburg: Metropolis-Verlag.

Jessen, Ralph. 1999. *Akademische Elite und kommunistische Diktatur. Die ostdeutsche Hochschullehrerschaft in der Ulbricht-Ära.* Vandenhoeck und Ruprecht: Göttingen.

Kielmansegg, Peter. 2000. *Nach der Katastrophe. Eine Geschichte des geteilten Deutschland.* Berlin: Siedler.

Lepsius, M.R. 1993. Das Erbe des Nationalsozialismus und die politische Kultur der Nachfolgestaaten des "Großdeutschen Reiches". In *Demokratie in Deutschland. Soziologisch-historische Konstellationsanalysen.* Ausgewählte Aufsätze, 229–245. Göttingen: Vandenhoeck & Ruprecht.

Lübbe, Hermann. 1983. Der Nationalsozialismus im deutschen Nachkriegsbewußtsein. *Historische Zeitschrift* 236: 579–599.

Lück, Helmut E., und Susanne Guski-Leinwand. 2014. *Geschichte der Psychologie. Strömungen, Schulen, Entwicklungen.* Stuttgart: Kohlhammer.

Lutter, Mark, und Martin Schröder. 2016. Who becomes a tenured professor, and why? Panel data evidence from German sociology, 1980–2013. *Research Policy* 45 (5): 999–1013.

Maffeis, Stefania. 2007. *Zwischen Wissenschaft und Politik: Transformation der DDR-Philosophie 1945–1993*. Frankfurt am Main and New York: Campus.

Münch, Richard. 2007. *Die akademische Elite. Zur sozialen Konstruktion wissenschaftlicher Exzellenz*. Frankfurt am Main: Suhrkamp.

Nitsch, Wolfgang, Uta Gerhardt, Claus Offe, Ulrich K. Preuß, Heinz Großmann, and Peter Müller. [1961] 1965. *Hochschule in der Demokratie. Kritische Beiträge zur Erbschaft und Reform der deutschen Universität*. Berlin and Spandau: Luchterhand.

OECD. 2016. *Education at a glance 2016. OECD indicators*. Paris: OECD.

Petermann, Werner. 2004. *Die Geschichte der Ethnologie*. Wuppertal: Hammer.

Ringer, Fritz K. 1969. *The decline of the German mandarins the German academic community, 1890–1933*. Cambridge, MA: Harvard University Press.

Rompeltien, Bärbel. 1994. *Germanistik als Wissenschaft. Zur Ausdifferenzierung und Integration einer Fachdisziplin*. Wiesbaden: VS Verlag für Sozialwissenschaften.

Schnädelbach, Herbert. 1994. German philosophy since 1945. *Cogito* 8 (3): 205–212.

Sohn, Alexander. 2016. Poor university professors? The relative earnings decline of German professors during the twentieth century. *Scandinavian Economic History Review* 64 (2): 84–102.

Sontheimer, Kurt. 1962. *Antidemokratisches Denken in der Weimarer Republik. Die politischen Ideen des deutschen Nationalismus zwischen 1918 und 1933*. München: Nymphenburger Verl.-hdlg.

Sozialistischer Deutscher Studentenbund. 1961. *Hochschule in der Demokratie. Denkschrift des Sozialistischen Deutschen Studentenbundes zur Hochschulreform*. Frankfurt am Main: Sozialistischer Deutscher Studentenbund, Bundesvorstand.

Statistisches Bundesamt. 1953. *Statistische Berichte. Die Lehrpersonen und das wissenschaftliche Hilfspersonal an den wissenschaftlichen Hochschulen des Bundesgebiets und West-Berlins im Wintersemester 1952/53*. Wiesbaden: Statistisches Bundesamt.

Wehler, Hans-Ulrich. 2008. *Deutsche Gesellschaftsgeschichte. 5. Bd.: Bundesrepublik und DDR 1949–1990*. München: C.H. Beck.

Weinreich, Max. [1946] 1999. *Hitler's professors. The part of scholarship in Germany's crimes against the Jewish people*. New Haven: Yale University Press.

Westemeier, Jens. 2016. *Hans Robert Jauß. Jugend, Krieg und Internierung*. Paderborn: Konstanz University Press.

Winkler, Heinrich A. 2014. *Der lange Weg nach Westen. Deutsche Geschichte II; Vom 'Dritten Reich' bis zur Wiedervereinigung*. München: C.H. Beck.
Winkler, Heinrich A. 2015. *Geschichte des Westens. Bd. 4: Die Zeit der Gegenwart*. München: C.H. Beck.
Wolf-Gazo, Ernest. 1980. Trends in German philosophy. A critical view. *Tijdschrift voor Filosofie* 42 (4): 777–780.

4

The Post-war Institutional Development of the SSH in the UK

Marcus Morgan

This chapter provides a broad overview of the institutional and demographic development of our seven social sciences and humanities (SSH) in the UK since the end of WWII. Given the enormous breadth of these concerns, a great deal will inevitably be left out. The purpose is to give a general impression of their institutional evolution rather than to approach an authoritative or exhaustive account of that process. Much of the quantitative data presented here should be handled with great care, and notes have been included on the various problems with these data where relevant.

The first part of the chapter provides a chronological sketch of key policy developments in British higher education over the last 70 or so years and their effects upon the SSH disciplines of interest. It also offers a brief account of the research funding landscape in the UK. The second part of the chapter goes on to suggest certain developmental trends that have taken place across the seven disciplines under study.

M. Morgan (✉)
University of Bristol, Bristol, UK
e-mail: marcus.morgan@bristol.ac.uk

C. Fleck et al. (eds.), *Shaping Human Science Disciplines*, Socio-Historical Studies of the Social and Human Sciences, https://doi.org/10.1007/978-3-319-92780-0_4

Key Policy Developments Since 1945

WWII had certain direct effects on both the philosophical underpinnings and practical institutionalisation of many of the disciplines under consideration. In terms of philosophical underpinnings, in English studies for instance, the war experience prompted scepticism towards the antebellum belief in the humanising forces of an education in English literature since it had become clear how little humanistic acculturation had done to avert much of the barbarity that had taken place. This scepticism led to changes in the self-conception of the discipline and its sense of moral mission following the war. In terms of direct effects on institutionalisation, psychology provides an example of a discipline that benefited directly from the war through psychologically devised intelligence and aptitude tests becoming mandatory requirements for entering the armed forces, a development which subsequently spread out into the civilian population following the war in the form of psychometric testing in educational and job assessment. Nevertheless, a very large part of both the philosophical and institutional shape of the post-war SSH in the UK has come about as an effect of governmental policy intervention.

A moderate increase in student numbers occurred following WWII due to government schemes providing support in entering university to those who had served in the war. However, the first major post-war policy development with special relevance to the social sciences was the Report of the Clapham Committee, chaired by the Cambridge economic historian Sir John Clapham. Published shortly after Clapham's death in 1946, the report was concerned with 'whether additional provision was necessary for research into social and economic questions' (Clapham 1946). The report argued that although more provision was needed for 'routine' social research, a Research Council was not, and in fact funding should instead be focussed on developing teaching departments in order to address the inadequate number of trained social scientists in the UK.

Other policy reforms which influenced the social sciences indirectly included the founding of the National Health Service (NHS), which was opened in 1948 in response to the wartime Beveridge Report

(1942). This Report, which lay the foundations for the British welfare state, was also authored by an academic social scientist, William Beveridge, who had been both the Director of the LSE and Master of University College, Oxford, and was an expert on unemployment and national insurance. The establishment of the NHS coincided with psychology becoming recognised as an established profession in England and from this point onwards although its clinical and academic manifestations developed in divergent ways, the two strands have consistently reinforced each other's legitimacy.

During the 1960s, and reflecting the wider rapid cultural change and global economic growth that was occurring as post-war austerity receded and the baby-boomers came of age, British higher education policy experienced major transformations. Within the period of interest, this was in many ways the 'golden age' for the British SSH, during which many such disciplines were creating more academic posts than there were doctorally trained candidates to fill them, a situation that has been entirely reversed in the current climate across the SSH. Although most Local Education Authorities had been paying fees and providing means-tested maintenance grants to university students for at least 20 or so years, the 1962 Education Act obliged them to do so for all full-time university students. This allowed students to attend university without the worry of financial burden, either at the time of their admission, or indeed in the future, since repayment was not required, leading to many students who would previously have been financially unable to enjoy the benefits of a higher education now being able to do so.

By far the most significant policy development during this decade occurred the year following the 1962 Education Act, with the publication of the Robbins Report of the Committee for Higher Education (Robbins 1963). This report had been commissioned by the Conservative Prime Minister Harold Macmillan in 1961 in order to investigate the need for university expansion, and if a need was found, to determine a plan for implementing such expansion. The committee was once again chaired by an academic social scientist, the LSE-based economist Lionel Robbins, and its primary recommendation was a massive increase in the provision of higher education in order to enable all those suitably qualified to engage. This expansion was to be achieved largely through the transformation of former Colleges of Advanced Technology (which had themselves been established as such in 1956) into Universities. The social sciences, and to a lesser extent the humanities too, grew enormously during this period and the quantity and diversity of new students dealt a definite blow to the pre-war image of a university education as the exclusive preserve of a white male ruling class. Such changes inevitably had an effect on the curriculum itself, and many subsequent developments—such as the rise of feminism in SSH departments during the 1970s, or the 'canon wars' that broke out in English Literature faculties during the 1980s[1]—cannot be adequately understood without factoring in the demographic changes ushered in by the so-called 'Robbins expansion'.

The report also suggested that as well as providing technical skills, universities should also be concerned 'to produce not mere specialists but rather cultivated men and women', and contained interesting reflections on the importance of balancing teaching and research, and their mutual implication within one another. For instance, following the nineteenth century Humboldtian ideal, the report stated that:

> there is no borderline between teaching and research, they are complementary and overlapping activities. A teacher who is advancing his [sic] general knowledge of his subject is both improving himself and laying foundations for his research. The researcher often finds that his personal work provides him with fresh and apt illustration which helps him to set a subject in a new light when he turns to prepare a lecture. (Robbins 1963: para 557)

As we shall see, this ideal of the rounded teacher-scholar has been gradually eroded over the last 54 years, and the separation and specialisation of teaching and research has grown. When Harold Wilson's Labour government came to power in 1964 the majority of the recommendations of the Robbins Report were implemented as policy.

---

[1]The 'canon wars' were (and to some extent continue to be) a series of politicised disputes over which body of works are to be taken as canonical of the Western tradition, and on what basis exclusions are to be made.

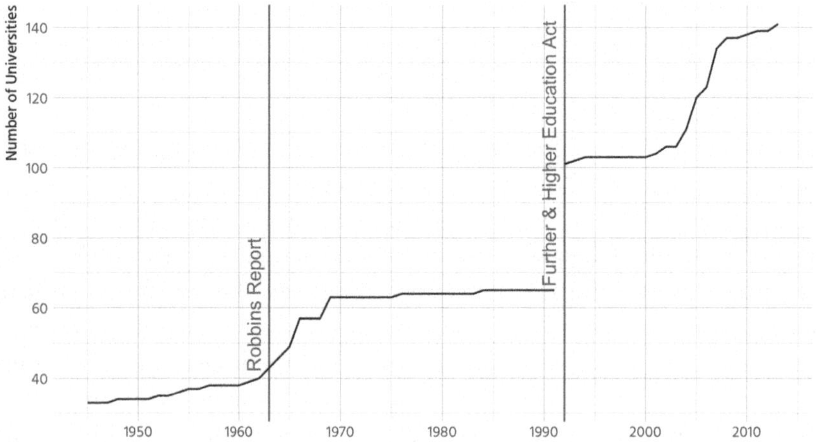

Fig. 4.1 British University expansion from 1945 (*Source* HESA)

Figure 4.1 shows the increase in the number of universities in the UK since 1945. Most of the jump that occurs in the mid-1960s was a direct consequence of the recommendations of the 1963 Robbins Report, though it should be noted that some of the so-called 'plate-glass' universities[2] that were established during this period in fact opened prior to the report's publication.

In 1965, following the 'Haldane principle' (a UK research policy principle established by Richard Haldane positing that academics rather than politicians should determine where research funds are to be spent), the Social Science Research Council (SSRC) was founded, and initially chaired by yet another social scientist, this time the sociologist Michael Young. Another significant development under the Harold Wilson government was the establishment of the Open University in 1969, a project that the same Michael Young was also centrally involved in. The Open University grew out of the earlier 'University in the Sky' and is a public correspondence or 'distance-learning', university that relies

[2]Named after the material with which many were built, in accordance with (and in distinction from) the informal naming of the so-called 'red-brick' civic universities built around the turn of the nineteenth century.

upon mailed, broadcast, and now digitally communicated material as its primary teaching media. Many significant social scientists have been employed there, most notably perhaps, the pioneer of British cultural studies, Stuart Hall.

During the 1970s, successive cuts to the public funding of universities occurred. In most SSH disciplines, the effect of the cuts and the reduction in targets for student numbers, was a slowing down of the growth of those engaged in higher education, rather than an actual contraction (Figs. 4.3, 4.4, 4.5).[3] Many of these cuts were motivated by economic difficulties, but with the election of the Conservative government of Margaret Thatcher in 1979, the economic issues found themselves quickly aligned with an ideological programme of rolling-back public expenditure, and the introduction of quasi-market competition through funding being selectively allocated on the basis of competitive research assessment.

The effect of Thatcherism on the SSH was felt particularly sharply, and whilst it is important to note that there are various limitations with the data presented,[4] Figs. 4.4 and 4.5 suggest that some of the SSH disciplines did contract slightly during this period. Thatcher was openly sceptical of the value of social scientific research, which was of course often critical of government policy, and she was reluctant to continue funding these disciplines to the extent they had enjoyed since the establishment of the SSRC. From 1979 to 1981, the funding to the SSRC was reduced

[3]In 2011, the UK had the lowest drop-out rate for university students in Europe (Schnepf 2014), and for this reason engagement in HE (as shown in Fig. 4.3) can be taken as a better indicator of actual degrees completed than it can elsewhere in the continent.

[4]Mills et al. (2006) point out that each of the different funders of HE research in the UK, from which many of the statistics come, uses a different definition of what constitutes a discipline, making national datasets unreliable, especially for the largest and smallest of disciplines. Different time periods have been taken for different data sources, and one of the major recent sources—the Higher Education Statistic Agency—has also changed its coding of disciplines (see footnote 6). Another set of issues arises from the fact that many subjects are studied as composite degrees, and yet another from the problems of extracting the specifically *social* scientific elements of subjects like psychology, which increasingly overlaps with natural science. The problems of the way in which disciplines are counted is not of course merely a methodological problem of data categorisation, it is also a philosophical problem of disciplinarity itself. Issues with the HESA series that is used from 1994 onwards can be found under the various footnotes for the Student and Staff tables on the HESA website (www.hesa.ac.uk). A useful alternative method of measuring student numbers is through the University and Colleges Admissions Service (UCAS) acceptance data, but since UCAS was only founded in 1992, such data can only be used for relatively recent time periods.

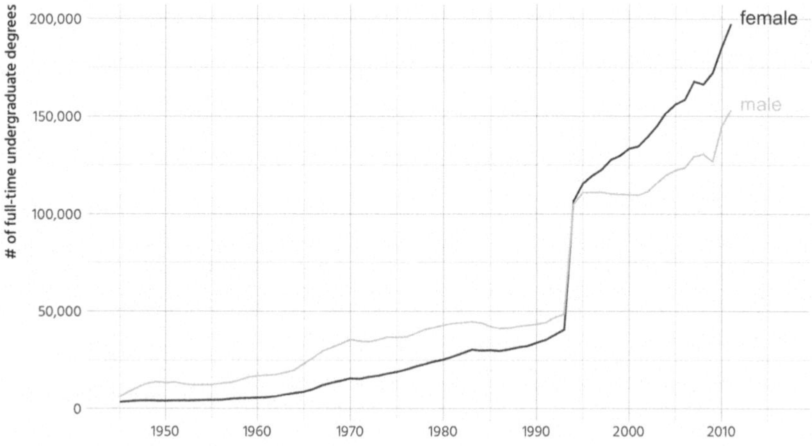

Fig. 4.2 Total full-time undergraduate degrees obtained in all subjects 1945–2012 (*Sources* Annual Abstract of Statistics, ONS/OPCS/CSA; Students in Higher Education, HESA)

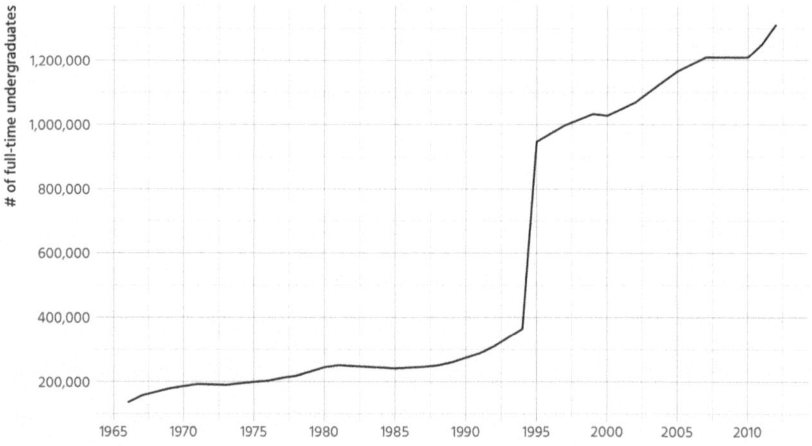

Fig. 4.3 Total full-time undergraduates engaged in HE 1966–2012 (*Sources* 1948/9–1965/6: Annual Returns to the University Grants Committee from Universities and University Colleges in Receipt of Treasury Grant; 1966–1994: Annual "Statistics of Education: Volume 6" Reports from Universities Grants Committee; 1994–2012: HESA)

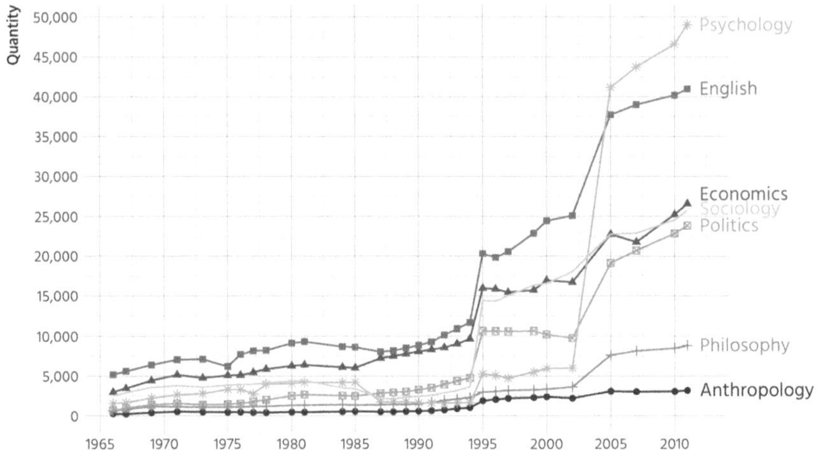

Fig. 4.4 Full-time undergraduates engaged in HE by subject 1966–2012 (*Source* 1966–1994: Annual "Statistics of Education: Volume 6" Reports from Universities Grants Committee)

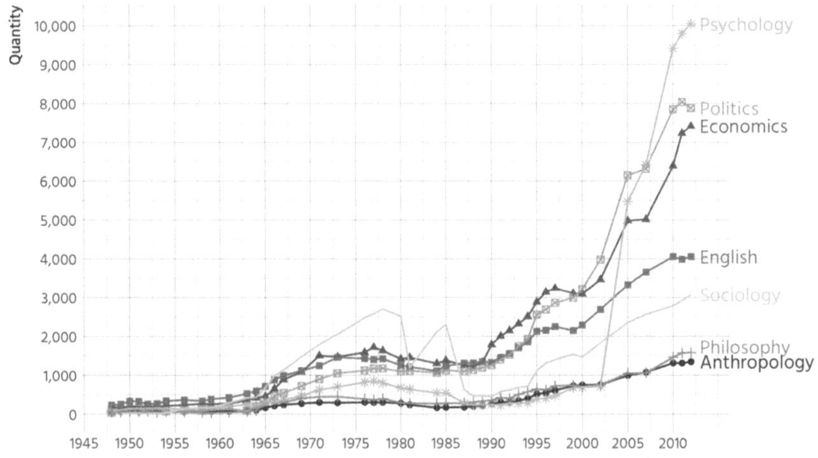

Fig. 4.5 Full-time postgraduates engaged in HE by subject 1948–2012 (*Sources* 1948/9–1965/6: Annual Returns to the University Grants Committee from Universities and University Colleges in Receipt of Treasury Grant; 1966–1994: Annual "Statistics of Education: Volume 6" Reports from Universities Grants Committee; 1994–2012: HESA)

step by step (Halsey 2004: 137). Then, in 1981, Sir Keith Joseph, one of Thatcher's key ideologues and at the time Education Secretary, commissioned an Enquiry into the SSRC. In spite of fears that it might be abolished entirely (Posner 2002), the Research Council survived, though its remit was altered to focus more strongly on empirical research with public relevance. This lead to its renaming in 1983 as the Economic and Social Research Council (ESRC), semantically significant as much in its foregrounding 'Economic', as in its excluding 'Science'. Also in 1983, the first private university—the University of Buckingham—gained university status by Royal Charter, in part due to Thatcher's influence, having previously been established ten years earlier under Thatcher's tenure as Education Secretary as the University College at Buckingham.[5]

In 1986, also during Thatcher's premiership, the first Research Assessment Exercise (RAE) appeared—a programme aimed at assessing the quality of university research in order to determine the variable allocation of government research funding. The proportion of funding allocated under this system was term 'quality-related' (QR) funding, and was originally conducted by the University Grants Committee, which had been set up in 1909. This audit was relatively light-touch at first, but with each successive round of assessment its remit, and the consequences of its results, have increased.

A notable blow to higher education staff came a little later under Thatcher's government with the Education Reform Act of 1988. This Act abolished tenure for all academics appointed after December 1987. Prior to this, there had been 90% tenure for UK university staff (Finkenstaedt 2011: 174). As time has gone on further casualisation of the sector has occurred, and in the academic year 2013–2014, Higher Education Statistics Agency (HESA) data shows that less than two-thirds of academics had permanent contracts.

In 1989, (the year of the second RAE), the University Grants Committee (UGC) was replaced by the Universities Funding Council (UFC), also under the Education Reform Act of 1988. The UFC was

[5]Thatcher's close relationship to the UK's first private university did not end when her time in politics concluded. After she retired, she became the Chancellor of Buckingham University for a decade, and in her honour, the institution has recently announced the establishment of the 'Margret Thatcher Centre' on the campus: a 'museum and library devoted to [her] life and ideals'.

responsible for funding all UK universities with relative autonomy from the Department of Education and Science. The Student Loans Company (SLC) was also established in 1989, before Thatcher's resignation and replacement by John Major the following year. The purpose of the SLC was to provide low-interest loans to students in order to assist in covering living costs. The prospect of debt was thereby introduced as a critical factor for prospective students considering the possibility of a university education.

As Figs. 4.4 and 4.5 show, sociology in particular declined under Thatcher, culminating in an enquiry into the discipline by the UGC in 1989. In 1992, under John Major's Conservative government, the Further and Higher Education Act replaced the UFC with the Higher Education Funding Council for England (HEFCE), the Scottish Higher Education Funding Council (SHEFC) (now the Scottish Funding Council (SFC)), and the Higher Education Funding Council of Wales (HEFCW). The Department of Education for Northern Ireland (DENI) distributed funds in Northern Ireland. Most significantly however, the Further and Higher Education Act also initiated by far the largest and most sudden expansion of higher education in the UK via the announcement that former Polytechnics (in England and Wales) and Central Institutions (in Scotland), which had originally been established as alternatives to universities, would gain University status. Although in many ways this was a nominal change—i.e. one of a re-classification of students to being counted as engaged in Higher Education, it effectively more than doubled the size of the university population in one fell swoop, and its effects can be seen in Figs. 4.1, 4.2, 4.3, and 4.4.[6] Since the vast majority of former Polytechnic and Central Institution students became undergraduates, its influence upon the postgraduate student population, though present, was less pronounced (Fig. 4.5).

[6]The meteoric rise in student numbers that can be seen in Fig. 4.4 from academic year 2002–2003 is on the whole misleading and artificial. This is especially the case for psychology, even though this discipline has certainly become one of the most popular undergraduate degree programmes (Collins 2012: 23–24). The huge jump seen in the graph is largely the result of a change in the Higher Education Statistics Agency (HESA)'s coding of subjects from 'HESA' to 'JACS' coding. For more on this change, please see https://www.hesa.ac.uk/content/view/158/233/.

Whilst they are legally identical, certain characteristics between the older and newer institutions continue to exist, with a larger emphasis on research or research combined with teaching in the pre-1992 universities, whereas the proportion of teaching-only staff is relatively higher in the post-1992 universities. Mills et al. (2006), note that another key difference is the less clearly defined 'disciplinarity' of the SSH in the post-1992 camp, which are often organised around interdisciplinary schools, rather than single-discipline departments. They also suggest that this feature may in part explain the relative failure of post-1992 institutions to access ESRC funding, especially for training, since this funding is predominantly disciplinary in structure (during the 2004/5 academic year, only 5% of research expenditure and 3% of training expenditure was received by such institutions).

One effect of the enormous growth within a relatively short period of time in the number of institutions operating within the HE sector, has been that academics trained in a handful of older institutions have ended up spreading out to teach, and often lead, SSH departments in the large number of new institutions that emerged during the second half of the century. Halsey (2004) and Fourcade (2009: 161) describe this situation with respect to LSE-trained sociologists[7] and economists respectively, and Kuper (1996) offers a similar picture with respect to the spread of Oxbridge, London, and Manchester-trained anthropologists. With respect to psychology, Hearnshaw (1964: 219) notes that in 1960, over half of the chairs in Britain were held by former students of Bartlett who had been trained at Cambridge. Figure 4.2 also suggests

[7]Institutionalised British Sociology was heavily monopolised by the LSE until the end of the two World Wars, although Liverpool University, where a small social science department had been established in 1909, also played a part. Oxford and Cambridge held out a long-standing resistance to the upstart discipline, which was only introduced at Cambridge into the Economics Tripos in 1961 (and properly as a department only in the 1980s by the efforts of Anthony Giddens) and at Oxford into the Philosophy, Politics, and Economics (PPE) degree only in 1962. During the 1950s a strong and steady expansion of the discipline took place, initially still predominantly at the LSE, but then at Bedford College, University of London, and in the civic 'redbrick' universities such as Birmingham, the University of Sheffield, and a little later, Leicester University (where the German sociologist Norbert Elias had moved in 1954). Nottingham had created a department of social administration in 1948, but sociology wasn't taught in earnest until 1964, when it was established a sub-department.

that a fairly large part of the increased representation of women in British Higher Education was, at least initially (the situation has now changed) due to their inclusion via post-1992 institutions.

In spite of the significant problems with the data, Figs. 4.4 and 4.5 still reveal something of the variable degree of expansion amongst the seven disciplines under study. If we compare anthropology with sociology for instance, we can see how anthropology has always been a numerically small discipline, and although it experienced restrained growth during similar moments to the other disciplines over the past 70 years, the extent of this growth has been comparatively limited, making it in some ways the most exceptional of the disciplines under study. This limited growth is in part an effect of the fact that it remained a discipline located primarily within the older institutions and therefore since the 1960s expansion was carried out mainly via the creation of new institutions, rather than the enlargement of pre-existent ones, such expansion had far less of an effect upon it (Evans-Pritchard 1967).[8] A similar story can be told of the second major expansion that took place in the early 1990s, and Mills et al. (2006: 8) note that in 2001 only 5% of anthropology staff were working in post-1992 institutions. In the academic year 2003–2004, anthropology had only approximately 300 staff members, compared with 1400 for sociology, and almost 2900 in psychology. Although many undergraduate programs now exist, for the vast majority of its history, British anthropology has also been above all a graduate pursuit, and it is rarely taught as an A-level in schools and offered by only one of the six exam boards (see Appendix 4.1).

On the other hand, sociology, which had been mostly neglected in the ancient universities and has long been both an undergraduate as well as a graduate pursuit, was expanding faster than it could fill its teaching posts during the 1960s, leading to many trained anthropologists moving into posts in sociology departments at this time. The Robbins Report spurred sociology's expansion on dramatically, and 28 new departments were established during the 1960s, predominantly within the new so-called 'plate-glass' universities, with particularly

[8]The significant role played by Sussex University as a training ground for anthropology graduates is a notable exception to this rule (Spencer 2000).

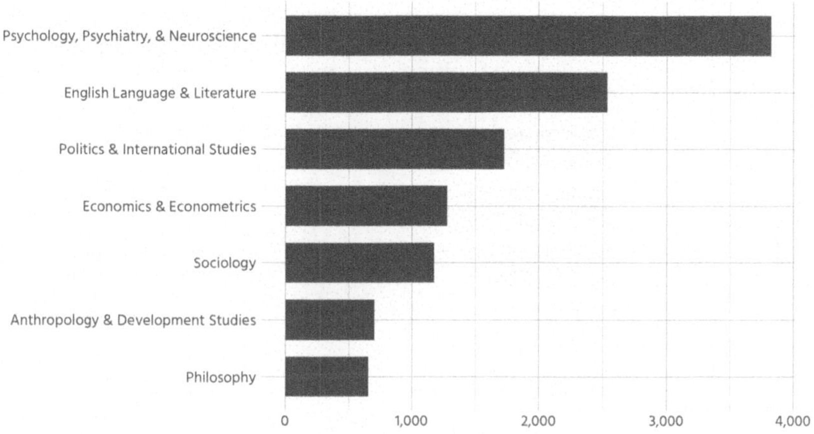

Fig. 4.6 FTE staff eligible for REF 2014 submission, by unit of assessment (*Source* HESA/REF)

important centres opening in places such as Essex and York. In 2006, a relatively large proportion (43%) of sociology's staff were over 50 years old, bearing testament to the 1960s demographic bulge working its way through (Mills et al. 2006: 69). Wakeling (2009) also notes that in 2009, approximately one-third of full-time sociology staff were employed in post-1992 institutions, again demonstrating that a large proportion of the growth that has occurred has done so through such institutions.

Figure 4.6 shows the number of staff operating in the different disciplines who were eligible to submit to the Research Excellence Framework in 2014.[9] Due to its small size, anthropology was counted

[9]The Higher Education Statistics Agency only collect staff data on the basis of 'cost centre' ('an accounting practice that can be used as a proxy for academic departments') or, since the introduction of the 'Joint Academic Coding System' (JACS), on the basis of the discipline of the staff member's highest qualification. Both of these methods present potential problems. The former has the tendency to over-define categories, whereas the latter does not necessarily reflect the discipline in which the staff member is now working. Figure 4.6 has therefore used the designated Units of Assessment (UoA) for submission to the Research Excellence Framework as its categories for locating staff members. Using this data as a proxy for staff numbers operating within separate disciplines is itself not without its own problems however, many of which are listed here: https://www.hesa.ac.uk/ref2014.

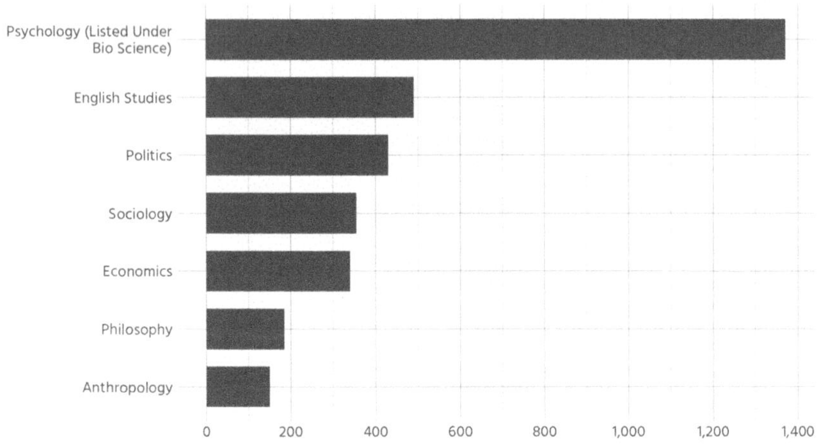

Fig. 4.7 Number of doctorates awarded 2013–2014 (*Source* HESA)

together with Development Studies as a single Unit of Assessment in the last REF, which of course boosted its figures substantially. Given that there were around 300 staff working in anthropology in 2003–2004, it is likely that the figure is still below 400 at the present time, making it significantly smaller staff-wise than philosophy. At the higher end, psychology also experiences 'over-counting' in that it includes the less social scientific psychiatry and neuroscience trends.

Figure 4.7 shows the number of doctorates awarded in the 2013–2014 academic year. Psychology is again by far the largest producer of doctorates amongst the disciplines under review, though numbers for this category are again inflated by the inclusion of the more biological side of the discipline. Collins (2012) argues that psychology's capacity to be applied to social problems—in particular those of education, industry, medicine, and the law—are of key importance to explaining its rise in Britain. Psychology has also enjoyed a large degree of sponsorship from British industry during the twentieth century, for instance through the National Institute of Industrial Psychology (NIIP)'s collaboration with the Rowntree Corporation (Tasker and MacLoed 2000: 1199), or through funding from large companies such as Cadbury's and Debenhams (Collins 2012: 16). It has also frequently been contracted by business to help in determining effective advertising. In 1972, A-level psychology appeared, and proved to be a popular choice for secondary school and college students. It is now offered by all but one of the national exam boards (Appendix 4.1), though it is still rarely a requirement for admission to degree level courses.

In 1997, the year that Tony Blair's 'New Labour' government came to power on a landslide victory, the Dearing Report was published, which amounted to perhaps the most thorough review of higher education since the Robbins Report some 34 years previously. The Report was chaired by Sir Ronald Dearing, a civil servant—rather than academic—who would later go on to become the Chancellor of the University of Nottingham. Its main recommendations were even further expansion of the sector, the creation of an interdisciplinary research council for the arts and humanities (which the following year became the AHRB—the Arts and Humanities Research Board), and reforms to the way in which students were to fund their studies (Dearing 1997). Concerning the latter, the report recommended a move away from grants for payment of tuition fees to a system more dependent upon low-interest loans provided by the SLC (loans had previously been introduced in 1990 as a combined system of 50% loan and 50% grant). Also in 1997, the Quality Assurance Agency for Higher Education (QAA) was established, in order to monitor and advise on standards and quality in higher education teaching.

In 1998, in response to the Dearing Report and amidst great public opposition, upfront tuition fees of up to £1000 per year were introduced for the vast majority of students. Against Dearing's recommendation the maintenance grant was cut, and then abolished in 1999, and replaced by student loans. Northern Ireland and Scotland (which had held a successful referendum on devolution and the creation of a Scottish Parliament in 1997) did not follow this model. In 2000, Scotland diverged even further form England in abolishing tuition fees and reimplementing grants.

Following the results of the 2001 RAE, the Roberts Review on research assessment, chaired by the academic physicist Gareth Roberts, was published in 2003. The Review recommended various revisions to research assessment that were integrated into the 2008 RAE.

In spite of the fact that in their 2001 Manifesto the Labour Party had pledged not to introduce so-called 'top-up fees', the controversial 2004 Higher Education Act went ahead and did precisely that, raising the cap that universities were able to charge in fees for tuition threefold, to up to £3000 per year. These fees were not to be paid upfront (as they had been previously) however, but deferred to the point at which a graduate was earning over £15,000 of income per annum, and this system was implemented in the academic year beginning in 2006. The university sector was split on the issue—the Russell Group[10] of universities (then a slightly smaller grouping than now) being in support of raising tuition fees, and some to a level higher than that adopted by the government, whilst others, including of course the vast majority of students, were in strong opposition. The 2004 Act also reintroduced maintenance grants for students from poorer backgrounds, and created the Arts and Humanities Research Council (AHRC) out of the preceding AHRB in 2005. Also in 2005, the National Student Survey began, asking for feedback from all final year undergraduate students on their university experience. This development in some ways related to the introduction of tuition fees and the reframing of the student as a (variably satisfied) market consumer of education. In 2004, a previous rule stipulating that institutions needed to award research degrees prior to gaining university status was dropped, ushering yet another wave of institutions into the university sector (Fig. 4.1).

In 2006, it was announced that the 2008 RAE would be the last of its kind. It was subsequently announced that it would be replaced by the new Research Excellence Framework (REF), with the intent of providing accountability for public research funding, and selectively distributing this funding across departments in 36 'Units of Assessment' (roughly disciplines). In 2009, regulations were introduced via the UK Borders Agency for student migrants coming from outside the EU in a government drive to reduce net migration into the country. These regulations have faced resistance due to the demonstrable cultural and economic benefits that international students bring to the

---

[10]The Russell Group is an association which represents the interests of a powerful group of research-intensive universities in the UK.

United Kingdom, combined with the universities' own economic interest in capturing the direct revenue gained via the much higher tuition fees they are able to charge such students relative to their UK or EU counterparts.

In 2010, under the new Conservative-Liberal Democrat coalition government, the Browne Review was published (Browne 2010).[11] Unlike many of the aforementioned Reviews, this Review was carried out not by an academic, but by a figure from the business world—the former CEO of British Petroleum, Lord Browne—and was commissioned by the Labour government in 2009. This decision could be taken to indicate a gradual move away from entrusting academics with formulating policy proposals for the institutions in which they work, which has occurred alongside the growth of a managerial class taking responsibility for administering universities. These shifts are also not unrelated to the repositioning of universities as service providers, and students as consumers of such services. The Browne Report recommended a variety of significant changes, most controversially, another massive increase of funding to be gathered from students through abolishing the fee-cap. It also recommended changes to student loans, suggesting that they should now only be repaid once graduates earned over £21,000 per annum. Any debt not repaid after 30 years, it was suggested, should be written off. Most of these proposed reforms were carried through in 2011, though in the face of much opposition the fee-cap was not entirely abolished. Instead the fee-cap was once more tripled, allowing universities to charge up to £9000 per annum for tuition.[12] In Scotland, it was decided that Scottish and EU students would not be required to pay tuition fees, while in Northern Ireland and Wales, EU and home students were only required to pay part of their tuition fees.

In 2012, largely through the initiative of the philosopher A. C. Grayling, another private university, the New College of the Humanities was founded, again generating much controversy, not least due to its

---

[11]Collini (2010) offers an excellent critical overview.

[12]The Liberal Democrats had, during their election campaign, vowed to scrap tuition fees entirely, so their subsequent tripling under the coalition government was understood as a betrayal by many students who had offered their votes to them for precisely this reason.

piggybacking upon the resources of the publically-funded University of London. The results of the first REF were published in 2014, and it was announced that the SSH had fared relatively well overall.

The recently published White Paper on Higher Education (BIS 2016) confirmed that, alongside various other market-oriented proposals such as 'creating a competitive market', promoting 'value for money', and easing both 'market entry' and (most alarmingly) 'market exit' for HE providers, a Teaching Excellence Framework (TEF) will be introduced alongside the current REF during the next round of assessment. A new 'Office for Students' will also be created from a merger of HEFCE and the Office for Fair Access, possessing a much wider remit which includes the granting of degree awarding powers and university title. Opposition to these proposals from various quarters within the higher education sector is ongoing at the time of writing and with the result of the recent referendum on EU membership, the UK SSH may be entering yet another period of uncertainty.

Funding

Public higher education research funding in the UK, i.e. research funding ultimately coming from Her Majesty's Treasury, is organised into what is known as a 'dual support' system, which separates the individual Research Councils' funding, from what is known as the 'block grant' (Hughes et al. 2013). Funding is based around the 'Haldane principle' mentioned above, that researchers, rather than politicians should determine where research funds should be spent. The Higher Education funding bodies—the HEFCE, the Higher Education Funding Council for Wales (HEFCW), and the SFC—provide 'block grants' to universities for research infrastructure and in order to support their strategic priorities. The majority of this funding is QR and based upon a formula which factors in both the volume of research activity taking place, the quality of that research, the attraction of business and charity funding, and the provision of doctoral training within the particular department. Research quality is assessed via the REF (formerly the RAE) every five years or so.

The second element of government support comes from the Research Councils, funded via the governmental Department for Business, Innovation, and Skills (BIS). The relevant research councils for the SSH are the AHRC and the ESRC. These research councils award funding on the basis of peer-review for specific projects proposed by individual researchers as well as proposals submitted in response to particular calls (Bastow et al. 2014). Additional public SSH funding comes from government departments other than BIS, as well as local authorities that commission research directly from universities.

Funding from both of these sources has increasingly come under the influence of the so-called 'impact agenda', whereby both during research assessment, and when applying for money through the Research Councils, the 'impact' of projects is considered as a significant part of the assessment criteria. The idea is that this should act as a mechanism to improve the relevance of research and ensure public accountability. Much of the controversy over 'impact' within the academy has naturally centred around the indefinite meaning of the concept, with critics arguing that much actual 'impact' of research on 'stakeholders' or 'beneficiaries' is difficult to measure (and especially so at the outset of a research project), whilst much apparently measurable impact, is arguably nothing of the sort. 'Impact', its detractors claim, is at best a red herring, and at worst an aspiration that risks distorting the actual merits of research practice (Fig. 4.8).

Additional non-governmental sources of funding for SSH research come from private industry, the EU, and charities and trusts, such as the Wellcome Trust, the Leverhulme Trust, the Joseph Rowntree Foundation, and the Nuffield Foundation. The national academy for the SSH—the British Academy—also provides research funding, and receives an annual grant from BIS (£27 million in 2014/15). Since student fees were introduced, universities have also received a notable proportion of funding directly from students in the form of loans, and HESA data shows that the proportion of funding coming from this source has risen progressively since the early twenty-first century (HESA 2015). However, this is a trend that began long before 2008. From the post-war period right up until 1990, the entire cost of tuition was covered by the taxpayer, but since 1990 this proportion has gradually

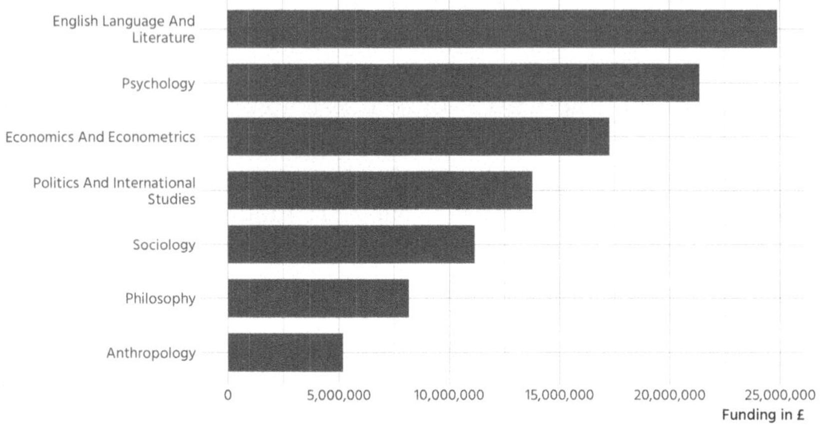

Fig. 4.8 HEFCE mainstream QR funding by unit of assessment 2014–2015 (*Source* HEFCE)

diminished as students have paid a greater share (Wyness 2010). The SSH disciplines also generally require less costly equipment for teaching and research than more technologically oriented disciplines, which effectively means that SSH students receive a lower proportional public subsidy for their education than STEM students. Wyness (2010) also shows how as students numbers have increased, per-head funding levels have simultaneously declined. As well as the creeping proportion of the cost of university being burdened by graduated students, a recent OECD survey also found that England now has the highest average undergraduate tuition fees in the world, higher on average even than those in the United States (OECD 2015).

The increasing reliance upon student fees combined with the arrival of private providers and the competition for resources that is encouraged through research assessment has led to the sector being steadily reorganised around market principles. This fact has become increasingly apparent with recent government policy documents (BIS 2016) and has involved reframing students as private consumers, and academics as service providers in explicit market competition with one another. Ironically perhaps, since the marketisation of higher education has in large part been introduced and established through assessment—and will be even more so once departments are able to raise tuition fees differentially depending upon their relative TEF performance—both regulation and marketisation have increased in tandem, as Foucault (2008) argued is typical of neoliberal forms of government.

Finally, it is worth mentioning that an important historical element of funding for the social sciences in the UK has come from philanthropic organisations based in the the United States, especially during the period of many of these disciplines' initial institutionalisation before and following WWII. The Rockefeller Foundation has been particularly important in this respect, helping finance many key social science institutions. Amongst others, notable recipients have included the London School of Economics, Nuffield College, Oxford, the Oxford Institute of Statistics, the Oxford Economists Research Group, the Economics Research Section at Manchester University, the Royal Anthropological Institute, the National Institute for Economic and Social Research, Chatham House (the Royal Institute for International Relations), the Cambridge Department of Applied economics and the National Institute for Industrial Psychology (Backhouse and Fontaine 2010: 46; Fleck 2011: 75–111; Fosdick 1989: 217; Fourcade 2009: 134–135; 140–141). Fleck (2011: 78) notes that Britain received approximately 40% of the Rockefeller funds dedicated to Europe between 1924 and 1941, the highest proportion by a considerable way.

Trends

The first, and perhaps most obvious trend in the British SSH over the past 70 years has been their massification. As we have seen above, this process has, however, occurred unevenly across the disciplines and the majority of expansion took place in two major waves following the exogenous logic of government policy intervention (in part in response to upturns in the global and national economy), rather than the endogenous logic of the 'natural' evolution of the disciplines themselves.

A second important trend has been the gradual feminisation of many of the SSH disciplines. Overall there are now more female undergraduate students (approx. 55%) at universities in the UK than there are male undergraduates, again however, as we will see, individual subjects differ hugely. In terms of staff, Mills et al. (2006) show that in comparison to their counterparts in the natural sciences, social science academics are on the whole older, and more likely to be female. Whilst outnumbering men in non-academic positions (63% in 2013–2014), women are however still in a minority in terms of academic staff, but the proportion has increased steadily from 40% in the academic year 2003–2004 up to 45% in 2013–2014 (HESA 2014). On the basis of 'cost centre' allocation, women made up only 40.9% of staff in the academic social sciences in the academic year 2013–2014, whereas in the humanities, the figure rises to 44.6%. The increased representation of women in the humanities and social sciences, to the extent that it has happened, has by no means been merely a demographic change, but has also transformed the substance of the disciplines themselves. Much of the change coincided with, and of course contributed to, the rise of feminism within the academy, especially from the mid-1960s.

According to the HESA staff record for the academic year 2013–2014, women constituted 45% of the academic workforce in the UK, though they only occupied 22% of the professoriate, demonstrating a classic 'glass ceiling' effect. In almost all disciplines the proportion of women declines steadily from undergraduate level, through postgraduate degrees, and up the academic rank, what is sometimes called the 'leaky pipeline' effect. Although the change has been gradual and there is still a very long way to go, the overall trend signals an improvement in the position of women: whilst making up just 15% of the professoriate in the academic year 2003–2004, a decade later their proportion had increased to 22%.

On the whole, the humanities have a more equalised gender-balance than the social sciences, which are in turn more balanced than the proportions found in Science, Technology, Engineering, and Mathematics (STEM) subjects.[13] However, examining gender balance through looking at overall groupings of disciplines in this way can be misleading,

[13]In certain STEM subjects the imbalance is quite remarkable. For instance, undergraduate Engineering students were 84% male in 2015 (Universities UK 2015).

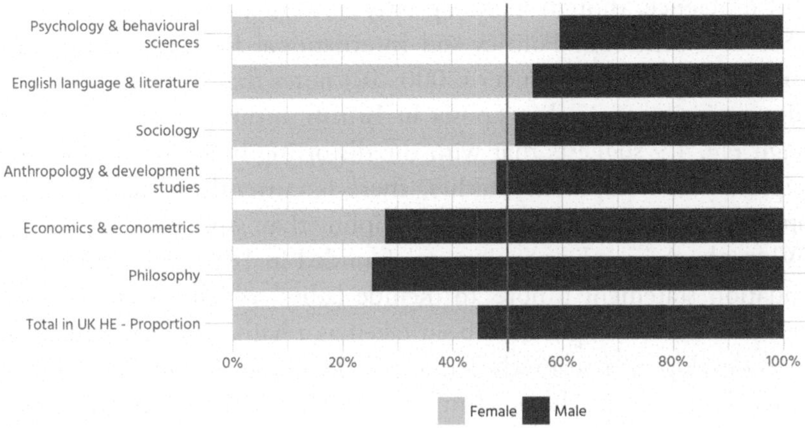

Fig. 4.9 Gender balance in UK SSH by 'cost centre', 2013–2014 (*Source* HESA)

since it masks the stark exceptions found within particular disciplines. For example, as Fig. 4.9 shows, the humanistic discipline of philosophy was the least feminised discipline out of those under review in 2013–2014, whereas Psychology, including behavioural sciences, was the most. In the academic year 2013–2014 women made up almost 60% of academic staff in the cost centre 'Psychology and Behavioural Sciences', up from just over 50% ten years earlier.

English studies is another strongly feminised discipline, and in the academic year 2008–2009 women made up 73% of undergraduates, 70% of Masters-level students, and 61% of Ph.D.-level students. Economics is one of the least feminised outside of STEM subjects, and although there has been some improvement in recent years, women made up only 32% of research students, 17% of full-time permanent lecturers, and only 4% of professors in 1998 (Booth and Burton 1999). By 2006, the situation had scarcely improved with women making up 5% economics professors. By the academic year 2013–2014, women still only made up 28% of all staff allocated to the cost centre 'economics and econometrics' (Fig. 4.9). Grant (2010: 153) notes how the proportion of women in the Political Studies Association actually declined from just 7.3% in 1953 to 5.6% in 1963. In the academic year 2012–2013, women represented 43% of the 2013 undergraduate intake into

Political Science, though made up only 35% of those employed in the HESA 'cost centre' of Politics and International Relations (Briggs and Harrison 2015: 107). Spencer (2000: 10) notes that women have occupied a majority of teaching posts in British anthropology since 1989, though Fig. 4.9 suggests that with the distorting inclusion of the interdisciplinary Development Studies, there is currently a slightly higher proportion of men working in the discipline than women.

When the Aristotelian Society was founded in 1880, it included in its foundation statement a note 'to include ladies' which, given the social mores of the time, might have been taken as a harbinger that philosophy was to develop as a relatively feminised discipline during the following century. However, in the academic year 2013/2014 women made up a meagre 25% of overall philosophy staff, making the most male-dominated of the seven disciplines under study, with women representing only 275 out of a total of 1085 jobs. Barriers to access have included the usual obstacles of implicit bias, stereotype threat and alternative life course models (SWIP/BPA 2011). Prominent British female philosophers who have defied the odds and made successful careers have included the Oxford-based philosopher and novelist, Iris Murdoch, the moral philosopher and scourge of the public atheists, Mary Midgley and the Hegelian philosopher and critic of post-modernism, Gillian Rose (although it may not be insignificant to note that the latter held university posts in European studies and social and political thought, rather than philosophy). Like other disciplines, female philosophers are also more poorly represented at the higher career and education levels (SWIP/BPA 2011), demonstrating the classic 'glass ceiling' and 'leaky pipeline' effects found across the academic sector. For instance, the percentage of female philosophy undergraduates was 46% in 2010/11, but declined to 37% for Masters students, 31% for Ph.D. students, 28% for temporary staff, 24% of permanent staff and only 14% of professors (HESA).

Although they make up a majority of undergraduates, women are still in a slight minority of professional sociologists (approximately 47%), measured either by those whose highest degree was in sociology, or by those who submitted to the 'sociology' Unit of Assessment in the latest REF. As in philosophy, this is especially the case the further one ascends the academic hierarchy, though there are signals of improvement in gender balance among more recent entrants into the discipline (Wakeling 2009). There is also data suggesting that many women chose to take higher degrees in sociology and join the profession following second-wave feminism in the late 1960s and 1970s, and therefore after the first great wave of expansion had taken place. This meant that they were caught in what Platt (2000) has described as a 'demographic promotion blockage', i.e. they were trying to stake out careers during a period in which many of the more senior positions were already taken by men who had secured their posts before second-wave feminism had taken hold. The fact that men make up the majority of the older age groups supports this idea.

A third trend has been the emergence of the USA as perhaps the leading national centre of SSH activity. In many disciplines in which the UK or other European countries may have been able to submit a claim to pre-eminence prior to WWII, such claims have often lost their feasibility during the second half of the twentieth century during which the centre of gravity of many disciplines shifted across the Atlantic. This trend is noted by various authors (Ash 2010; Bunn, n.d.; Fourcade 2009: 163; Fuller 2008; Hammerstein 2011: 407; Kuper 1996; Magee 1971; Mills et al. 2006: 10), and has encompassed the so-called 'brain drain' phenomenon whereby many British and other European academics and students have moved to better-resourced American universities. As was noted in the earlier discussion of funding, a fair proportion of the initial institutionalisation of the British social sciences was in fact made possible due to American philanthropy, as of course was much post-war European reconstruction in general. The UK has itself however, also benefited from American students and academics over this period, and Fourcade, for instance, points out that between '1945 and 1995, 40 percent of the LSE staff with a doctorate were U.S.-trained, including many Americans' (2009: 161).[14]

The rise of America as a global hegemon and the post-war decline of the United Kingdom is of course part of a far broader historical shift associated with the loss of the British empire and the weakening of the UK's international global and economic power. Decolonisation—the vast

---

[14]Mills et al. note that in the academic year 2003/4 '38% of new permanent appointments in economics did their highest degrees in the US' (2006: 10).

majority of which occurred following WWII, and particularly during the 1960s—of course had a more direct effect on certain disciplines, such as anthropology (the early fieldwork of which was conducted almost exclusively in British colonial territories) than it did on others. It remains to be seen what the effect of Brexit on the British SSH will be, although the vast majority of commentators within the UK academy have warned of its likely adverse consequences, in terms of the potential loss of students, staff, access to research funds, and effects on research collaboration.

Whereas work carried out in early British analytic philosophy, especially in centres such as Cambridge, was particularly influential prior to the two world wars, and ordinary-language philosophy continued to exert great influence, especially from centres such as Oxford, for some time following WWII, there is little doubt of the growing influence of American philosophy during the second half of the twentieth century. This was especially the case with the work of key figures such as Willard Van Orman Quine in the analytic tradition and Thomas Kuhn in the philosophy of science. By 1971, Brian Magee wrote that British philosophy 'seems on the way to becoming a chief province in a territory whose capital is elsewhere' (1971: ix) and many successful British philosophers moved to American universities during the twentieth century. Whitehead was an early representative of this migration and was followed later by figures such as Derek Parfit, Colin McGinn, and Alasdair MacIntyre. Many well-known British sociologists, such as Michael Mann and Michael Burawoy, have also enjoyed successful careers in the US academy.

Ash (2010) also identifies a trend of 'internationalisation as Americanisation' in psychology. One indicator of this is that in 1983 British heads of psychology departments voted the American B. F. Skinner to be the most important living psychologist (Bunn, n.d.). In 2012, using H-index measures, the International Benchmarking Review of UK Psychology ranked the UK second in terms of citation, with the USA coming out on top (ESRC/BPS/EPS/AHPD 2011). Likewise in anthropology, especially during the 1980s, America rose to prominence, a situation exacerbated by the financial strains on the discipline in the UK, freezes on appointments, and in many cases encouraged early retirement.

Of course the story is more complex than simple relative decline, and a recent government report notes that relative to research and development expenditure, UK academia still performs very well on a global plane and continues to act as a strong node for research collaboration and international researcher mobility (BIS 2013).[15] Accrued historical reputation (symbolic capital) combined with the English language acting as the global academic lingua franca, no doubt explains part of the UK's continuing success in attracting funding and quality international students and staff. The story is also more complex in that many countries outside the US maintain leadership in particular sub-disciplinary fields, specialisations, or schools. In 2011, British social psychology, for instance, enjoyed a higher citation index than its American counterpart (ESRC/BPS/EPS/AHPD 2011).

A fourth trend has been the specialisation of the disciplines, their differentiation from one another, their gradual professionalisation, and the proliferation of sub-disciplines as they have expanded in size.

Prior to their institutionalisation, the precursors to today's social sciences were often studied within universities in Britain under the vague rubric of the 'moral sciences'. In Victorian Britain, they were often pursued by so-called 'renaissance men' (and indeed they were typically *men*) such as Herbert Spencer, who whilst being a key British sociologist, also authored a central early text in the development of academic psychology (Spencer 1855). W.H.R. Rivers was a similar figure, setting off in 1898 to Melanesia on the famous Torres Straits Expedition not as an anthropologist, but as a psychologist primarily interested in understanding the islanders' psyches. Once there he became fascinated by issues of genealogy, descent and martial relationships, pioneering the anthropological study of kinship.

Processes of disciplinary differentiation can be illustrated by journals such as *Mind*, which began life in 1876 interested in 'this question as to the scientific standing of psychology', yet eventually developed away from psychology into one of the major journals specialising in analytic psychology (the subtitle accordingly changed from *A Quarterly Review of Psychology and Philosophy* to *A Quarterly Review of Philosophy*). The manner in which

[15]The report points out that 'While the UK represents just 0.9% of global population, 3.2% of R&D expenditure, and 4.1% of researchers, it accounts for 9.5% of downloads, 11.6% of citations and 15.9% of the world's most highly-cited articles' (BIS 2013: 2).

'political psychology' or 'political economy' evolved into the disciplines of political science and economics track comparable courses.

Sub-disciplinary proliferation also usually occurred, and all of the seven disciplines examined here can be defined as 'exporter', rather than 'importer' disciplines (Mills et al. 2006), in that they train academics who may then go on to work in other, more specialised fields, or in the production of so-called 'mode 2 knowledge', in which multi-disciplinary teams are brought to bear upon specific research problems or areas. Anthropology is an extremely strong 'exporter discipline', in that out of 230 anthropologists trained from 1994 to 2004 and working within UK universities '46% [are] now employed outside anthropology departments, both in permanent and temporary positions' (ibid.: 55). 'Cultural studies', by contrast, might serve as an example of an 'importer' discipline, in that it has frequently received academics trained elsewhere, as exemplified by the English literature critic Richard Hoggart who founded the influential Centre for Contemporary Cultural Studies at Birmingham University in 1964.

Whilst specialisation helps develop depth, it simultaneously runs the risks of both anti-intellectualism (Said 1994), and the slowing down of innovation through disciplinary cross-fertilisation (more of which below). Specialisation has also occurred within the SSH through the often informal separation of teaching from research and the increase in teaching-only or research-only contracts, and therefore the dissolving of Lord Robbins's aforementioned Humboltian ideal.

Professionalisation has occurred in the SSH both in the looser sense that many disciplines have become more formalised and mono-paradigmatic over time (for instance, economics, with its 'neo-classical consensus'), and also in the stricter sense that practitioners of some social sciences, have become accredited professionals, such as occurred in psychology with the introduction of the publication of a 'Code of Conduct for Psychologists' in 1985 and the 'Official Register of Chartered Psychologists' two years later. It is of interest however, that unlike the relationship between academic and clinical psychology, sociology has usually been institutionally separated from the applied disciplines of social work or social policy. Following the model provided by the natural sciences, the increased importance of publishing in peer-reviewed journals has also taken place across the SSH, especially in disciplines like economics, although the monograph does still hold a strong place, and especially so in the humanities.

A fifth trend in the post-war British SSH is that innovation (and occasionally resultant controversy) has often taken place as the result of the importation of ideas or individuals both from other disciplines and from other countries. In terms of the latter, this often occurred through the arrival on British shores of émigré intellectuals, particularly those that came following the rise of National Socialism in continental Europe.

For example in philosophy, just prior to the outbreak of WWII, the Vienna Circle's ideas of 'logical positivism' were imported from Austria to Britain in large part via A.J. Ayer (1936) who had been encouraged to visit Vienna by his mentor Gilbert Ryle, and who effectively synthesised these ideas with British Humean empiricism. The Vienna Circle were themselves strongly influenced by Russell and Wittgenstein's early works, and one way of understanding Ayer's book is therefore as the re-importation of British ideas, in part devised by an Austrian thinker (Wittgenstein), back into Britain following their development in Austria, with innovation occurring at each step along the path of international knowledge-exchange. What's more, the Vienna Circle case itself also shows the strong influence of the disciplines of mathematics and physics (in particular Einstein's work) on the development of analytic philosophy.[16]

The Austrian philosopher Karl Popper likewise reinvigorated British Philosophy of Science during his time at the LSE following the war, as did his fellow countryman Friedrich Hayek a little earlier in the Economics Department. The Canadian G.A. Cohen helped innovate British political philosophy from a Marxist perspective, and the Polish sociologist Zygmunt Bauman breathed fresh life into British sociology after his arrival at Leeds University in the late 1960s. British psychoanalysis benefited hugely from the arrival of Melanie Klein who came from Austria at the invitation of Ernest Jones in the mid-1920s, and of course Sigmund Freud himself, and his daughter Anna, who fled from Nazi anti-Semitism and settled in London in 1938.

---

[16]In the US, an idea originally developed by the British philosopher Gilbert Ryle would similarly have a innovating effect on cultural anthropology through Clifford Geertz's development of the notion of 'thick description'.

In English studies the emergence of 'Theory' was an innovation which was largely imported both from France, and also to some extent from the social sciences, especially via thinkers such as Barthes, Althusser, Lacan, Deleuze, Foucault, Kristeva and Derrida. Backlashes to this development in the UK occurred during the 1970s and 1980s in what came to be known as the 'Theory Wars' (Morgan and Baert 2015). Levi-Strauss's structuralism (itself developed out of Swiss and Russian linguistic theory) was another Gallic force that had a similarly innovative influence on British anthropology during the late 1950s and 1960s, especially via the figures of Mary Douglas at UCL, Edmund Leach at Cambridge, and Rodney Needham at Oxford. Again, the effects of Brexit on the UK benefiting from all this cross-cultural innovation remains to be seen.

By way of conclusion, it seems clear that the institutionalisation of the British SSH disciplines has advanced a great deal in the past 70 or so years. As well as rooting themselves within universities, government, and civil society research institutions, their influence has also percolated out to wider British society via their applied professionalisation, as in psychology, or their role in public and private consultancy, as for instance frequently takes place in economics. Another way in which this has occurred has been through the diffusion of SSH terminology—and this applies especially to the social sciences—into everyday language, with non-specialists, journalists and laypersons routinely employing so-called 'psychobabble', sociological jargon, or economic metaphor. Unfortunately this diffusion into the everyday has often been less than faithful to the original meaning of many of the terms, as for instance has been the case with British politicians' routine and sincere promotion of Michael Young's (1958) critically- and satirically-conceived term 'meritocracy'. Whilst policy reform enabled the growth of the SSH following the war, it also threatened the institutionalisation of these disciplines during the 1980s, and from this period to the present day, such policy reform has also arguably fundamentally restructured the very character of what the British university is, and the role that the SSH are to play within it.

Finally, it is important to stress that much more systematic data collection is needed in order to provide more accurate commentary on longitudinal trends within the British SSH. Tying data collection processes in the UK to methods used in other European countries would of course also enable much more confident cross-country comparison too.

Appendix 4.1: Examination Boards Offering SSH Subjects at School-Level

A-Level								
Examination board	Anthropology	Sociology	Politics	Economics	English language	English literature	Psychology	Psychology
AQA	✓	✓	✓	✓	✓	✓	✓	✓
OCR		✓	✓	✓	✓	✓	✓	
Edexcel			✓	✓	✓	✓	✓	
WJEC		✓	✓	✓	✓	✓	✓	
CCEA			✓			✓		
Scottish highers		✓	✓	✓	✓	✓	✓	✓
GCSE								
Examination board	**Anthropology**	**Sociology**	**Politics**	**Economics**	**English**		**Psychology**	**Psychology**
AQA		✓		✓ (Business studies)	✓		✓	
OCR		✓		✓	✓		✓	
Edexcel				✓ (Applied business)	✓		✓	
WJEC		✓		✓ (Business)	✓		✓	
CCEA				✓ (Business)	✓			

Bibliography

Ash, M.G. 2010. Psychology. In *The history of the social sciences since 1945*, ed. Roger E. Backhouse and Philippe Fontaine, 16–37. Cambridge: Cambridge University Press.

Ayer, A.J. 1936. *Language, truth, and logic*. London: Victor Gollnacz.

Backhouse, R.E., and P. Fontaine (eds.). 2010. *The history of the social sciences since 1945*. Cambridge: Cambridge University Press.

Bastow, S., P. Dunleavy, and J. Tinkler. 2014. *The impact of the social sciences: How academics and their research make a difference*. London: Sage.

Bell, D. 1982. *Social sciences since the Second World War*. New Brunswick, NJ: Transaction Publishers.

Beveridge, W. 1942. *Social insurance and allied services*. London: His Majesty's Stationary Office.

BIS. 2013. *International comparative performance of the UK research base*. Elsevier.

BIS. 2016. *Success as a knowledge economy*, May 16.

Bolton, P. 2012. Education: Historical statistics. *House of Commons Library*, SN/SG/4252.

Booth, A., and Burton, J. 1999. *The position of women in UK academic economics*. ISER, University of Essex.

Briggs, J., and L. Harrison. 2015. The status of women in UK political science. *European Consortium for Political Research* 14: 105–115.

Browne, J. 2010. *Securing a sustainable future for higher education: An independent review of*.

Bunn, G. n.d. A chronology of psychology in Britain. Published online at http://www.bps.org.uk.

Clapham, S.J.H. 1946. *Report of the committee on the provision for social and economic research ….* London: H.M. Stationery Office.

Collini, S. 2010. Browne's gamble. *London Review of Books* 32 (21): 23–25.

Collins, A. 2012. England. In *The Oxford handbook of the history of psychology: Global perspectives*, 1st ed., ed. D.B. Baker. Oxford: Oxford University Press.

Commission on the Social Sciences. 2003. *Great expectations: The social sciences in Britain*.

Dearing, R. 1997. *National committee of inquiry into higher education*. London: Her Majesty's Stationery Office.

Education Reform Act. 1988. London: Her Majesty's Stationary Office.

ESRC/BPS/EPS/AHPD. 2011. *International benchmarking review of UK psychology*. Swindon: RCUK.

Evans-Pritchard, E.E. 1967. Social anthropology and the universities in Great Britain. *Higher Education Quarterly* 21 (2): 167–181.

Finkenstaedt, T. 2011. Teachers. In *A history of the university in Europe: Volume IV: Universities since 1945*, ed. W. Rüegg. Cambridge: Cambridge University Press.

Fleck, C. 2011. *A transatlantic history of the social sciences*. London: Bloomsbury.

Fosdick, R.B. 1989. *The story of the Rockefeller Foundation*. Oxford: Transaction Publishers.

Foucault, M. 2008. *The birth of biopolitics: Lectures at the collège de France, 1978–1979*. Basingstoke: Palgrave.

Fourcade, M. 2009. *Economists and societies: Discipline and profession in the United States, Britain, and France, 1890s to 1990s*. Princeton, NJ: Princeton University Press.

Fuller, S. 2008. The coroner is not for turning. *Philosophy of the Social Sciences* 38 (3): 383–387.

Grant, W. 2010. *The development of a discipline: The history of the Political Studies Association*, 1st ed. Chichester: Wiley-Blackwell.

Hammerstein, N. 2011. Social sciences, history, and law. In *A history of the university in Europe: Volume IV: Universities since 1945*, ed. W. Rüegg. Cambridge: Cambridge University Press.

Halsey, A.H. 1992. *Decline of donnish dominion: British academic professions in the twentieth century*, 1st ed. Oxford, England and New York: Clarendon Press.

Halsey, A.H. 2004. *A history of sociology in Britain: Science, literature, and society*. Oxford: Oxford University Press.

HESA. 2014. *Staff in higher education, 2013–14*.

HESA. 2015. *Finance record: 2015/6*.

Hearnshaw, L.S. 1964. *A short history of British psychology, 1840–1940*. London: Methuen.

Higher Education funding & Student finance, 12 October 2010.

Hughes, A., M. Kitson, A. Bullock, M. Isoble. 2013. *The dual funding structure for research in the UK: Research Council and Funding Council allocation methods and the pathways to impact of UK academics*, February 2013.

Knorr Cetina, K. 1999. *Epistemic cultures: How the sciences make knowledge*. Cambridge, MA: Harvard University Press.

Kuper, A. 1996. *Anthropology and anthropologists: The modern British school*, 3rd ed. London and New York: Routledge.
Magee, B. 1971. *Modern British philosophy*. London: Martin Secker & Warburg Ltd.
Mills, D., A. Jepson, A. Coxon, M. Easterby-Smith, P. Hawkins, and J. Spencer. 2006. *ESRC demographic review of the social sciences*. Economic and Social Research Council.
Morgan, M., and P. Baert. 2015. *Conflict in the academy: A study in the sociology of intellectuals*. London: Palgrave Macmillan.
OECD. 2015. *Education at a glance: OECD indicators*. Paris: OECD.
Platt, J. 2000. Women in the British sociological labour market, 1960–1995. *Sociological Research Online* 4 (4): 1–12.
Posner, M. 2002. The social sciences under attack in the UK (1981–1983). *La revue pour l'histoire du CNRS* [En ligne], 7. http://histoire-cnrs.revues.org/547. Accessed 7 Dec 2015.
Robbins, L.R.B. 1963. *Higher education: Report of the committee appointed by the Prime Minister under the chairmanship of Lord Robbins, 1961–63. Appendix*. H. M. Stationery Off.
Rüegg, W. (ed.). 2011. *A history of the university in Europe: Volume IV: Universities since 1945*. Cambridge: Cambridge University Press.
Said, E. 1994. *Representations of the intellectual: The 1993 Reith lectures*. New York: Vintage Books.
Schnepf, S.V. 2014. Do tertiary dropout students really not succeed in European labour markets? IZA Discussion Papers, no. 8015.
Spencer, H. 1855. *The principles of psychology*. London: Longman Brown, Green & Longmans.
Spencer, J. 2000. British social anthropology: A retrospective. *Annual Review of Anthropology* 29: 1–24.
SWIP/BPA. 2011. *Society for women in philosophy/British Philosophical Association UK report on women in philosophy in the UK*, September 2011.
Tasker, F., and MacLoed, A. 2000. A brief history of British psychology. *The Lancet* 356: 1199–1200.
UCU. 2007. *Use of fixed-term contracts in the employment of UK academic staff 1995–6 to 2005–6*.
Universities UK. 2015. *Higher education in facts and figures*.
Wakeling, P. 2009. *International benchmarking review of sociology: Statistical overview and commentary*. ESRC.

Wakeling, P. 2010. *International benchmarking review of psychology: Statistical overview and commentary*. BPS/ESRC.
Wyness, G. 2010. Policy changes in UK higher education funding: 1963–2009. Institute of Education Department of Quantitative Social Science Working Papers, no. 10-15, July 2010.
Young, M. 1958. *The rise of the meritocracy, 1870–2033: An essay on education and society*. London: Thames & Hudson.

5

Discipline and (Academic) Tribe: Humanities and the Social Sciences in Italy

Barbara Grüning, Marco Santoro and Andrea Gallelli

Introduction

Italy is the country where the ancient Roman legacy has deep roots and where Renaissance humanism found its major cradle. It has many reasons to claim a central position in the contemporary field of humanistic studies. As the country of Vilfredo Pareto, Gaetano Mosca and Antonio Gramsci, Italy might appear fertile soil for the rise and the development of the social sciences in the twentieth century. This was not the case, however.

B. Grüning (✉) · M. Santoro · A. Gallelli
University of Bologna, Bologna, Italy
e-mail: barbara.gruning2@unibo.it

M. Santoro
e-mail: marco.santoro@unibo.it

C. Fleck et al. (eds.), *Shaping Human Science Disciplines*, Socio-Historical Studies of the Social and Human Sciences, https://doi.org/10.1007/978-3-319-92780-0_5

While Italy is a relatively well-established place to cultivate classical and social studies, it does not figure in the top level of the usual rankings of academic institutions. Italian institutions generally do not figure in these ranking until about the 100th position.[1] There is nothing comparable to Oxford or Cambridge in Italy, not to mention Harvard or Yale. It would be difficult to find an Italian equivalent of the French *Écoles* system or the German Max Planck network. This chapter, however, is not the place to investigate the historical, social and political reasons for this.[2] Suffice to say that Italy—one of the countries where the University originally took shape in the Middle Ages (Collins 1981; Verger and Charle 1994)—has already lost its privileged position in the burgeoning transnational higher education system. In the twentieth century, it has possibly become one of the less promising places in contemporary Western society to pursue a scholarly career. It might not be possible to argue that neither Gramsci nor Benedetto Croce—probably the two most famous Italian intellectuals in the twentieth century—had no academic footing in Italy, while Vilfredo Pareto pursued his academic career in Switzerland, not Italy.

When Italy was unified in 1861, it hosted the largest concentration of academic institutions in Europe, if not the world, with 20 universities spread across the national territory (including ancient ones such as Bologna and Padua, and small, local ones such as the Camerino or Sassari). Currently, the system includes 103 institutions of higher education (56 of which were funded after 1945). The structure of the academic system comprises 68 state universities (including 3 Polytechnics), 18 'free' (i.e. private or non-state, including Catholic) universities, 11 private 'telematic' universities (since 2004), 3 'higher schools' (including the *Scuola Normale Superiore di Pisa*) and 3 graduate institutes. Compared to those in other countries, the Italian system 'enjoys' strong

[1]For an idea of the impact these rankings may have on current debates on the Italian university, see Graziosi (2010).

[2]For a general history of Italian universities, see Miozzi (1993), Brizzi et al. (1999), and Moretti and Porciani (1997). A specialized journal has been founded in 1997 on this topic, the «Annali di storia delle università italiane», edited by the Centro interuniversitario per la storia dell'Università, created in 1996 and based at the University of Bologna.

recognition from the centralized state—an institutional benefit that has a cost in terms of self-regulation and professional autonomy (Giglioli 1979; Clark 1987; Capano 1998; Capano and Regini 2015). At present, the system works on a dual organizational principle. The state Ministry of Education, University and Research (MIUR), maintains the formal power to regulate the entire system, managing the (legal) classification systems underpinning curricula and disciplines and the organizational structures (Departments and Schools, called 'faculties' until 2010), as well as the evaluation systems governing academic recruitment, career progression and the distribution of research funding. Local departments make up the cornerstone of the everyday academic profession, enjoying some degree of political autonomy with respect to the practical organization of teaching and research activities (there is no distinction between research-only and teaching-only departments in Italy).[3]

A crucial feature of the Italian academic system is that all professors, researchers, post-doc research fellows (even Ph.D. programmes, and therefore graduate students) are rigidly classified according to so-called Scientific Disciplinary Sectors (SSD) (Santoro and Pisati 1996). Since its formal institution in 1996, the classification system of SSD functions as the institutional matrix, establishing formal boundaries for each discipline and sub-discipline (see Appendix 5.2 for an example). SSD are further aggregated by disciplinary macro-areas, which set the limits of each (sub)discipline and regulate its relationships with other (sub)disciplines. For example, the subdiscipline of political philosophy—intellectually a partition of philosophy—is located in the macro-area of political and social sciences, along with sociology, political science and the history of political ideas. It is, therefore, severed from

[3]The current academic hierarchy is the outcome of a series of institutional reforms temporally dislocated and far from being coordinated: full and associate professors (since 1980), tenured researchers (a position however abolished with the reform n. 240/2010), temporary researchers without tenure track (created by a law in 2005 also abolished in 2010), senior temporary researchers with tenure track, and junior temporary researchers without tenure track. The figure of senior temporary researcher should represent the new first step for academic career (and recruitment), but in the last five years only about 500 positions have been created for all the disciplines.

the macro-area of philosophy. The main effect of this rigid classification system is strong academic segregation along disciplinary lines, a sort of *tribalization* according to a disciplinary principle that makes it very difficult for both teachers and researchers to move within the academic space. 'Academic tribes' are far from being an Italian peculiarity (Becher and Trowler 2001; see also Abbott 2001). We contend that Italian academic tribes enjoy a degree of institutionalization and formalization difficult to find in other places, which enforces the *formal* institutionalization of disciplines at the price of scientific communication and interdisciplinary exchange. Importantly, it also affords little opportunity to adapt or change intellectual or knowledge conditions.[4]

Bachelor (B.A.) and Master's (M.A.) degree courses are classified and aggregated according to 'classes' that comprise a different system of classification with respect to SDD. The correspondence between teaching classes and SSD is far from perfect since often classes have an interdisciplinary character, something begun, and empowered, by the Bologna process in 2001. Still more complex is the classification of the graduate (Ph.D.) programmes, especially since 2001. The 'disciplinary identity' of each graduate programme depends on the interplay of the SSDs of the professors involved (calculated as a percentage). The integration by law of many small departments (enacted in 2010) and the consequent increase of interdisciplinary departments, together with a parallel decrease in public funding, exacerbated the existing trend towards interdisciplinary graduate programmes; a trend that contrasts with the persistence, if not strengthening, of disciplinary boundaries within the academic system.[5]

[4]For making just an example, no chair currently exists in Italy in nowadays relatively well-established intellectual fields as cultural studies or STS—and this not because of lack of practitioners but because of a missing recognition of these fields as such at the national, administrative level. There are some teachings in this field of course, but from scholars institutionally identified according to other labels, as sociology or literary studies or even philosophy. It is this institutional identification of membership to a disciplinary sector which sets the rules of the play.

[5]As a consequence also the number of Ph.D. students as a whole decreased in the last years: 37,550 Ph.D. students enrolled in 2005; 35,492 in 2010; 33,037 in 2013. Source: http://statistica.miur.it/scripts/postlaurea/vdottori_isc1.asp.

Finally, it is worth remembering that, in Italy, scientific research is carried out almost exclusively within universities, especially in the social sciences and humanities (SSH).[6] Of the eight national research centres under MIUR's control, only one—the CNR, *Centro nazionale di ricerca*, or National research centre—provides space for (applied) social sciences, but out of 11 departments and 103 institutes, only one department and six institutes are devoted to SSH-research areas (for a total of 157 researchers).[7]

Continuities and Changes in the Italian Academic System (1861–2015)

Even a brief history of the seven disciplines under investigation is useful, not only to identify the different degrees of institutionalization reached at the end of the Second World War, but also to pinpoint their different trajectories in the post-war phase and highlight the current structure of their symbolic boundaries. For this, in what follows we will describe the institutionalization process of our seven disciplines, focusing on the legal and social transformations of the academic field in which they are embedded. We will pay particular attention to three crucial historico-political breakpoints: Italian unification (1861), the advent of the Fascist Regime (1922) and the foundation of the Italian Republic (1946). Our aim is not to offer a sketch of Italian history, but to identify the major points of entanglement between this general history and the more particular histories of our disciplines.

At the time of Italian unification in 1861, only philosophy, Italian literature and economics enjoyed some degree of academic institutionalization. At the beginning of the nineteenth century, a first step towards the academic autonomization of philosophy occurred in the northern

[6] The separation between the two careers is strictly regulated: researchers who work in public research institutes have a national employment contract whereas academic researchers and professors are public officials comparable to diplomats and judges.

[7] Data collected on November 20, 2015. Sources: website of ILIESI, ISPF, CERIS, IRPPS, ISTC, ISEM.

part of Italy with a Napoleonic reform law (1810) establishing the new faculty of '*Belle lettere*'. Since then, both Medicine and the Natural Sciences were severed from philosophy, to which they had been attached since the Middle Ages. This innovation was soon followed by the creation of two faculties of '*Belle lettere*' in Rome and Perugia (1810) by the Papal State and two separate degree courses in 1826, one in philosophy and the other in philology.[8] A similar move occurred two decades later in the Kingdom of Savoy with the institution of a faculty of '*Belle lettere*', distinguished from that of 'Natural Sciences'. In the process of political unification, first with the Casati reform of 1959 and later the Coppino reform of 1877, philosophy eventually gained its full institutional autonomy as a discipline: the formerly established the Faculty of '*Lettere e filosofia*', and the latter distinguished two degree courses: one of 'Letters' (including both classical studies and Italian literature) and one of 'philosophy'.

Before 1861, the teaching of *Italian literature* was relatively neglected because of the cultural supremacy of Latin and Greek studies, as well as the weak identity of Italy as a cultural space. With the exception of some failed experiments in the University of Pavia at the beginning of the nineteenth century, the first chairs in Italian poetry and rhetoric were created in the Papal State and included in several academic curricula as preparatory teaching (in arts, theology, law, medicine, for example). With the University reforms in 1877, these chairs were renamed as 'Italian literature' (Formica 2000; Ferraresi and Signori 2012; Archivio storico della camera dei deputati, online). In 1882, the institution of the Faculty of Magistero (alias 'Pedagogy'), which aimed to prepare students to teach in secondary schools, paved the way for an increase of chairs in both 'Italian literature' and 'philosophy'. The same phenomenon occurred during the Fascist regime when Magistero was transformed into a full—albeit socially stigmatized—faculty (Di Bello 2006).

While the first chair in *Economics* in the Italian peninsula was established in Naples in 1754, a tangible academic establishment of the

[8] The denomination of 'philology' followed the German tradition in studying national and foreign literature.

discipline came only after the Casati reform, which made the teaching of *economia politica* (the Italian name for 'economics') and 'financial science' (corresponding to 'public economy') compulsory within the Faculty of Law. The first chair in economics was created in Turin in 1859 and the first '*Laboratorio di economia politica*' (Laboratory of Economics) was founded was founded in Turin, within the Law faculty. Other steps towards the institutionalization of the discipline were the foundation of Schools[9] of Commerce after unification, the first in Venice in 1869, following the German Model of the *Realschule* and originally principally concerned with applied economics. The political decision in 1920 to rename these as 'high schools of economic studies' marked a gradual shift towards the 'pure' field of economics with a consequent gain in prestige. At the *Istituto Superiore di Scienze Sociali* ('Higher Institute of the Social Sciences'), founded in Florence in 1875, economics was taught alongside the 'science of administration', with the aim of educating the new Italian political class. From 1902, this was also the case in the newly founded *Istituto superiore commerciale* Luigi Bocconi in Milan, which in the following decades became the most prestigious academic research institution in the field of economics in Italy (i.e. Rabbeno 1891; Falkner 1891; Loria 1891; Ritrovato 2013; Garofalo and Graziani 2004; Cagiani 2006).

Anthropology and *Psychology* started their route toward academic institutionalization just after the birth of the Kingdom of Italy. The first chair in 'Anthropology and Ethnology' was created in Florence in 1869. Initially, the teaching of anthropology was included in the degree courses of philology and philosophy, but the University reforms in 1877 relocated it to the natural sciences. Both the first Anthropological Institute—founded in 1884 in Rome—and the first chair in Criminal anthropology—held since 1905 by Cesare Lombroso in Turin (formerly a professor of Legal Medicine)—were attached to the Faculty of natural sciences. It wasn't until 1914 that the first chair exclusively devoted to Ethnography was created, three years after the first Italian Congress

[9]Venice (1869); Genua (1884); Bari (1886); Milan, Bocconi (1902); Rome (1906); Palermo (1918); Catania (1919); and Naples (1920).

of ethnography and, arguably, in its wake (Clemente 1985; Fedele and Baldi 1988). The institutionalization of *Psychology* also began in Florence at the end of the nineteenth century and, like anthropology, the discipline was located at the boundary between medicine, biology and humanities. A natural scientific orientation was dominated the first phase, indeed various laboratories of experimental psychology were created—the first in 1889 by the Institute of Anthropology in Rome—and the first chair in experimental psychology was created in 1906 at the Faculty of Medicine in Rome (i.e. Chiabra 1904; Ferrari 1905; Lombardo and Foschi 1997; Cimino and Dazzi 1998; Marhaba 1981; Guarnieri 2014).

Even if *Political Science* and *Sociology* took their first steps in the same period as psychology and anthropology, their trajectories were quite different. Their development can be seen as more intellectual than academic. Sociology entered the academic system in the 1870s, but only as a subject for free courses usually taught by professors with chairs in law or philosophy. There was a serious attempt to establish sociology in the academic field in 1906 with a formal request to the Minister of Education signed by more than 20 professors, but the attempt failed mainly because of the resistance of more established disciplines as philosophy (including Croce) and law. Many other attempts have been made to give some institutional footing to the discipline of sociology, but these were generally impractical and doomed to failure (see Fiamingo 1895; Michels and Hartsough 1930; Ferrarotti 1957; Burgalassi 1996; Scaglia 2007; Santoro 2011, 2013).

While strongly rooted in Italian culture, at least since Machiavelli and Botero, *Political science* was formally excluded from academic curricula as a fully fledged discipline. Even Gaetano Mosca—the author of *Elementi di Scienza Politica* (1896), a book considered a founding text in the discipline, never taught the discipline, devoting himself instead to the teaching of 'constitutional law' and the 'history of political ideas'.[10]

[10]Indeed, Mosca had the opportunity to teach 'Political Science' between 1918 and 1923 at the (private) University Bocconi in Milan. In 1926, Roberto Michels held a free course in sociology at the Università La Sapienza in Rome. See *Giornale degli economisti* (1981, vol. 40, p. 54) and Michels (1927).

The greatest obstacles for the acceptance of (positivistic) sociology in the academic system were related to the strength of traditional philosophy and the increasing influence of neo-idealism. However, the study of political science as an autonomous field, and politics more generally, was chiefly hindered by the hegemony of legal studies, as is evidenced by the two failed attempts (in 1862 and 1875) to create an autonomous degree course in politics (Villari 1902).

In order to follow the institutionalization processes of SSH disciplines during the fascist regime, three aspects must be taken into account: the hegemony of the 'neoidealistic thought' favoured by the new minister—and philosopher—Giovanni Gentile in 1923, the penetration of political goals within the academic field, and the rigid definition of curricula set forth by the Gentile Reform. Gentile's 1923 reform principally reinforced the supremacy of the humanities within the academic hierarchy of the SSH disciplines by stressing the elite character of the gymnasium and strengthening the distinction between leading Universities (those with all the main faculties, i.e. letters and philosophy, law, mathematics and medicine), secondary universities and High schools in applied and technical sciences, the latter under the jurisdiction of the Ministry for the economic development (i.e. Ostenc 2012). It was an unintended consequence that this measure favoured the expansion of these schools including the High schools in Economical and commercial studies, a process further developed in 1924 with the equalization of their diplomas with academic degrees and, in 1935, with their transformation into Faculties of Economics and Commerce.

A second crucial aspect of the Gentile Reform was the administrative and teaching autonomy conferred to the University, albeit within a framework of political control from the (Fascist) state. This led to a 'two-faced' process. On the one hand, the Fascist regime pushed the institution of new Schools in Political sciences (1924) aimed at educating a new fascist ruling class. These were already being gradually transformed into Faculties by 1925. The study of politics was, however, still embedded within other studies—historical, statistical, economic, legal and so on—and was not a discipline in its own right (De Vivo and Genovesi 1984; Comparato et al. 2011). As a consequence, and paradoxically, political science didn't exist as a separate field of study and research during the fascist regime. The only teaching called 'political science' was at the Institute of Social sciences in Florence, but was suppressed in 1938 (i.e. Spreafico 1964). In Pavia and Rome, 'political science' was taught under the name of 'State doctrine' instead. It was left to legal theorists, and strongly informed by neo-idealistic thought (Bartolé 1991; Mongardini 2003). Greater autonomy in the organization of political studies was conceded to the Catholic University of Milan, a private university linked to the Catholic Church. In 1926, under the direction of Father Agostino Gemelli, a 'School of political, economic social sciences' was funded. It taught two degree courses, one in economic studies and one in political and social sciences.[11] The school was transformed into a Faculty in 1932, still preserving its Catholic origins and a philosophical orientation (Palano 2005).

Meanwhile, in both the liberal and the socialist milieu, sociology was losing the original attractiveness that guaranteed some footing in both the academy and the public sphere in the last decades of the nineteenth century. In 1921, the *Rivista Italiana di Sociologia*—a journal founded in 1897 that, for two decades, offered a venue for discussion in the social sciences and in sociology more specifically—ceased publication. According to an oft-repeated and well-established interpretation, sociology suffered during fascism, almost unto death. However, it was *during* fascism that sociology was eventually introduced—even if only as complementary teaching—in political science degrees and as compulsory teaching in demography and statistics. In demography and statistics, this was largely due to the entrepreneurship of Corrado Gini, professor of statistics at the University of Rome and one of the leading intellectuals and technocrats of the regime (Barbano 1987; Losito and Segre 1992; Padovan 2006; Cassata 2006; Santoro 2013).[12]

During the fascist regime, anthropology began to be increasingly understood as a natural science discipline, partly due to the growing

[11]The original faculty of social sciences was transformed in 1923 in faculty of law.

[12]Close to fascism were both Roberto Michels, author of the influential *Sociology of Political Party* (1911), and Camillo Pellizzi, holder since 1938 of a chair in *Doctrine of Fascism* in Florence, who would be the first chaired professor in sociology in Italy, after the fall of Fascism, in 1950 (see above).

political relevance of racial studies at the time (Bottai 1939). Cultural anthropology, however, was marginalized. This was partly due to its subordination to history of popular traditions and history of religions and paleography, under the influence of a Catholic culture that held views contrary to evolution, and partly because of compatibility, or otherwise, with the neo-idealistic historicism of Croce. So, for example, in 1937 ethnology was taught, 'unofficially', with 'historical sciences'. In 1942, an institute devoted to ethnological studies was founded but was called the 'Institute of Primitive societies' (Grottanelli 1977; Testa 1996).

The institutionalization of psychology was also partially hindered by Gentile's hostility towards positivist thought; he restricted the number of chairs in psychology. Research in experimental psychology increased, however, with the addition of two laboratories at the University of Padua and Milan's Catholic University. Further, in the 1930s, Musatti successfully introduced teaching of psychoanalysis at the University of Rome. A more crucial role, however, for the furthering of psychology after the Second World War was played by the school of psychology of Agostino Gemelli at the Catholic University of Milan (Lombardo and Foschi 1997; Proietto and Lombardo 2015).

The early post-war period mostly continued concerns of the previous phase as regards chairs, degree courses and faculties in Italian literature, philosophy, economics and, partially, psychology. Conversely, the institutionalization process of the less established disciplines did not begin until the 1960s. The first chair in sociology was created in 1950 by converting Camillo Pellizzi's existing chair of 'State and doctrine of fascism'. Until 1963, it was the only chair for the discipline. The institutionalization of political science was hindered by the hegemonic model of political sciences (i.e. Sartori 1965). Cultural anthropology was still subordinate to the history of popular traditions and of religions.[13]

The situation changed in the 1960s in connection with student protest. In 1962, a higher academic Institute of social sciences opened in Trento. By 1967, it had been transformed into a faculty and had the

[13]In 1952 there was still a chair in 'Colonial ethnography' held by Raffaele Corso that will be later transformed in 'ethnography'.

first degree course in sociology (Balbo et al. 1975). The first chair in political science was established in 1966 in Florence and in 1967, three chairs in 'ethnology' were created at the universities of Rome, Bari and Genoa. With the curriculum reforms in 1969, the teachings of sociology and political science became compulsory within political science degrees (cf. Lotti and Pasquino 1984). In 1971, two degree courses in psychology were inaugurated in Padua and Rome. Degree courses in economics and commerce were flanked by more specific degree courses in either political or management economics (cf. Castellino and Zanetti 1980; Garofalo and Graziani 2004). Degree courses in political science and cultural anthropology were introduced only after the reform of '3+2' with the Bologna process (2001) and then only for the master's degrees.

Patterns of Institutionalization

If the institutionalization of an intellectual activity can be taken to mean: 'the relatively dense interaction of persons who conduct that activity within a social arrangement which has boundaries, endurance, and a name' (Shils 1980, 168), the previous section shows that only a few disciplines were institutionalized by the end of the Second World War. Neither sociology nor political science, for example, had chairs, nor it was possible to grant degrees in sociology or psychology, even the subjects were taught under these names. No courses called political science existed until the 1960s. So, what about chairs and teaching in the period after 1945?

Disciplines and People

The first set of indicators we consider in order to assess the institutionalization of the social sciences and humanities in Italy concerns academic careers. As shown above, since its origins, university teaching in Italy has taken on several forms, usually in response to reform processes and different recruitment strategies. It goes without saying that

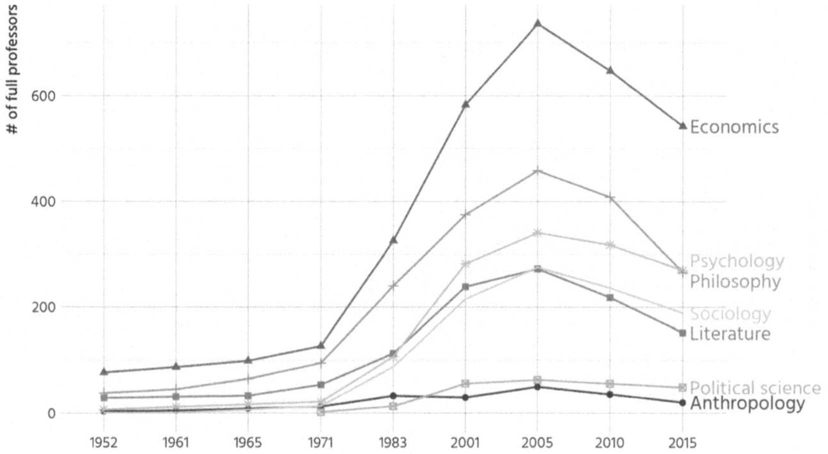

Fig. 5.1 Full professors in the seven selected disciplines

the endogenous transformations in the academic field follow global complex societal change and therefore trends in the number of scholars are related, inevitably, to the expansion of educational opportunities, changes in the labour market, the strengthening of the welfare state and so on.

There is both a period effect and disciplinary trend in the number of full professors in Italy. A similar evolution for all the disciplines is observed, shown in Fig. 5.1. The first 20 years of the period of observation are characterized by a low number of full professors, with any tendency towards growth scarce, if not absent. Across the whole period, there are three significant turning points. From 1971, there is a strong increase in the number of professors in all disciplines. From 2001, the growth becomes much weaker and, in some cases, negative. From 2005 onwards, there is a constant and strong reduction in the number of full professors. Political science is the only exception showing a stable, if low, trend in the last 15 years.

The 'hierarchy' among the disciplines is relatively stable over the whole period. Economics consistently has the highest number of professors, followed by philosophers. Nevertheless, there are some interesting points to note in our evidence. Literature was the third discipline

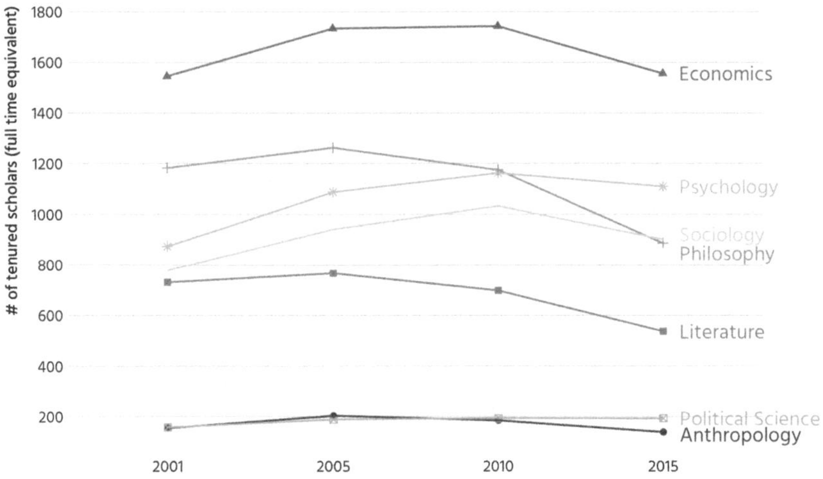

Fig. 5.2 Number of tenured scholars (full time equivalent) in the seven Interco disciplines, 2001, 2005, 2010, 2015

for the first 30 years, but in 2015 it is fifth (out of seven), with fewer professors than psychology and sociology. Prior to 2015, the difference between the number of psychology and philosophy professors was considerable, with fewer in psychology. In 2015, however, this position was reversed.

All of these changes must also be analysed within the framework of internal institutional processes. One that is relevant to our analysis was the introduction in 1980 of the faculty classification according to full professors, associate professors and researchers. For comparability within the whole period of observation, Fig. 5.1 considers only full professors. Figure 5.2 gives a more complex image of the last 15 years, including all tenured faculty. In general, between 2000 and 2010, the number of scholars in the INTERCO disciplines increased slightly, followed by a drastic decrease in 2015. In 2015, there were 5320 scholars, including full professors, associate professors and researchers; 100 fewer than 15 years earlier.

In 2015, no disciplines show more faculty members than in 2010. The reduction was most certainly a general one, though some disciplines lost fewer staff than others. One exception is political science, which counts the same number of professors in 2010 as it did in 2015; the only discipline showing no reduction. Despite the reduction over the last five years, in 2015 sociology has 901 professors, compared to 778 in 2001. Economics and psychology, in spite of a fluctuating trend, also have more academic staff in 2015 than in 2001. Literature and philosophy are the disciplines in which a decrease is evident.

As mentioned above, distinction among full professors, associate professors and tenured researchers was introduced in 1980. It is hard to say whether the evolution of the articulation of the academic staff caused some of the changes in the number of academics in the various disciplines. Nevertheless observing the internal composition of the disciplines in question allows an evaluation of any differentiated changes resulting from different career levels.

Table 5.1 contains data for the years 1983, 2001, 2005, 2010 and 2015. The values are expressed as the percentage of each academic class within each discipline.

It is not always possible to highlight precise trends in the evolution of the composition of the academic staff, but some evidence is noteworthy. The table begins at 1983. At this time the percentage of full professors was the lowest for almost all disciplines, except for economics and anthropology. This percentage reaches its highest in 2005. For associate professors, 2015 represents their highest percentage, even though the time period is not split evenly. In almost all disciplines, there has been a 'swap' between the proportion of researchers and associate professors over time. The exception is economics, which has the most homogeneous distribution of positions throughout the whole period. In 1983, researchers were the most consistent category, while in 2015 associate professors dominate. Political science is an interesting case. In 1983, it consisted of only 9% of associate professors, against 66% of researchers (Fig. 5.3).

This is also strong evidence for the two disciplines representing the humanities. For both literature and philosophy, 1983 is when the percentage of researchers is highest and those for professors (both full and associated) are lowest, while at the end of the period the trend is reversed. This data should be read in the context of the general reduction of literature and philosophy faculty, as seen early on. It is probable

Table 5.1 Ratio faculty/courses, 2005, 2010, 2015

	Total courses			Total faculty			Ratio faculty/courses		
Discipline	2005	2010	2015	2005	2010	2015	2005	2010	2015
Economics	171	130	129	1734	1743	1556	10,14	13,41	12,06
Sociology	47	48	37	941	1033	901	20,02	21,52	24,35
Anthropology	11	12	9	203	184	138	18,45	15,33	15,33
Psychology	104	118	124	1087	1162	1109	10,45	9,85	8,94
Political science + international relations	116	118	89	188	193	193	1,62	1,64	2,17
Literature	119	113	102	767	698	537	6,45	6,18	5,26
Philosophy	90	82	73	1262	1174	886	14,02	14,32	12,14
Total	*658*	*621*	*563*	*6182*	*6187*	*5320*	*9,40*	*9,96*	*9,45*

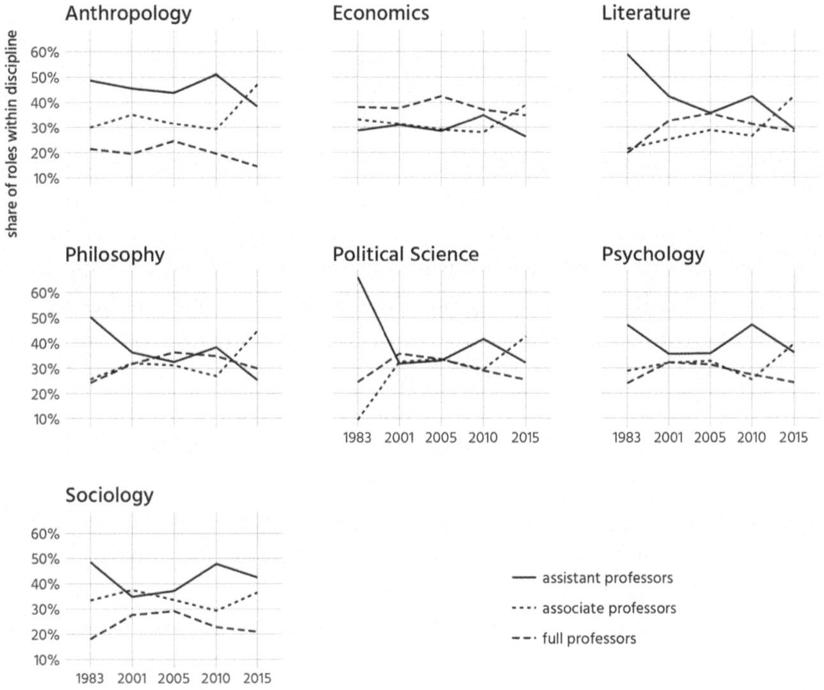

Fig. 5.3 Faculties by rank and discipline, 1983–2015, (in per cent)

that someone researching during the 1980s reached the rank of associate or full professor in the late 2000s, while fewer researchers have been recruited in the intervening years. Thus it can be argued that the overall decrease of scholars has been more significant at the lower end of the academic career.

Teaching

In order to evaluate the institutionalization processes of the social sciences and humanities in Italian academia, a further set of indicators concerns the educational supply and the rates of graduates. As outlined in previous sections, some important normative turning points steered the expansion of educational opportunities and impacted upon the general structure of the Italian academic system. In 1969, access to higher education was liberalized, regardless of the 'sending' school, which increased widely the opportunities for students to enroll for a university degree. In 1980, the third educational level (Ph.D.) was introduced by law in Italy, and in 1997, the government gave universities the power to create degree courses for themselves. Last, but not least, a 1999 university reform introduced the '3+2' years system. This was a big change as degree programmes in Italy had lasted four to six years traditionally, depending on the degree (six for medicine, five for engineering and architecture, four for humanities and the social sciences). As part of the Bologna process, the reform also introduced a distinction between bachelor's and master's degrees. Our analysis must take account of this breaking point, which marks a divide between two periods of the university institutional path.

Figure 5.4 shows the number of degree courses from 1950 to 2000 in the seven INTERCO disciplines. Over the whole period, literary studies is the discipline with the greatest number of courses provided by Italian universities, peaking at 55 in 2000. Just after the Second World War, economics and philosophy had the same number of courses, but from 1975 onwards, economics started to increase strongly with a peak of

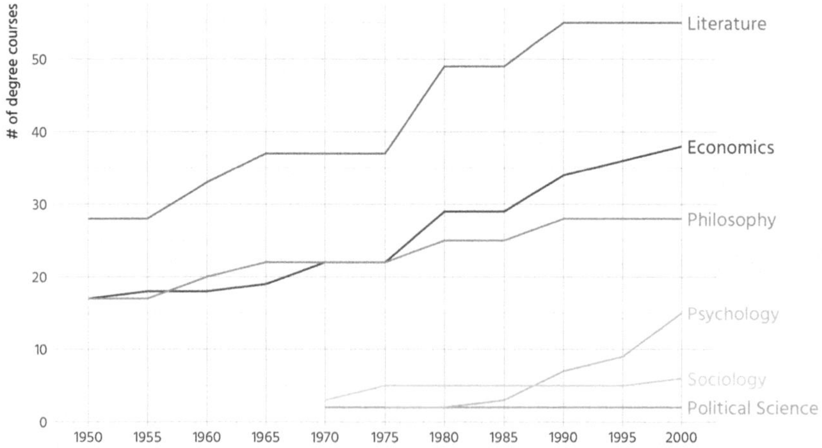

Fig. 5.4 Number of degree courses pre-Bologna process, 1950–2000

38 courses in 2000, against 28 from philosophy. Psychology also shows a strong increase, especially as before 1971 there were no psychology courses at all. Up to 2000, sociology presents a low number of courses, with almost no growth over time. For the whole period, political science counts only two courses, which are actually courses in 'economics and political science' (at universities on Modena and Teramo). Even if political sciences—note the plural, which we are using to denote multidisciplinary, rather than courses focussed only on political science—faculties existed in Italy, courses comprise disciplines such as law, economics, sociology, public administration; it doesn't really have a focus on political science as a proper discipline.

Anthropology is the only discipline among the seven with no degree courses until 2000. Under the faculties of letters and philosophy, there was specialization in anthropological studies, but it is not until the reform that introduced the master's degree that there was for a specialized curriculum dedicated to anthropology.

There was a general increase in the number of courses from 1950 to 2000, during which all of the disciplines doubled the number of degree programmes. But this increase is even more evident after this because of university reform; the introduction of a three-year degree generated an important growth of courses in all disciplines. In 2000, there were overall 144 courses in the INTERCO disciplines, by 2005, the bachelor programmes alone numbered 339. Between 2000 and 2005, then, the increase is evident. After 2005, a counter-process of general reduction of the number of courses began. In Fig. 5.5 we can see a clear drop in all courses. This reduction is especially notable for economics, which lost 38 courses between 2005 and 2015. The position of literature is also noteworthy; it is the only discipline that has a lower number of courses in 2015 than in 2000. In spite of this gradual decrease over the last 10 years, there is now a much wider educational offer than in the past. Political sciences courses are difficult to assess, at least at bachelor level, because it largely comprises multidisciplinary courses. Thus it is difficult, if not impossible, to gain insight on it as a single discipline.

Master degree courses (Fig. 5.6) show a slightly different trend. Firstly, the decrease in the number of courses is not as large as it is for bachelor programmes. In 2015, psychology counts almost 20 more

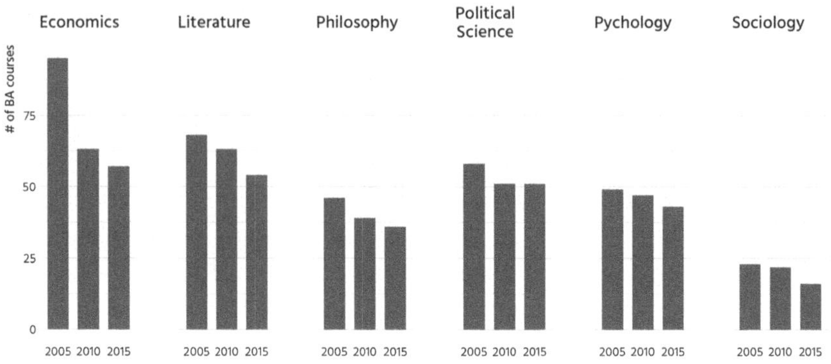

Fig. 5.5 Number of B.A. courses, 2005, 2010, 2015

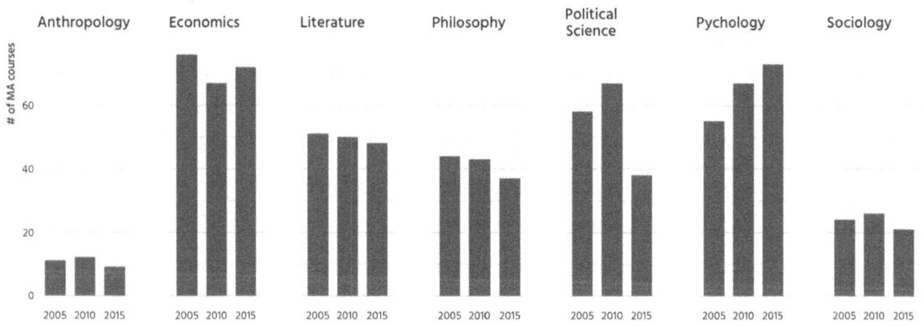

Fig. 5.6 Number of M.A. courses, 2005, 2010, 2015

M.A. courses than in 2005. More interestingly, disciplines such as anthropology and political science (singular) finally find a broad space for their teaching and acquire the possibility to be recognized as institutional degree programmes. Of course, neither the history of political science nor anthropology begins with the 1999 university reform but, as is already argued, normative changes have a deep impact on the institutionalization of disciplines, and this is the first time in the history of Italian academia that the two disciplines obtained full institutional recognition within the educational framework.

So far we have seen data about academic staff in the seven SSH disciplines and the evolution of education as a whole in Italy. In most cases, a deciding factor in the number of courses provided by a university is the number of professors available in specific topic or discipline. As a result, in order to evaluate the capacity of a discipline to spread and provide teaching, it is useful to observe both the data on the presence of professors and on degree courses. The figures in Table 5.1 show the average proportion of professors per course in the different disciplines. With such data, we aim to provide two types of information: how many scholars of that specific discipline there are on average for each course and, more generally, the level of fragmentation of the disciplines. In other words, to test the assumption that the lower the ratio of faculty/courses the weaker the presence of professors belonging to the discipline within a course. This would indicate that courses in that discipline have many professors that belong to other disciplines.

Sociology shows the greatest cohesion as a discipline. In Italy, degree programmes in sociology have, on average, about 20 sociology professors, and this ratio has increased over the past 10 years. We cannot say what the exact percentage of sociology professors is over the totality of the professors in sociology courses, but we can presume that 20 is quite a high number of representatives of the discipline. Anthropology also shows a good proportion between scholars and degree programmes. This was slightly higher in 2005, but remains stable at 15.33 in 2010 and 2015. Economics, psychology and philosophy show a medium ratio, ranging to almost 9 points in psychology, to 12 in economics and philosophy indicating that, even with good rates of disciplinary teaching, their programmes usually comprise other disciplines. Literature and political science have a very low ratio, showing a high multidisciplinary level. In particular, we see that approximately two political science professors are involved in degree programmes in political science and international relations, even if this shows an increase over the last 10 years. Except for sociology and political science, in all SSH disciplines, the proportion between faculty and degree programmes has been decreasing, showing both a greater multidisciplinary generally and the fact that fewer professors are employed for more than one course.

The next set of indicators concerns graduate rates in the seven disciplines from 1948 to 2010. As already mentioned, before 2000 the degree lasted four years, with medicine and law remaining exceptions. Figure 5.7 shows graduate rates in Italy according to the previous

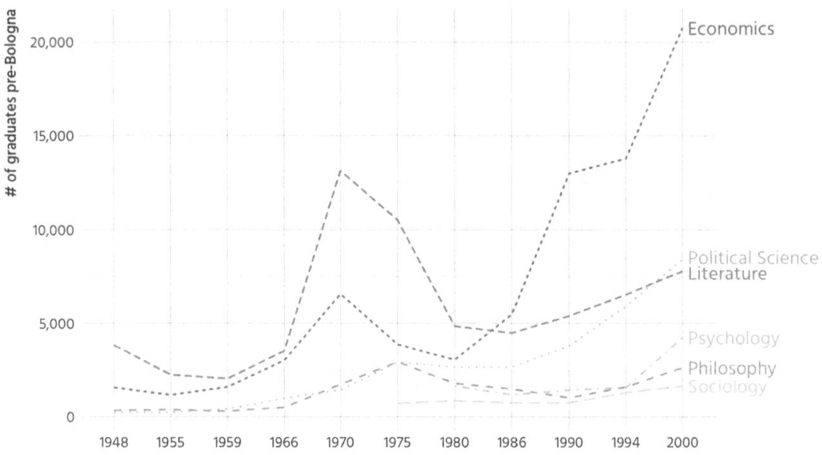

Fig. 5.7 Graduates in pre-Bologna process courses, 1948–2000

education system. Until 1966, the number of graduates is relatively low. In that period, the most popular disciplines were literature and economics, even though they had few graduates. From 1966 to 1980, literature has the most graduates at over 13,000, about twice as many as economics. Philosophy and political sciences are also present, but with almost negligible graduate rates. Two interesting trends concern these disciplines. From 1975, the number of philosophy graduates gradually decreases, with the number of graduates in 2000 almost the same as in the early 1960s. Political sciences, on the other hand, increase their graduates year on year, eventually exceeding literature in 2000.

While no anthropology degree shows up until the early 2000s, sociology and psychology start a gradual, if weak, increase in graduate rates from the 1980s onwards. It is psychology that shows the highest number of bachelor's and master's graduates in both 2005 and 2010.

Figure 5.8 shows graduate data after the introduction of the new 3+2 system for 2005 and 2010. Despite general increases, it is not easy to interpret the absolute values because 2005 remains close to the previous system. This might be because those who started at university in the late 1990s could be still enrolled in the pre-reform programme in 2005. The two systems overlapped for a number of years. In some cases, students

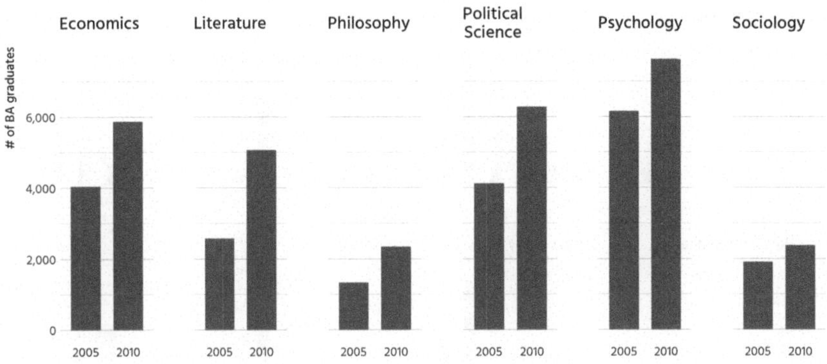

Fig. 5.8 B.A. graduates, 2005 and 2010

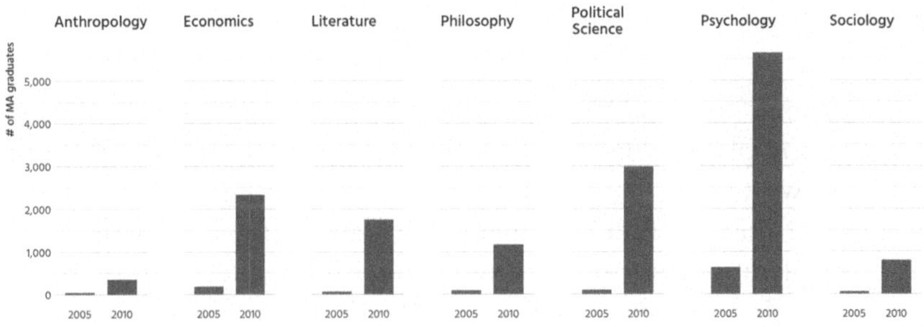

Fig. 5.9 M.A. graduates, 2005 and 2010

could choose to pass from the old to the new degree, or not. As a result, it is only the 2010 data that are reliable. What is nevertheless clear, beyond the absolute values of graduates rates, is the new 'hierarchy' of the disciplines, which is rather similar for bachelor's and master's degree.

As mentioned earlier, psychology is the best-represented discipline in 2010. It is followed by political sciences (plural), economics, literature, sociology and philosophy (in that order). For the M.A., the rank is basically the same, the only exception being philosophy, which has more graduates than sociology (Fig. 5.9).

As has already been mentioned, the *doctorate* was introduced late in Italy, dating back only to 1980, a singularly anomalous feature of this national system (Cesaratto et al. 1994). As for other aspects

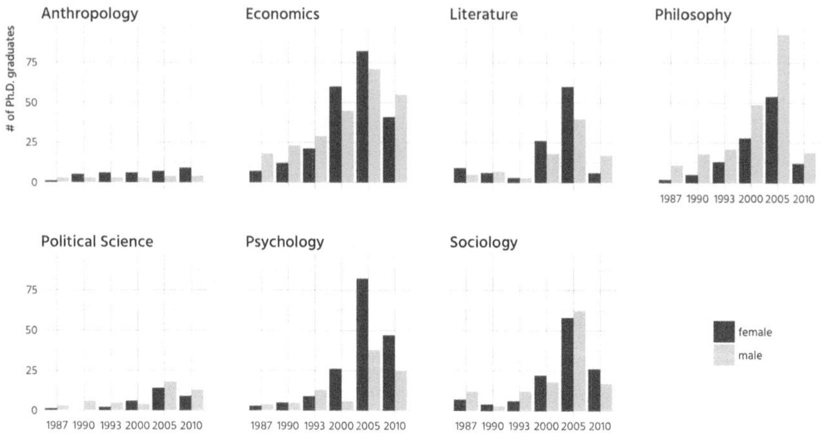

Fig. 5.10 Ph.D. graduates in the Interco disciplines, by gender

of the system, the Ph.D.'s aims and scopes are normatively defined, including the places where it is possible to pursue it and the positions associated with it. As a rule, there is a *numerus clausus* in the annual provision of posts as Ph.D. students (with an annexed fellowship; only a very small proportion of positions are without a fellowship and the numbers of these are fixed each year). The doctorate is conceived as (at least) a three-year period of training for academic research professionals. Indeed, only a few years after its introduction the academic career has been subordinated to the acquisition of a doctorate. Figure 5.10 shows the evolution of the number of Ph.D. graduates in Italy in the seven SSH disciplines. It shows a general increase. From 1987 to 2010, the number of Ph.D. graduates goes from 86 to 303. Nevertheless, the growth is not constant over the observed period. In 2010, there is a strong decrease in the number of Ph.D. graduates. The total difference between 2010 and 2005 is 56%, with the number of graduates in the seven SSH disciplines even lower than 2000.

Beyond the overall values, the differences among disciplines present a slightly different picture when bachelor's and master's graduates are compared. Economics is the most represented discipline, in line with the previous data on courses, followed by psychology and, quite surprisingly, then by sociology. Since the beginning, sociology Ph.D. graduates have been over-represented compared to the other education levels. Thus recently, while disciplines like philosophy or literature seem to be gradually losing importance, sociology—together with psychology—seem to gain 'space' in academia.

Professional Devices: Associations and Journals

Disciplines may also be established outside of the university system. The academic establishment confers greater cultural legitimacy on a discipline in modern societies (e.g. Shils 1980; Collins 1979, 1981; Bourdieu 1984), but a discipline can also survive, and even flourish, outside of academic settings as well, psychoanalysis being a well-known case. Indeed, even when firmly rooted in the academic system, disciplines are usually eager to exploit links with other institutions.

Professional associations are among the most important institutions that ground disciplinary projects; their number and quality say a lot about the status of a discipline as a scientific community and professional group (see Cole 1972; Larson 1977). As Fig. 5.11 shows, the number of associations has grown substantially in the last 50 years. This growth is especially visible in the case of philosophy (reaching 11 associations between 2001 and 2010). The trend is not always progressive, as evidenced by economics, the first discipline to experience an intense associative life beyond traditional academies, inherited from the *ancient régime* (which was especially strong in the humanities) (Table 5.2).

Scientific journals are pivotal institutions around which disciplines coalesce and organize themselves as both scientific communities and professional groups. This is especially true when disciplines are in the early stages of academic institutionalization, at this point journals offer chances to organize and form an identity, at least for intellectual schools when they are not part of larger disciplinary projects. We have already referred to journals when describing the early histories of the SSH disciplines in Italy. Table 5.3 collates some relevant information about the journals that have contributed to the establishment of those disciplines in the decades following unification.

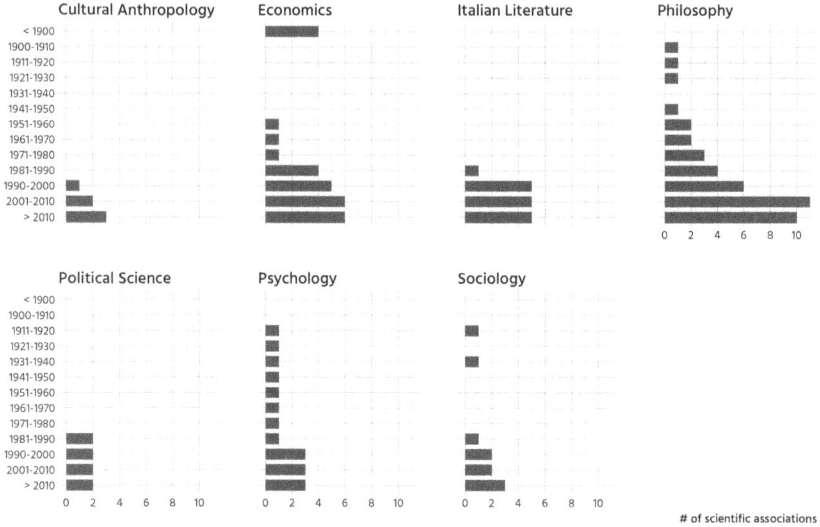

Fig. 5.11 Number of scientific associations by period and discipline

It should come as no surprise that it is the old humanities who first established their journals; philosophy in 1870 philosophers and literary scholars in 1883. More surprising is that anthropologists had their own journal since 1871, which still exists, and that a journal devoted to sociology was started in Italy before any in France or the US. This is a well-explored issue however in the history of the social sciences, and we can only refer to these specialized studies here (cf. Santoro 2013).

As already noted, the Italian academic system—and even academic research—are strongly affected by the existence of classificatory schemas, established by law, that classify knowledge. Labels such as 'sociology', 'political science', 'economics', etc., in Italy, refer to research and scholarly fields as they do elsewhere, but also specifically to formal administrative categories called *settori scientifico-disciplinari (SSD).* Since 1997, and earlier in some cases, almost everything in the academic system has been organized, managed, decided and assessed according to these SSD. While in some cases there is a correspondence between an established discipline and an administrative category—as in political science—in most cases there is no correspondence and the disciplinary

Table 5.2 Scientific associations in Italy, for seven disciplines

Discipline	Association	Year of foundation	Year of death	N members (2016)
Anthropology	AISEA—Associazione italiana per le scienze etno-antropologiche	1992		102
	ANUAC—Associazione nazionale degli universitari dell'antropologia culturale	2007		
	SIAA—Società italiana di antropologia applicata	2013		
Economics	SIE—Società italiana degli economisti	1950		1007
	AIEL—Associazione italiana economisti del lavoro	1985		30
	SIET—Società italiana di economia dei trasporti e della logistica	1986		
	SIEP—Società italiana di economia pubblica	1988		173
	AISPE—associazione italiana per la storia del pensiero economico	1992		60
	STOREP—Società italiana per la storia dell'economia politica	2003		153
Literature	SICL—Società italiana di comparatistica letteraria	1985		
	COMPALIT—Associazione per gli studi di teoria e storia comparata della letteratura	1993		
	ADI—Associazione degli italianisti	1996		552
	MODLET—Società italiana per lo studio della modernità letteraria	1998		
	SFLI—Società dei filologi della letteratura italiana	1999		102
Philosophy	SFI—Società filosofica italiana**	1953 (1906)		
	SILFS—Società italiana di logica e filosofia delle scienze	1972 (1952)		236
	AISS—Associazione italiana di studi semiotici	1972		104

(continued)

Table 5.2 continued

Discipline	Association	Year of foundation	Year of death	N members (2016)
	AISE—Associazione italiana per gli studi di estetica	1984	2011	
	SIFA—Società italiana di filosofia analitica	1992		101
	SIFL—società italiana di filosofia del linguaggio	1994		132
	SIE—Società italiana d'estetica	2001		227
	SIFP—società italiana di filosofia politica	2001		
	AIFR—Associazione italiana di filosofia della religione	2002		
	SIS—Società italiana storia della filosofia	2002		
	SISFA—Associazione italiana di storia della filosofia antica	2009		78
Political science	SISE—Società italiana di studi elettorali	1980		
	SISP—Società italiana di scienza politica*	1981		220
Psychology	Società italiana di psicologia scientifica	1911		
	Associazione italiana di psicologia	1992		
	Associazione italiana di psicologia di comunità	1994		
	AISC—Associazione italiana di scienze cognitive	2001		
Sociology	AIS—Associazione italiana di sociologia*	1983		529
	SPE—Sociologia per la persona	1995		106
	SISE—Società italiana di sociologia economica	2017		200

*Since 1951 an association for political and social sciences collected all sociologists and political scientists (but also some jurists and historians). In 1981, political scientists founded their own association, and two years later sociologists did the same, bringing this association to an end

**The SFI was dissolved in 1926 and refounded in 1931 by the fascist regime with the new name of 'Associazione Filosofica italiana'. It was included however in the Istituto di studi filosofici, ruled by the Fascist state. In 1953 it has been refounded as a free association

Table 5.3 First Italian academic journals, in seven disciplines

Discipline	Oldest journals	Founding year	Closing year
Anthropology	Archivio per l'antropologia e l'etnologia	1871	–
Economics	Giornale degli economisti e annali di economia	1875	2012
Literature	Giornale storico della letteratura italiana	1883	–
Philosophy	La filosofia delle scuola italiane	1870	
	Rivista italiana di filosofia	1886	1898
Political science	Rivista trimestrale di scienza politica e dell'amministrazione;	1968	
	Rivista italiana di scienza politica	1971	–
Psychology	Rivista di psicologia applicata*	1905	–
Sociology	Rivista Italiana di Sociologia	1897	1921
	Rivista di Sociologia	1894	1898

*Since 1912 'Rivista di psicologia'; suspended from 1933 to 1954

field is fragmented according to administrative criteria. 'Sociology', for example, is fragmented into 'General sociology', 'Sociology of culture', 'Economic sociology', 'Political sociology' and so on. There are precise lists establishing which journals are recognized as pertinent for a definite disciplinary category. These correspond to a class in recruitment and career administration. Interdisciplinary journals do exist, but are always in danger of not being acknowledged as relevant to any disciplinary sector.

In a sense, SSDs work as academic 'cage's, perhaps rationalizing the management of recruitment processes, but at the cost of constraining research within rigid categories and around frozen borders. This explains why well-bounded professional associations have increased in number in the last 20 years, becoming increasingly more specialized, which corresponds to the time period following the formal introduction of the device of the SSD. For journals, considerations of bureaucratic/political opportunity mesh with intellectual—and even generational—factors to produce the outcome represented in Fig. 5.12. This shows the distribution of journals per discipline, many of which would be little known even to the average practitioner, especially in disciplinary sectors marked by an overflow of titles, such as sociology and literary studies.

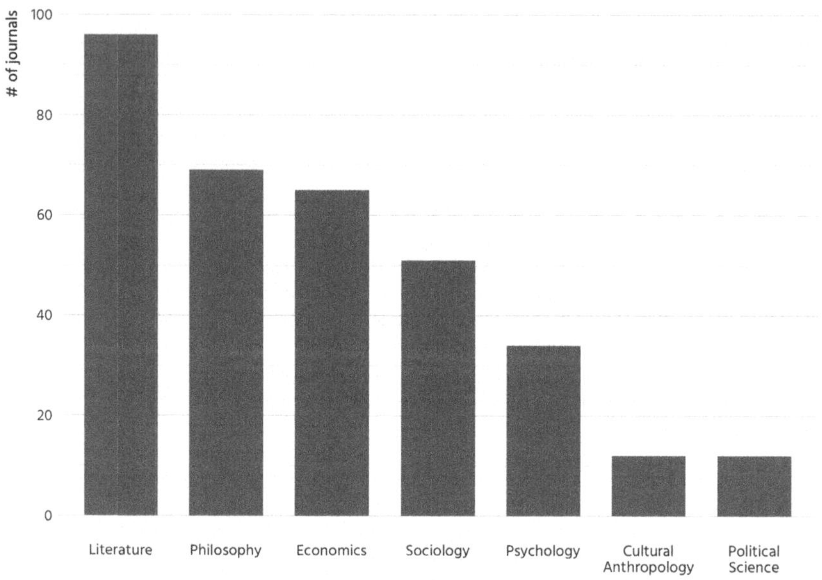

Fig. 5.12 Number of Italian journals by discipline, in 2013 (*Source* ANVUR [only journal recognized and assessed])

Conclusions

In this chapter, we have given an overview of the main developments of SSH in Italy. Instead of focusing on individual disciplines—sociology, economics and so on—we have chosen to adopt a comparative approach, trying to keep alive a view of the 'big picture'. We have tried to embed our data in a narrative that captures the essential trends and points of Italian academic history, highlighting its peculiarities and often contradictions. The '*academic tribe*' has been the (often implicit) category underpinning our narrative and data analysis. The assumption is that what our data represent are just the superficial manifestation of dynamics (alliances, struggles, arrangements, etc.) occurring at a deeper level, the level where academic communities organized according to officially defined disciplinary regions (or sectors) with their boundaries and internal structure. To go deeper would mean to focus on those often hidden and usually complicated 'dense interaction of persons' that for Shils forms the core of an intellectual activity (1980, 168).

Among the general patterns that emerged from our overview, we would conclude by underlining the ambivalent relationships academic men and women cultivate with political institutions, especially with the national government. In a sense, we can say the national government has always been involved in the field of higher education, setting the rules for almost everything that can happen in universities, from academic careers to curricula formation and research practices. Since Italy's unification, the state has massively affected scholarly work, to the point where it determines the contents and the boundaries of knowledge fields. In another sense, it seems that Italian academics have been very successful in maintaining a strong degree of autonomy and self-government, even corporate discipline, in spite of state intervention, both practical and legislative. Co-optation mechanisms, based on personal or group relationships, are the driving forces for recruitment, notwithstanding the many strictures set forth by the law on public concourse.

In order to understand this apparent paradox, it is useful to move to data we have not yet presented; the participation of academics in political life, including national government. The closeness of university and politics in Italian history is a well-researched topic in Italy (e.g. Santoni Rugiu 1991), including documented reference to practices of misconduct and 'mafia'-like arrangements (Froio 1973, 1996; Simone 1994; Perotti 2008). Less well known is that this co-penetration between the political and academic fields is growing, as Fig. 5.13 shows, and is especially relevant for some disciplines (such as economics) (Fig. 5.14).

In this, Italy is not unique. As elsewhere, the development of the social sciences and humanities has been highly influenced by parallel developments in other countries, especially France (still strong as a reference in the field of philosophy and, in part, anthropology) as well as the UK and US, common reference points for Italian sociologists, political scientists and economists. Italy does differ, however, from other, smaller, countries, such as the Netherlands or Denmark. During recent decades, local, indigenous traditions and legacies have been resisting the many pressures towards greater standardization (which usually means, at least since the Second World War, the adoption of models developed in the Anglophone world). This has produced a dual effect. On the one hand it has contributed to the persistence of diversity in intellectual life

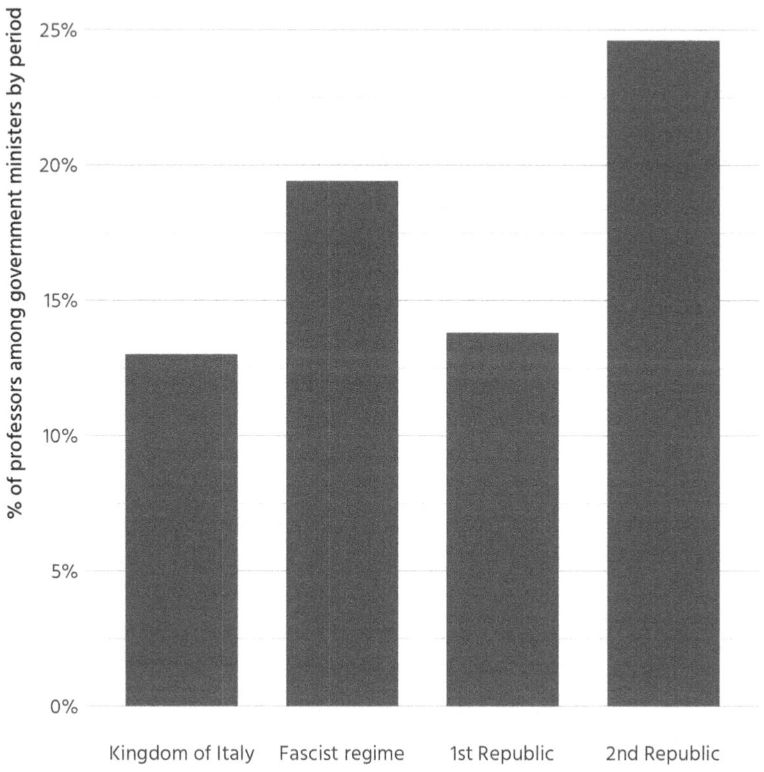

Fig. 5.13 University professors in governments, Italy 1861–2016 (share of ministers, by historical period)

and scholarly production, which is apparent in the persistence of local intellectual traditions (like the Gramscian, or what is currently named 'Italian Theory' [see Gentili 2012]), and in diffused practices of writing, publishing and teaching reflecting well established conventions and informal rules of academic work possibly dating back to the foundation of the national academic system (see e.g. Perotti 2008; Graziosi 2010). On the other, it results in complicated and often inconsistent blends of foreign models with local practices and indigenous inventions. This is especially apparent currently in the field of research evaluation, a practice totally unknown in Italy till a few years ago.

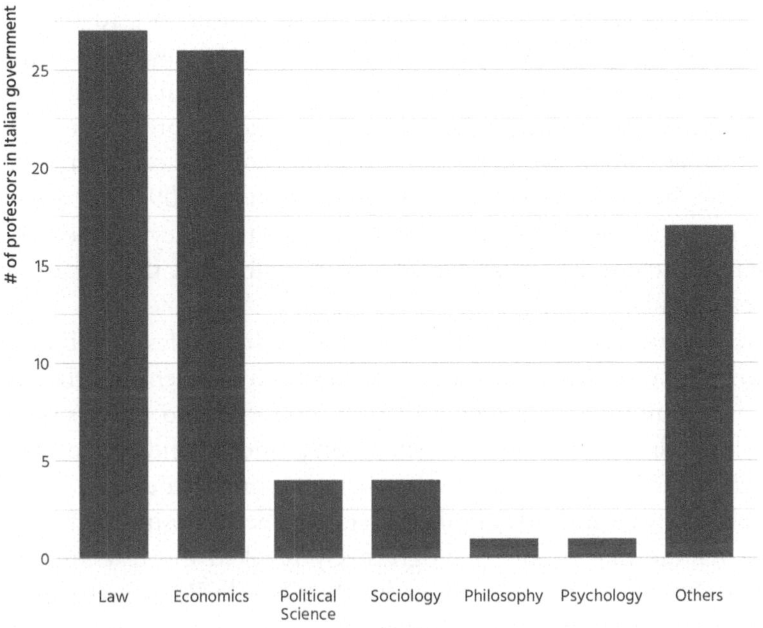

Fig. 5.14 Distribution of professors in governments, by discipline (1994–2016)

Appendix 5.1. An Historical-Institutional Profile of the Italian University

The first constitutive act of university politics in Unified Italy was the 1859 Casati law (n. 3725). The main disposition—which worked, substantially, until the University reform in 1969—was to set the number of universities, faculties and chairs in legislation.[14] It also established a three-level ranking of universities: leading; secondary; and private. This corresponded implicitly with a ranking of faculties. First came the faculty of law, followed by faculties of letters and philosophy, medicine and natural sciences, with the various High schools for applied sciences at

[14]It defines also the different stages of the academic hierarchy: full professor, appointed professor, extra-ordinary professor and 'libero docente', corresponding to the position of a teaching assistant who was waiting for the qualification to teach.

the bottom. Finally, it limited access to the universities to those who had attended the classical gymnasium. The Casati law, then, depicted an elitist university under government control, with an unclear mission; should it prepare graduates for a scientific career or a profession?

A further important step in the organization of academia is represented by the dispositions of the Minister Rava in 1909. They were also effective until the end of the twentieth century. They divided fundamental and complementary teaching and established that chairs could be created only by law.

The Gentile Reform of 1923 ratified a (de)professionalization of the university studies, granting more autonomy in defining the curricula and the number and denomination of teaching to universities, and abolishing the distinction between leading and secondary universities, though it did stress the distinction between universities and the High Schools oriented towards applied sciences. The dispositions during the fascist regime that followed progressively revoked the teaching autonomy conferred to the universities by the Gentile Reform. Of particular importance was the 1938 Bottai reform (n. 1652), which not only strictly redefined the curricula, but also correlated them to defined professional careers, effectively outclassing the scientific mission of the university.

After the Second World War and the fall of fascism, one of the few dispositions concerning higher education issued by the new 'democratic' government (law n. 312/1953) allowed universities to expand complementary teaching within the various existing degree courses. Interest in the university system arose once again in 1962 with the constitution of a parliamentary commission to examine the condition of public instruction. The commission highlighted two questions, which culminated in the 1969 reform (law n. 910): the possibility to access to all degree courses independently of the sending school; and the possibility for each student to arrange his/her curricula differently from the curricula set by law. As a consequence, the existing separation between main teaching and complementary teachings lost importance.

The increase of students and teaching, but not chairs, in the 1970s instigated a reform of the university staff. Decree 382 in 1980 subdivided the academic structure into three levels: full professors; associate professors; and tenured researchers. Further, it instituted the 'departments,' substitutes for the pre-existing 'Institutes,' and ratified that each University could establish 'Graduate schools' based on proposals of the faculties or of the departments (Capogrosso Colognesi and Cerulli 1981). From the end of the 1980s to 1999, a set of decrees and laws gradually gave greater autonomy to universities. In 1990, the law n. 341 ratified a new system of disciplinary classification for professors (by scientific disciplinary sectors, SSD). The application of SSD was not immediate and, at the end of the 1990s, it was extended to include tenured researchers, Ph.D. candidates and post-doctoral research fellows. In 1997, the law n. 127 conferred upon the university the power to institute degree courses. Finally, the ministerial decree of 1999 (n. 509) replaced the existing '*corso di laurea*' with three years of Bachelor's degree courses and two of Master's degree courses (the so called '3+2' system) that was redefined in 2004 (law n. 270). It also introduced a new classification system—the 'classes'—to classify the various degree courses that proliferated after the universities became autonomous (Gasparri 2005). The last University reform in 2010 (law n. 240) caused three main changes within the academic system, although it was only partially enacted because of the lack of funds. It introduced two new positions for temporary researchers, the national habilitation, and several and juxtaposed evaluation systems to evaluate the research activities of the permanent staff, Universities and departments that constituted new criteria for the distribution of public funding (Brollo and De Luca Tamajo 2011).

Appendix 5.2. Sociology as an Example of SSD

Where is sociology in the Italian SSD system? There are three issues to consider. The first concerns the curious classification criteria adopted by the Italian institutional system to cluster disciplines in 'disciplinary macro-area', numbered from 1 to 14. The first areas regard 'pure natural science disciplines (mathematics, physics) whereas sociological SSD are included in the last Area, the so-called 'Scienze politiche e sociali' (Political and social sciences), suggesting a hierarchy of relevance among natural, human and social sciences. The second issue concerns sociology itself as divided in six SSD: General sociology; Sociology of culture and communication; Economic sociology; Urban and environmental sociology; Political sociology; and Sociology of Law. Thus, within Area 14, sociological sectors are grouped with non-sociological disciplines, such as political philosophy, political science and the History of Africa, among others. The inclusion in an Area suggests the pertinent disciplines have both meaningful affinities and similar intellectual distances to others. In this regard, General sociology seems closer to History of Africa than to Anthropology, the latter being included in the humanistic Research Area 11 (an area comprising various historical and philosophical disciplines).

The third issue concerns how the sociological SSD have been constructed by clustering the sociological 'subdisciplines' that existed before the introduction of the SSD system (in 1994 there were 65 subdisciplines). It is not surprising that the decision to classify a subdiscipline in a SSD, rather than in another area, is the result of internal [institutional?] struggles among sociologists, alongside those between sociologists and scholars of other disciplines (i.e. for Sociology of law, between sociologists and jurists).[15] However, the main point is that these 'struggles' led the symbolic spaces and boundaries of SSD being fixed in official declaratory judgments (*declaratorie*). The *declaratorie* are relevant for identifying supposedly 'objectified' criteria to be used in competitive examinations (i.e. for academic positions or for the habilitation), or in order to establish if the scientific products (and activities) of a 'sociologist' are pertinent, or not, to the SSD as he competes for acceptance or for upgrading (see for example the list of topics included in SPS/07 'Sociologia Generale'). These *declaratorie* were last updated in 2001. It follows that potential new sociological areas developed at both national and international levels in the last 15 years are formally excluded from the official definition of the discipline, which itself is not really consistent with the interior articulation. In the end, we can compare this classification system based on Research Areas and SSD to the famous classification imagined by Jorge Luis Borges as pertaining to 'a certain Chinese encyclopedia: 1. those that belong to the Emperor,

[15]Interview with Alberto Febbrajo, Emeritus Professor of Sociology of law, April 2016.

2. embalmed ones, 3. those that are trained, 4. suckling pigs, 5. mermaids, 6. fabulous ones, 7. stray dogs, 8. those included in the present classification, 9. those that tremble as if they were mad, 10. innumerable ones, 11. those drawn with a very fine camelhair brush, 12. others, 13. those that have just broken a flower vase, 14. those that from a long way off look like flies'. The tragedy is that this arbitrary, and logically inconsistent, classificatory system is institutionally effective and impacts on the reproduction and transformation of the symbolic and social boundaries of sociology as practiced in Italy.

Area 14

1. SPS/01 Filosofia politica.
2. SPS/02 Storia Delle Dottrine Politiche.
3. SPS/03 Storia Delle Istituzioni Politiche.
4. SPS/04 Scienza Politica.
5. SPS/05 Storia E Istituzioni Delle Americhe.
6. SPS/06 Storia Delle Relazioni Internazionali.
7. SPS/07 Sociologia Generale.

    Metodologia Delle Scienze Sociali (Q05a)
    Metodologia E Tecnica Della Ricerca Sociale
    Politica Sociale
    Principi E Fondamenti Del Servizio Sociale
    Sistemi Sociali Comparati
    Sociologia
    Sociologia Dei Gruppi
    Sociologia Della Salute
    Sociologia Della Scienza
    Sociologia Della Sicurezza Sociale
    Sociologia Dello Sviluppo
    Storia Del Pensiero Sociologico
    Teoria E Metodi Della Pianificazione Sociale

8. SPS/08 Sociologia Dei Processi Culturali E Comunicativi.
9. SPS/09 Sociologia Dei Processi Economici E Del Lavoro.
10. SPS/10 Sociologia Dell'Ambiente E Del Territorio.
11. SPS/11 Sociologia Dei Fenomeni Politici.

12. SPS/12 Sociologia Giuridica, Della Devianza E Mutamento Sociale.
13. SPS/13 Storia E Istituzioni Dell'Africa.
14. SPS/14 Storia E Istituzioni Dell'Asia.

References

Abbott, A. 2001. *Chaos of disciplines*. Chicago and London: University of Chicago Press.

Archivio storico della camera dei deputati. http://archivio.camera.it/patrimonio/archivio_della_camera_regia_1848_1943/are01o/documento/CD1100020201.

Balbo, L., G. Chiaretti, and G. Massironi. 1975. *L'inferma scienza. Tre saggi sull'istituzionalizzazione della sociologia in Italia*. Bologna: Il Mulino.

Barbano, F. 1987. *Sociologia come scienza e scienza dei sociologi, '45–'85*. Torino: Giappichelli.

Bartolé, S. 1991. Scienza politica e diritto: commento. *Rassegna italiana di scienza politica* 1: 129–135.

Becher, T., and P. Trowler. 2001. *Academic tribes and territories: Intellectual enquiry and the culture of disciplines*. London: McGraw-Hill Education.

Bottai, G. 1939. *La carta della scuola*. Milano: Mondadori.

Bourdieu, P. 1984. *Homo academicus*. Paris: Editions de Minuit.

Brizzi, G.P., P. Del Negro, and A. Romano (eds.). 1999. *Storia delle Università*, 3 vols. Messina: Sicania.

Brollo, R., and R. De Luca Tamajo (eds.). 2011. *La riforma dell'università tra leggi e statuti*. Milano: Giuffré.

Burgalassi, M. 1996. *Itinerari di una scienza. la sociologia italiana tra ottocento e novecento*. Milano: Angeli.

Cagiani, R. (ed.). 2006. *La facoltà di economia. Cent'anni di storia*. Soverio Manelli: Rubettino.

Capano, G. 1998. *La politica universitaria*. Bologna: Il Mulino.

Capano, G., and M. Regini (eds.). 2015. *Come cambia la governance. Università italiane ed europee a confronto*. Fondazione Crui.

Capogrosso Colognesi, L., and I. Cerulli (eds.). 1981. *La riforma universitaria (d.P.R. n. 382 dell'11 luglio 1980)*. Milano: Giuffré.

Cassata, F. 2006. *Il fascismo razionale. Corrado Gini fra scienza e politica*. Roma: Carocci.

Castellino, O., and G. Zanetti (eds.). 1980. *Guida alla facoltà di economia e commercio*. Bologna: Il Mulino.

Cesaratto, S., S. Avveduto, M.C. Brandi, and A. Stirati. 1994. *Il brutto anatroccolo. Il dottorato di ricerca in Italia fra universitento/are01o/documento/CD1100020*. Milano: FrancoAngeli.

Chiabra, G. 1904. The tendencies of experimental psychology in Italy. *The American Journal of Psychology* 4: 515–525.

Cimino, G., and N. Dazzi. 1998. *La psicologia in Italia*. Milano: Led.

Clark, B.R. 1987. The problem of complexity in modern higher education. Working paper No. 9. Los Angeles: California University.

Clemente, P. 1985. Alcuni momenti della demologia storicista in Italia. In *L'antropologia italiana un secolo di storia*, AA.VV, 3–50. Bari: Laterza.

Cole, S. 1972. Continuity and institutionalization in science: A case study of failure. In *The establishment of empirical sociology*, ed. Anthony Oberschall, 73–129. New York: Harper & Row.

Collins, R. 1979. *The credential society: A historical sociology of education and stratification*. New York: Academic Press.

Collins, R. 1981. Crisis and declines in credential systems. In *Sociology since midcentury*, ed. R. Collins, 191–215. New York: Academic Press.

Comparato, V.I., R. Lupi, and G.E. Montanari (eds.). 2011. *Le scienze politiche. Modelli comparati*. Milano: Angeli.

De Vivo, F., and G. Genovesi (eds.). 1984. *Cento anni di università. L'istruzione superiore in Italia dall'Unità ai nostri giorni*. Napoli: Edizioni Scientifiche Italiane.

Di Bello, G. 2006. Dall'istituto superiore di Magistero alla facoltà di scienze della formazione. In *Formazione e società della conoscenza*, ed. G. Di Bello, 9–27. Firenze: Firenze University Press.

Falkner, R.P. 1891. Academic instruction in political and economic science in Italy. *Annals of the American Academy of Political and Social Science* 1: 635–661.

Fedele, F.G., and A. Baldi (eds.). 1988. *Alle origini dell'antropologia italiana. Giustiniano Nicolucci e il suo tempo*. Napoli: Guida.

Ferraresi, A., and E. Signori (eds.). 2012. *Le università e l'unità d'Italia*. Bologna: Clueb.

Ferrari, G.C. 1905. Experimental psychology in Italy. *The American Journal of Psychology* 2: 225–227.

Ferrarotti, F. 1957. Sociology in Italy, problems and perspectives. In *Modern sociological theory*, ed. H. Becker and A. Boskoff, 695–710. New York: Dryden Press.

Fiamingo, G. 1895. Sociology in Italy: The sociological tendency of today. *American Journal of Sociology* 3: 335–352.

Formica, M. 2000. Il secolo dei Lumi. In *Storia della facoltà di lettere e filosofia dell'Università della Sapienza*, ed. L. Capo and M.R. Di Simone, 305–339. Roma: Viella.

Froio, F. 1973. *Università: Mafia e potere*. Firenze: Vallecchi.

Froio, F. 1996. *Le mani sull'università. Cronaca di un'istituzione in crisi*. Roma: Editori Riuniti.

Garofalo, G., and A. Graziani (eds.). 2004. *La formazione degli economisti in Italia (1950–1975)*. Bologna: Il Mulino.

Gasparri, W. 2005. Università degli studi e autonomia didattica: la disciplina del D.M. n. 270/2004. In *Materiali sullo stato della riforma degli studi giutidici nelle facoltà di Giurisprudenza*, ed. L. Pietrolta, M.M. Procaccini, and O. Roselli, 163–177. Napoli: Edizioni scientifiche italiane.

Gentili, D. 2012. *Italian theory. Dall'operaismo alla biopolitica* Bologna: Il Mulino.

Giglioli, P.P. 1979. *Baroni e burocrati. Il ceto accademico italiano*. Bologna: Il Mulino.

Graziosi, A. 2010. *L A. (2010) per tutti*. Bologna: Il Mulino.

Grottanelli, V. 1977. Ethnology and/or cultural anthropology in Italy: Traditions and developments. *Current Anthropology* 4: 593–614.

Guarnieri, P. 2014. *Senza Cattedra. L'istituto di psicologia dell'università di Firenze tra idealismo e fascismo*. Firenze: Firenze University Press.

Larson, M.S. 1977. *The rise of professionalism: A sociological analysis*. Berkeley: University of California Press.

Lombardo, G.P., and R. Foschi. 1997. *La psicologia italiana e il novecento*. Milano: Angeli.

Loria, A. 1891. Economics in Italy. *Annals of the American Academy of Political and Social Science* 2: 59–80.

Losito, M., and S. Segre. 1992. Ambiguous influences: Italian sociology and the fascist regime. In *Sociology responds to fascism*, ed. D. Käsler and S. Turner, 42–87. London: Routledge.

Lotti, L., and G. Pasquino (eds.). 1984. *Guida alla facoltà di scienze politiche*. Bologna: Il Mulino.

Marhaba, S. 1981. *Lineamenti della psicologia italiana: 1870–1945*. Firenze: Giunti Editore.

Michels, R. 1927. *Corso di sociologia politica*. Rome: I.E.S.; English Translation *First lectures in political sociology*. Minneapolis: University of Minnesota Press, 1949.

Michels, R., and M. Hartsough. 1930. The status of sociology in Italy. *Social Forces* 1: 20–39.
Miozzi, U.M. 1993. *Lo sviluppo storico dell'università italiana*. Firenze: Le Monnier.
Miur (online source): attiministeriali.miur.it/.
Mongardini, C. 2003. Gli studi politologici e la facoltà di scienze politiche di Roma. In *Passato e presente della facoltà di scienze politiche*, ed. F. Lanchester, 205–224. Milano: Giuffrè.
Moretti, M., and I. Porciani. 1997. Il reclutamento accademico in Italia. uno sguardo retrospettivo. In *Annali di Storia delle università italiane* [online].
Ostenc, M. 2012. In memoria di Luigi Ambrosoli, storico dell'educazione. La riforma Gentile dell'insegnamento superiore del 1923. In *Luigi Ambrosoli e la storia d'Italia*, AA.VV, 69–78. Milano: FrancoAngeli.
Padovan, D. 2006. Organicismo sociologico, pianificazione e corporativismo in Italia durante il fascismo. *Rassegna italiana di sociologia* 4: 681–717.
Palano, D. 2005. *Geometrie del potere. Materiali per la storia della scienza politica italiana*. Milano: Vita e Pensiero.
Perotti, R. 2008. *L R. (2008) truccata*. Torino: Einaudi.
Proietto, M.G., and G.P. Lombardo. 2015. The 'crisis of psychology' between fragmentation and integration: The Italian case. *Theory and Psychology* 3: 313–327.
Rabbeno, U. 1891. The present condition of political economy in Italy. *Political Science Quarterly* 3: 439–473.
Ritrovato, E. 2013. Tra la scuola superiore di Commercio e la facoltà di economia: precedenti storici e sviluppi fino agli anni '70 del novecento. *Annali di Storia delle università italiane*, vol. 17, source online. www.cisui.unibo.it/annali/17/testi/12_Ritrovato_frameset.htm.
Santoni Rugiu, A. 1991. *Chiarissimi e Magnifici. Il professore nell'università italiana (dal 1700 al 2000)*. Firenze: La Nuova Italia.
Santoro, M. 2011. Esiste una comunità scientifica per la sociologia italiana? *Rassegna Italiana di Sociologia* 52: 253–282.
Santoro, M. 2013. Empire for the poor: Imperial dreams and the quest for an Italian sociology, 1870s–1950s. In *Sociology & empire*, ed. G. Steinmetz, 106–165. Durham/London: Duke University Press.
Santoro, M., and M. Pisati. 1996. *Dopo la laurea. Status sfide strategie*. Bologna: Il Mulino.
Sartori, G. 1965. Gli studi politici nelle facoltà di scienze politiche. In Atti del terzo congresso nazionale di scienze politiche e sociali: *Gli studi politici e sociali in Italia*, 41–55. Milano: Vita e Pensiero.

Scaglia, A. 2007. 25 anni dell'Associazione italiana di sociologia. materiali per scriverne la storia. *Quaderni del dipartimento di sociologia e ricerca sociale*, 39. Trento: Università di Trento.

Shils, E. 1980. *The calling of sociology and other essays on the pursuit of learning*. Chicago: University of Chicago Press.

Simone, R. 1994. *L'università dei tre tradimenti*. Roma-Bari: Laterza.

Spreafico, A. 1964. Studi politici e scienza politica in Italia. *Annuario politico italiano* 2 (1): 202–230.

Testa, M.F.-E. 1996. La présence des études de folklore, d'ethnologie et d'anthropologie dans l'enseignement universitaire italien: éléments pour une histoire. *Acta Ethnographica Hungarica* 40 (3–4): 403–408.

Verger, J., and C. Charle. 1994. *Histoire des universitts*. Paris: PUF.

Villari, P. 1902. *Le scuole di scienze sociali e le facoltà giuridiche, le scienze politiche; raccolta di scritti politici e sulla pubblica istruzione*. Discorso per l'inaugurazione dell'anno accademico 1901–1902 nel R. Istituto di Scienze Sociali "Cesare Alfieri".

6

The Institutionalization of SSH Disciplines in the Netherlands: 1945–2015

Rob Timans and Johan Heilbron

In this chapter, we outline the institutionalization of the SSH in the Netherlands shortly before and after the Second World War. We focus on economics, psychology, sociology, anthropology and political science, as well as on philosophy and the sciences of literature. It is not possible to provide a complete overview within in the space of one chapter; therefore, we will study this institutionalization process in relation to the development of state policies of higher education. In the period addressed here, the balance of power between the Dutch state and the academic field gradually shifted. The role of the state in the field of higher education increased, which finally resulted in a more centralized decision-making process regarding the structure of academic

R. Timans (✉)
Erasmus University Rotterdam, Rotterdam, The Netherlands

J. Heilbron
Centre Européen de Sociologie et de Science Politique (CESSP), Paris, France
e-mail: johan.heilbron@planet.nl

J. Heilbron
Centre National de La Recherche Scientifique (CNRS) – EHESS, Paris, France

C. Fleck et al. (eds.), *Shaping Human Science Disciplines*, Socio-Historical Studies of the Social and Human Sciences, https://doi.org/10.1007/978-3-319-92780-0_6

education and the content of research. The academic field thus became less autonomous, which affected the SSH disproportionally, especially the more weakly institutionalized disciplines. Furthermore, this rearrangement of the position of Dutch academic institutions *vis-à-vis* the state was accompanied by the institutional convergence of the universities, resulting in a less diverse field. Combined with the increasing intellectual orientation towards the US of most of the disciplines this led to less diversity in the intellectual position taking of Dutch SSH scholars.

In the first section, we will describe briefly the principal historical stages in the increasing role of the state bureaucracy regarding higher education. The second section outlines the main developments in individual disciplines against the backdrop of the process described in section one. We summarize the main developments in the conclusion of the chapter.

General Developments

The historical development of the involvement of the state with the field of higher education in the Netherlands can be characterized by four distinct stages. The first stage, lasting from the end of the nineteenth century until the end of the Second World War, is marked by the introduction of legislation that shaped the modern university. At roughly the same time most of the disciplines with which we are concerned began a process of institutionalization within the traditional faculties. The second stage, lasting approximately from 1945 until 1964, saw the first attempts of the state to control the costs of higher education. This is when the effects of the rapid increase in student numbers became clear and terms such as 'efficiency' entered the discourse. In the third phase, from 1965 until around 1981, the temporal pressures on the academic field increased, eventually materializing in the form of new regulation aimed at cost control and reshaping curricula. The fourth, and final, stage we examine, from around 1982—when a major law concerning the structure of higher education came into effect—until the present day, is characterized by new rounds of budget cuts.

Phase 1: Early developments: until 1945

The Higher Education Law of 1876 laid the foundation of the institution of the modern university and had a distinct function: to provide an education suitable for careers in science and 'positions in society requiring a scientific education'.[1] When the law was introduced, the Netherlands had three so-called *Rijksuniversiteiten* (state universities)—Leiden, Groningen and Utrecht—a municipal university (the municipal University of Amsterdam), a technical vocational school in Delft and a School for higher education in agriculture in Wageningen (without the right to grant Ph.D. titles). Fifty years later, the field of universities looked quite different. In 1918, a separate Ministry of Education, Arts and Science was created, followed by an advisory committee on all educational matters; the *Onderwijsraad* (Council for Education; 1919). In Amsterdam, a second university was founded on Protestant-Calvinistic principles (the Free University of Amsterdam), while in Rotterdam and Tilburg two schools of commerce were established, which specialized in economics. A Catholic confessional University was also established in Nijmegen (1923). All of these institutes were founded by non-state actors, although the state increasingly contributed to their funding over the years (Roelink 1955). Further changes were that the Jus Promovendi had been granted to the School of Agriculture in Wageningen (1917) and the technical School in Delft (1905). Table 6.1 lists all the relevant academic institutes in the Netherlands.

The absence of elite universities was a distinct feature of the Dutch field of higher education at this time. While each institute has its distinct profile, there was not an obvious hierarchy among the universities. Further, these universities dominate the production of scientific knowledge, holding a virtual monopoly in this area. At the disciplinary level, the social sciences became distinguished from the humanities (*geesteswetenschappen*, a direct translation of the German *Geisteswissenschaften*)—such as philosophy, history and language and literature studies—at an early stage. This situation is different from the French system, for example, in which the *Sciences humaines* or *Sciences de l'homme et de la société* also encompass disciplines like sociology.

[1]Wet tot Regeling van het Hooger Onderwijs, April 18, 1876.

Table 6.1 Institutes of higher education in the Netherlands

Institute	Year established	Remarks	Changes after year of foundation
University of Leiden	1575	–	State University of Leiden (since 1876)
University of Groningen	1614	–	State University of Groningen (since 1876)
University of Utrecht	1636	–	State University of Utrecht (since 1876)
Polytechnical School of Delft (Royal Academy from 1842 until 1864)	1864	Technical school	• Technical vocational School (1905) with Jus Promovendi • Technical University of Delft (1986)
Municipal University of Amsterdam (Atheneum Illustre from 1632 until 1877)	1877		University of Amsterdam (1961)
State vocational School of Agriculture Wageningen	1876	Agricultural school	• State School of Agriculture, Horticulture and Forrestry (1918 with Jus promovendi) • Agricultural University of Wageningen (1986)
Free University of Amsterdatm	1880	Private university with Confessional-Calvinistic charter	Jus Promovendi in 1905
Dutch vocational School of Commerce Rotterdam	1913	Initially private school	• Dutch School of Economics (1938, acknowledged as University) • Erasmus University Rotterdam (1973)
Catholic University of Nijmegen	1923	Confessional Catholic charter	Radboud University of Nijmegen (2004)
Catholic vocational School of Commerce Tilburg	1927	Business and economics, confessional: Catholic	• Catholic School of Economics (1938, acknowledged as University) • University of Tilburg (1986/2001)

(continued)

Table 6.1 (continued)

Institute	Year established	Remarks	Changes after year of foundation
Dutch education institute for abroad (NOIB)	1946	Private Business School founded by large companies	Nijenrode Business University (1982 academic status)
Technical vocational School of Eindhoven	1956	Technical	Technical University of Eindhoven (1986)
Technical vocational School of Twente	1964	Technical	University of Twente (1986)
Rijksuniversiteit Limburg	1976	State university	Maastricht University (1996)

Source For pre-1930 situation: Jensma and De Vries (1997: 14)

From 1815 to 1921, research and education in these disciplines at state universities was organized in the 5 'traditional' faculties: Letters and Philosophy, Mathematics and Physics, Law, Medicine, and Theology.[2] In 1921, the government introduced a major revision of the so-called 'academic statute' (*Academisch Statuut*).[3] This was a governmental decree that regulated the subjects in which students could graduate and specified exam requirements. The main changes introduced in 1921 were as follows. Firstly, the barriers to enter into academia were lowered; a gymnasium secondary education (including Greek and Latin) was no longer the only degree that allowed access to an academic education. After the decree, an HBS (*Hogere Burgerschool*, or Higher Civilian School) degree also gave a student the right to enter university directly, marking the beginning of the unravelling of an education system highly stratified along class lines. Secondly, it provided for the creation of two so-called inter-faculties; a structure within which faculties could cooperate in research and education organization in overlapping subjects. The 1921 decree recognized two of these 'joint ventures': the interfaculty of Law and Letters (in which the subject of social geography was located), and the interfaculty of Mathematics and Letters

[2]Prior to 1815, philosophy, letters, physics and mathematics were united in the 'Artes Faculty'.

[3]Academisch Statuut, Koninklijk Besluit of June 15, 1921. Published in *Staatsblad* no. 800.

(for physical geography). The first university to establish the former was the University of Leiden in 1925 (followed by the University of Utrecht in 1930 and the University of Amsterdam in 1931; Jensma and De Vries 1997: 200).[4] Thirdly, the decree recognized psychology and sociography as separate subjects within philosophy and geography, respectively.

By 1939 there were a total of 11,251 students at the various institutions of higher education in the Netherlands. Of these, 865 studied in the traditional faculties of Philosophy and Letters, 190 in the economics faculty in Amsterdam, 375 at the interfaculties of Law and Literature, and 711 at the Schools of Commerce in Rotterdam and Tilburg (Jensma and De Vries 1997: 210). On the eve of the Second World War, then, roughly 20% of the student population was enrolled in SSH or SSH-related subjects. Within the faculties of Philosophy and Letters, students could obtain a degree in philosophy with sociology, psychology or pedagogy as an elective major. They could also obtain a degree in indology at the United Faculty of Law and Letters in Leiden or Utrecht, or a degree in economics in Rotterdam, Amsterdam or Tilburg.

Phase 2: 1945–1964: expansion

During, and shortly after, the War a number of Dutch scholars problematized the position of the nation's intellectuals during the rise of national socialism in the 1930s and the German occupation of the country from 1940 to 1945. They concluded that a lack of political sophistication and naiveté concerning international politics among the intellectual elite was one of the main reasons for underestimating the fascist threat before, and a passive stance during, the occupation (Brandt 1998; Knegtmans and Keman 1998). A small group of academic entrepreneurs, consisting of scholars like historian Nicolaas Posthumus, historian Jan Romein, polymath/pedagogist Philip Kohnstamm and historian Jacques Presser discussed the idea of a 'seventh faculty' at the University of Amsterdam. The educational programme at this faculty would aim to educate the future elite of the Netherlands in the

[4]In Leiden, the interfaculty Law and Letters already welcomed 199 students in its first year (Jensma and De Vries 1997: 203).

social sciences, with an emphasis on the political sciences. The establishment of such a faculty, with the study of society and political processes at its centre, would also serve to forge a stronger bond between academia and the public. This was also to be achieved by lowering the entry requirements to study at the new faculty. The academic entrepreneurs felt that the academic elite had been impervious to social developments too many times in the past, while the resistance effort during the war had shown that true leadership did not require an elite background (Knegtmans and Keman 1998: 16).[5] Politically, momentum was building towards the kind of programme envisioned by the initiators of 'the seventh', as it became known. After the War, the Dutch government prioritized rebuilding the nation's industry and its damaged infrastructure, but there were also concerns about social cohesion in this rapidly modernizing society. The social sciences were seen as an important element of the rational planning of the rebuilding effort; they could provide the scientific basis for government policies intending to restore pre-war culture (Bijleveld 1996: 21). Similarly, a broad commission installed by the Secretary of Education in 1946 to assess the state of higher education in the country concluded in its section report on the social and political sciences that there was an increasing societal need for academically trained professionals in these disciplines. Although various universities offered courses in social scientific disciplines, it was felt that there was a lack of a common structure. The Commission believed this issue had to be addressed by the nation's universities.

The University of Amsterdam did not wait for the publication of the final report to act. It opened its 'seventh faculty' of Political and Social Sciences in 1947. This consisted of three sections: political and social sciences ('section A'), sociology and sociography ('section B'), and psychology and pedagogy ('section C'). Being a municipal university, the University of Amsterdam was more agile than the state universities, although the appointment of the first occupants of the newly established chairs in sociology, political sciences, contemporary history, economics and press, propaganda and public opinion was mired

[5]The proposal to lower entry requirements did not make it into the final charter of the Faculty.

in controversy over the alleged socialist and communist sympathies of various candidates (Daalder 1995: 41; Brandt 1998). The Faculty was perceived by some members of the municipal council as a bulwark of Marxism and 'red sympathies' that could not be trusted to educate future politicians (Brandt 1998; Knegtmans 1998: 279).[6]

The establishment of the Amsterdam 7th faculty had a major impact on the field of higher education in the SSH in the Netherlands. It brought the long lingering competition with the University of Leiden to the fore, which traditionally had been close to the state and ministries, such as the department of foreign affairs. The establishment of a curriculum in political science sharpened suspicion of the 'leftist faculty'. In addition, the organizational structure raised questions about the viable organizational form of any faculty for the social sciences. At other universities, chairs in social science subjects were still scattered across the faculties of Law, Economics and various 'United Faculties', depending on the local circumstances. The Amsterdam constellation, with psychological sciences as part of the faculty, was a source of heated debate within the social sciences section of a national Commission installed to discuss educational reform (Rupp 1997: 155ff.; Coumou 2013: 12).[7] Eventually, in their final report (1949), the Commission recommended a different model than the one used in Amsterdam: the establishment of a 'central interfaculty' for sociology and political sciences, organized between the existing faculties of Law, Economics and Letters. They reasoned that, since these subjects transcended traditional disciplinary boundaries, the disciplines would be best served by a faculty that connected all other faculties (Groen 1989: 173; Rupp 1997: 156; Coumou 2013: 25). According to the 1960 law on Scientific Education (*Wet Wetenschappelijk Onderwijs*, WWO), every university had to establish such an interfaculty, but that proposed for social and political sciences never came to fruition. Instead, a separate Faculty

[6]This controversy was also brought to the national stage: the secretary of Education had to confirm the appointments, but the newly appointed Catholic-conservative secretary Gielen also opposed the candidates forwarded by the university, as well as the reduced entrance requirements.

[7]The *Staatscommissie inzake de Vernieuwing van het Hoger Onderwijs*.

of Social Sciences was proposed that would include three sub faculties: one for Western Sociology, Non-Western Sociology and Cultural Anthropology, one for Psychology and a third for Pedagogy. Thus, after 20 years, the US-inspired idea to organize the social sciences into a separate faculty finally prevailed.[8]

The humanities were less visible in the academic-political debate in this phase. Indeed, in 1960, the board of the department of studies of literature (*letterkunde*) of the Royal Netherlands Academy of Arts and Sciences (KNAW), together with the board of the ZWO (the predecessor of the Netherlands organization for scientific research, NWO), wrote a letter to the secretary of education in which they expressed their concern over the one-sided focus of national science policy to the natural sciences (Haas 1975: 3ff.). In 1961, the secretary duly installed a commission to recommend "coherent policies" for a "balanced execution" of the humanities (Uhlenbeck 1975: 22). One of its recommendations was the foundation of a Dutch institute comparable to the American (sic) Institutes of Advanced Study in Princeton and Stanford (ibid.: 23), which led to the founding of the Netherlands Institute for Advanced Study in the Humanities and Social Sciences (NIAS) in 1970. This is an example of the steadily increasing influence of the US-style of higher education on developments in the Netherlands at this time. American conceptualizations of social science were introduced in the Netherlands partly through the Fulbright programme, which allowed European researchers to spend time in the US and vice versa (Rupp 1997: 201ff.). It also influenced the content of curricula and style of teaching. There is a marked, early influence on the content of the political science curriculum, which was modelled after epistemic practices in the US. In economics and, somewhat later, psychology and sociology, practices based on US-style science replaced national and European-influenced approaches like phenomenological psychology and sociography.

During this stage, the first initiatives were taken to reorganize the anachronistic structure of the faculties at the Dutch universities. Rising student numbers (see Fig. 6.1) put a strain on the existing

[8]Rupp (1997: 146ff.) provides a detailed overview of the discussions in the commission.

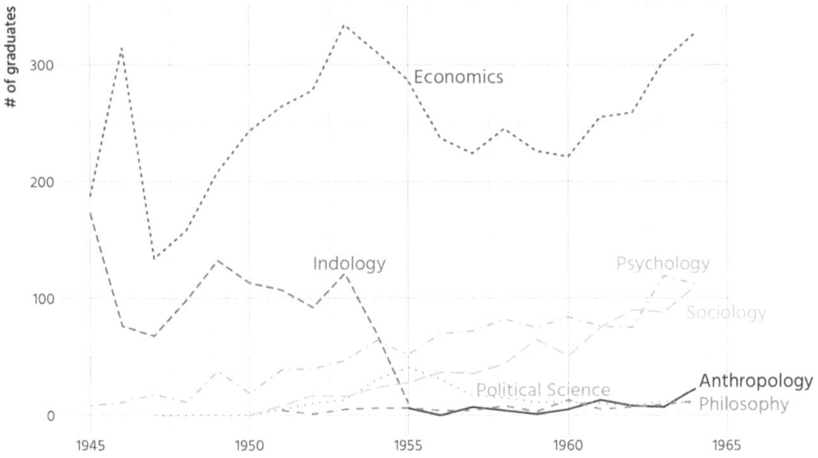

Fig. 6.1 Number of students who graduated 1945–1964 in philosophy, economics, political and social sciences (University of Amsterdam and, later, University of Nijmegen), sociology (recognized as a subject in 1950 under the name 'Social Sciences'), psychology, anthropology and indology. Indology is included because of its importance for anthropology (*Source* CBS, various statistical publications)

arrangements, but disciplines like sociology and psychology finally found an institutional home in the Faculty of Social Sciences, while philosophy was based in a central interfaculty. Although the final settlement was contested, this at least brought an end to the plethora of curricula spread over faculties and united faculties (see Table 6.2). Moreover, the Ministry of Education engaged in a series of initiatives to tackle the issue of the long time it took students to complete their studies. One of the first measures was introduced in the *Wet Wetenschappelijk Onderwijs* (WWO) of 1960 and comprised the transfer of the responsibility to determine the duration of curricula to the faculties. Up to that point, professors were solely responsible for the content of curricula and the research agenda (Foppen 1989: 59, 86).

Phase 3: Initial reforms: 1965–1981

Discussions on how to reform the structure of academic curricula continued into the second half of the 1960s. By and large, the participants in this discussion (mostly the Academic Council and the Ministry of Education) felt that curriculum reform should aim to further reduce

Table 6.2 SSH disciplines and major events in the establishment of curricula

SSH discipline	First chair	Initial situation	Major changes (based on academic statute and otherwise)
Philosophy		Subject within Faculty of Letters and Philosophy (since 1876)	1921: Doctoraal 1926: Kandidaats/Doctoraal 1963: CIF Wijsbegeerte 1986: F of Philosophy
Economics	1831 (Utrecht)	Required course within Faculty of Law (since 1876, *Kandidaats*)	1913: NHH (Commerce sciences) 1921: Major within 'vrije studierichting rechtswetenschappen' 1922: F of Economics UvA 1927: KHHT (Commerce sciences) 1938: Subject of Economics within F of Economics
Cultural Anthropology	1877 (Leiden)	Subject within Faculty of Letters and Philosophy (since 1929)	1953: Subject within UF Letters and Philosophy, Law and/or Economics 1963: Subject in F of Social Sciences 1980: Merger of Cultural Anthropology and Non-Western Sociology
Psychology	1890 (Groningen)	Faculty of Letters and Philosophy: major within subject of Philosophy	1921: Psychology as a major within subject of Philosophy 1941: Psychology as a subject within F of Letters and Philosophy 1946: Psychology subjects in UF Mathematics and Physics, Letters and Philosophy and/or Medicine 1952: Psychology (Letters and Philosophy) into UF 1963: Subject in F of Social Sciences

(continued)

Table 6.2 (continued)

SSH discipline	First chair	Initial situation	Major changes (based on academic statute and otherwise)
Sociology	1922 (Amsterdam)	Faculty of Law: major within law studies	1921: sociography as separate subject 1950: Subject ('sociale wetenschappen') within UF Law, Letters and Philosophy 1952: Subjects 'Westerse sw' and 'niet-westerse sw' within UF Letters and Philosophy, Law and/or Economics 1963: Subject in F of Social Sciences (western/non-western sociology)
Political Sciences	1948 (Amsterdam) [1612 Leiden: Daniel Heinsius]	Faculty of Law	1948: Subject within F of Political and Social Sciences (UvA) 1950: Subject within UF Law and Letters and Philosophy 1967: Subject in F of Social Sciences
Dutch Letters/Sciences of Literature	1797 (Franeker and Leiden)/1956 (Utrecht)	Subject Nederlandse Letterkunde within Faculty of Letters and Philosophy (since 1876)[a]	1921: Nederlandse Taal- en Letterkunde 1950: minor Algemene Literatuurwetenschap for students in Leiden 1967: subject Algemene Literatuurwetenschap

[a]In 1876, the subject of history was detached from the designation of chairs in *Nederlandse Letterkunde*. History became a separate subject in 1921

the required time it took to complete studies and obtain an academic degree, as well as the effective duration of study.[9] Changing government coalitions and the prioritizing of governance reform (cumulating in the WUB, see below) meant that the process of drafting a law to that effect dragged on for years. However, by the mid-1970s there were signs of a growing political will to address the rising costs of higher education. The rising number of students meant that higher education, especially academic, was using up ever more financial resources. An alarming report by the Statistical Agency (*Centraal Bureau voor de Statistiek*, CBS) that argued that the expected influx of students would be unsustainable under the prevalent regime featured heavily in the discussions (CBS 1956). The Law on the restructuring of higher education (WHHO) was eventually passed by parliament in 1975. However, this law failed to realize the chief objective of reducing the maximum study duration. Earlier requests by the Ministry of Education to the faculties to submit proposals to reduce the length of the curriculum to four years failed as almost all submitted proposals used the exemption clause to allow 5 years of study.[10] According to the Secretary of Education, Van Kemenade, moves like these led to the forfeit of the academic institutes' opportunity to exert subsequent significant influence over the development of the new plans (Van Kemenade 1982). It strengthened the state's resolve to act increasingly unilaterally, rather than relying on commissions composed of stakeholders to prepare legislation (Groen 1983: 2).

Implementation of the 1975 law was postponed until 1980. In addition to tough negotiations between the parties, the slow progress of discussions was due to the urgency of addressing governance issues. A series of conflicts, mostly between students and professors, took centre stage. At the end of the 1960s, calls for more democratization of academic governance structures led to student protests, often concentrated in the newly formed social science faculties. The protests focused

[9]The Academic Council (*Academische Raad*) was a new body established by the Law on scientific education (*Wet op het Wetenschappelijk Onderwijs*, see below) of 1960 to function as a liaison between universities and colleges.

[10]Nationaal Archief, Archiefinventaris 2.14.35. For examples for sociology, anthropology, political sciences and psychology, see Inventarisnummer 171–172.

on broad political issues—political science students in Amsterdam rebelled against the teaching of political science that they saw as too focused on mainstream theories, and that this amounted to a tacit support for the Vietnam war and the use of (social) science as a means of waging war, for example—but also on a much bigger role for students and staff in the governance of the faculties. In 1970, the government introduced the *Wet op de Universitaire Bestruursstructuur* (WUB, the law on the governance structures of universities) that sought to fundamentally democratize decision-making processes. Gone was the dominant position of professors and in came faculty councils composed of scientific and non-scientific staff, as well as students. So-called competence groups (*vakgroepen*) democratized the existing structure based on chairs of professors. However, the increased power of the competence groups had the adverse effect, further fragmenting research activities and decision-making. Competence groups for sociology, political sciences, cultural anthropology etc., although formally part of the same sub-faculty, often went their separate ways. The anticipated integration under the umbrella of the faculty of Social Sciences and its sub-faculties proved, in most cases, to be illusory (e.g. Bak 2013: 169ff.) (Fig. 6.2).

Phase 4: 1981 and beyond: Restructuring and consolidation

While the 1970s can largely be characterized by the process of democratization of education, the 1980s are undoubtedly marked by fiscal restructuring and budget cuts. As mentioned above, in the 1960s and 70s, discussions regarding regulation of the duration of curricula and study advanced slowly. However, the economic crisis of the early 1980s and the ensuing need for budget cuts accelerated matters substantially. After a new administration (a coalition of Christian-democrats and right wing liberals) took office, a new bill revitalizing the WHHO was sent to parliament. In 1981, parliament passed this new law that organized education curricula into a two-phase structure. This structure, implemented in 1986, shared important elements with the scheme contained in the dormant WHHO. The law introduced a first doctoral phase of education lasting four years, to be concluded with a *doctorandus* (master's) degree, and a second phase of another four years to complete a Ph.D. This model introduced important elements of the

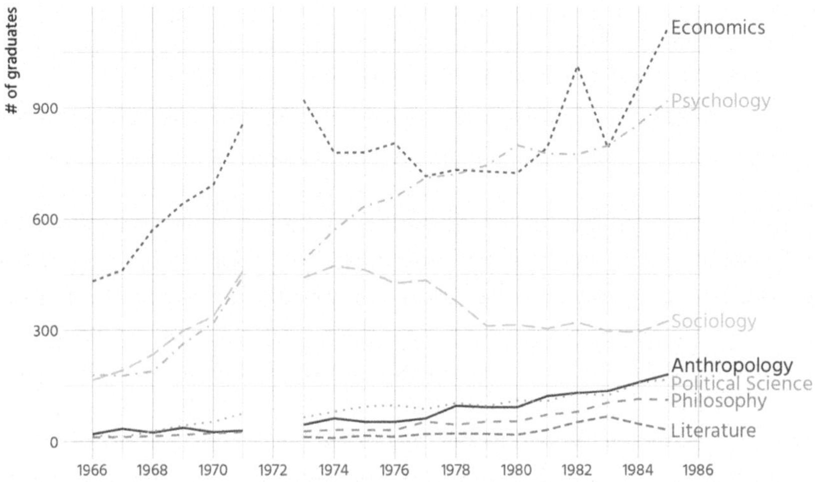

Fig. 6.2 Number of students who graduated 1966–1985 in economics, political sciences, sociology (including Western sociology and sociography), anthropology (cultural anthropology and non-Western sociology), psychology, sciences of literature and philosophy (CIF). The year 1972 is missing because of insufficient data (*Source* CBS, various statistical publications)

US higher education system into the Netherlands, splitting the original dual goals of academic education (preparing for the practising of science and for positions in society requiring an academic training). Although the integration between the two stages never became as seamless as was originally envisioned, the new organization of the educational trajectory resulted in the formalization of research education through the establishment of graduate schools in the social sciences.

Also in the early 1980s, the Ministry of Education announced a large scale review process of allocation of specializations (TVC).[11] In October 1982, the '*Nota taakverdeling en concentratie hoger onderwijs*' was presented, which introduced cutbacks of 258 million Dutch Guilders (roughly 4% of the annual budget for higher education) for

[11] The official term was 'Taakverdeling en Concentratie' (TVC; division of labour and specialisation; later the operation became known as 'Selectieve Krimp en Groei' (SKG; selective shrinkage and growth), which included proposals for personnel cutbacks.

the period 1983–1987. The aim of the policies was simple and clear: to increase efficiency. Universities were required to submit proposals stipulating which curricula would be discontinued as part of a general plan to achieve concentration and specialization. While the economics faculties remained largely untouched by the restructuring, other social sciences, alongside large parts of the humanities, were affected. In Utrecht for example, a new curriculum of General Social Sciences was introduced in 1985 (Selten 2005), while in the humanities cooperation between different competence groups resulted in new curricula like General Language Sciences and Medieval Studies (Overbeek 2010). The anthropology department in Groningen was closed, as was non-Western sociology at the Free University of Amsterdam as well as the entire andragogy curriculum nationwide. Research was also affected. A new system of conditional finance for research activities (*Voorwaardelijke Financiering*, or *VF programma's*) introduced the need to submit compact research proposals with a maximum duration of 5 years. The new budget structure aimed to halt the trend of increasing fragmentation of the social sciences indifferent specializations and studies (the so-called *kundes*) that, in the eyes of the Ministry, had led to 'little islands' of research groups that jealously guarded their autonomy. The proposals included initiatives to end this fragmentation. The new approach resulted in the financing of research becoming more independent from education and student numbers, and, by increasing the chances of a successful application for research funds, it rewarded cooperation between competence groups. It also benefited research groups that already were well organized and had gained a strong position in their disciplinary field (e.g. Van El 2002: 75).

A second round of budget cuts, the so-called Selective Contraction and Growth operation (SKG), was announced in 1987. Again an aim was to allocate disciplines and specializations among the Dutch universities. This time the cutbacks mainly targeted education. Curricula were to be combined and integrated and students had to take more interdisciplinary and 'modular classes'. Secretary Deetman's proposals to limit a full philosophy curriculum to three universities (Utrecht, the Free University of Amsterdam and Nijmegen) and to terminate Tilburg's curriculum in the theory of literature were not maintained in the final version of the TVC operation, but the sociology departments in Leiden and political science departments in Rotterdam eventually closed. In 1988, the sociology curriculum at the Free University of Amsterdam was also discontinued, although it was later reinstated.

Just as in the TVC operation previously, the faculties of Letters and Social Sciences bore the brunt of the cutbacks, but it was the social sciences that most suffered. Again, the rationale offered for the reorganization was efficiency, and the whole operation reflected an increasingly economized view of higher education. The disciplines singled out were those with the highest unemployment rates among graduates (i.e., sociology, psychology and anthropology; *NRC Handelsblad* 1993).

To summarize, in the last phase of the modern history of higher education policies in the Netherlands, central government increased its grip on the universities. Their status as relatively independent institutes was significantly diminished over this period, both in education and research. The Ministry of Education tackled organizational fragmentation and the replication of curricula by allocating specializations and imposing increasingly strict financial and operational constraints. In general, the smaller departments (such as anthropology and sociology) suffered most, often unable to muster the critical mass needed to stand up in the face of the academic bureaucracy charged with the implementation of the new measures.

Institutionalization of the Disciplines

The process of institutionalization of the social sciences in the Netherlands differed across disciplines. In chronological terms, economics and anthropology were institutionalized in academia well before sociology and political sciences, with psychology occupying a middle position, although psychology had a slower process of autonomization. Philosophy teaching preceded that of the other disciplines, but it held a rather weak position in that it was dependent on other disciplines, most notably theology. It would take until the 1960s for philosophy to get its own faculty. And although the first chairs devoted to the sciences of literature did not appear until the 1950s, the discipline quickly established itself as a small, but distinct, field.

The first chairs in the social sciences were created within one of the traditional faculties mentioned above, next to existing chairs in law, mathematics, philosophy and letters. Economics was first, with a chair in political economics (*Staathuishoudkunde*) established in 1831 in Utrecht within the Law faculty. Initial chairs in ethnology (1877; Leiden, at the Faculty of Law), psychology (1890; Groningen, at the Faculty of Philosophy and Letters) and sociology (1922; University of Amsterdam, at the Faculty of Law) followed.[12] Figure 6.4 provides an overview of the development of the number of ordinary chairs according to discipline. Dutch Literary studies is included to provide a comparison with sciences of literature.

Philosophy: A Central Discipline?

The first university founded in the territory of what is now Belgium and the Netherlands was the University of Leuven (in present-day Belgium) in 1425, which would remain the sole university in the Low Countries until the University of Leiden was founded in 1575, which was the centre of academic life in the fifteenth and sixteenth centuries. Philosophy was taught as an introductory subject for students in the traditional faculties of theology, law and medicine in a 'preparatory faculty'.[13] Students of medicine, theology and law first had to pass a propaedeutic exam in this faculty before they could continue their studies at one of the other faculties. In addition, most of the academic discourse in philosophy actually took place in the context of the faculties of theology and medicine, permitting philosophy only the status of an auxiliary subject. Moreover, a professorship in philosophy was often seen as a preparation for a chair in theology (Sassen 1959: 78, 80). Initially, this was also the case in Leiden.

In the middle of the seventeenth century, philosophy became increasingly differentiated as a separate discipline from theology and, later, the natural sciences and mathematics. The establishment of a Faculty of

[12]In 1877, the University of Amsterdam also established a chair in pedagogy.

[13]*Voorbereidende faculteit der bespiegelende wijsbegeerte en letteren.*

Letters and Philosophy in 1876 was a further milestone in this development, elevating the subject of philosophy from its status as a preparatory subject. According to some, however, this arrangement also resulted in a weaker connection between philosophy and the sciences, thus impeding the development of the philosophy of science in the Netherlands (McAllister 1997). For others, it led to an era of general and intellectually superficial philosophy (Peijnenburg 2009).

Although some welcomed the new institutional home for philosophy, according to Weijers (1995) there was a notable group of Dutch philosophers who continued to make the case for re-establishing philosophy as a central discipline, not least in order to provide a classical ideal of *Bildung* in the moral education of students. Outside of academia, an early result of this line of thought was the establishment of *the Internationale School voor Wijsbegeerte* in 1916. The School was a private initiative with a strong idealistic agenda aimed at international cooperation with emphasis on the unity of the sciences—with a central place for philosophy—and advancement of the education in philosophy (see Van den Hoven 1986). Initially, the significs movement played a role in its foundation, but after a conflict, the significs separated to found the Signific Circle (*Signifische Kring*) in 1921. The mathematicians Gerrit Mannoury (1867–1956) and Luitzen (Jan) Brouwer (1881–1966) were among its members. These two mathematicians made significant contributions to the philosophy of the foundations of mathematics with Brouwer's philosophy of intuitionism especially gaining international recognition in this respect. The Signific Circle, however, never fulfilled its ambition of socio-linguistic reform to eliminate conflicts arising from misunderstandings in human communication. It was dissolved in 1926.

Although various smaller associations for philosophers existed prior to 1933—most notably the Dutch chapter of the International Kant association, which existed from 1921 until 1933—the first broad association for philosophers in the Netherlands—the General Dutch Association for Philosophy (*Algemene Nederlandse Vereniging voor Wijsbegeerte*, ANVvWB) was founded then. The journal *Algemeen Tijdschrift voor Wijsbegeerte* pre-exists the Association (see Table 6.3). It was established in 1907 but was rekindled by the ANVvWB in 1933, published from 1933 to 1970 as *Algemeen Nederlands Tijdschrift voor Wijsbegeerte en Psychologie*. Thus, although the contributions from psychologists to the journal remained feeble, it recognized the growing importance of psychology.

After the Second World War, the perceived failings of the academic system to educate a critical political and cultural elite (Weijers 1995: 81ff.) was added to the arguments to give philosophy a central position in the institutional landscape of Dutch academia.[14] University students should all be able to acquire philosophical knowledge and learn to reason beyond the disciplinary boundaries of their primary subject (Krop 2006: 40). Such discussions formed the basis of the so-called *Centrale Interfaculteit* (Central Interfaculty of Philosophy, CIF), which was established in a section of the Law on Scientific Education (WWO 1960),[15] and was conceived as a supportive faculty. Philosophy was to function as a common element to the different academic disciplines and provide a common foundation, allowing for the philosophical reflection on disciplinary knowledge (Langendorff 1990: 20). Under the WWO, each Dutch university was required to instal a CIF. In 1963, this was achieved, but the interfaculty did not last; because of the weak position of philosophy as a discipline, the ambition of it acting as a unifying force was never truly achieved. The CIF was discontinued in 1986, under the regime of 1980s budget cuts.[16] Instead, philosophy was transferred to a separate faculty or to a department within the faculty of Letters. Although plans were laid out to substantially reduce the number of philosophy departments—philosophy would only be taught at the confessional universities and

[14]This argument was also important in the discussions about the seventh faculty in Amsterdam.

[15]The *Wet op het Wetenschappelijk Onderwijs* of 1960 defined the goals of an academic education as preparing for the independent practising of science, preparation for societal positions for which an academic education would be required or beneficial, and the integration of the sciences. The provision for the integration of the sciences was added to the tasks of universities compared to the previous law of 1876. Moreover, the goal of encouraging a sense of societal responsibility with students was also explicitly mentioned in the law.

[16]Before that year the Rijksuniversiteit Limburg (founded in 1976) was already no longer required to have a CIF.

the University of Utrecht—the other universities successfully resisted the proposed cuts. More recently, the discipline has enjoyed a steady rise in the number of students (see Fig. 6.3). At various universities 'philosophy of a discipline' programs are offered for students whose main study is in another discipline. The number of chairs in philosophy also rose relatively fast, more than doubling between 1969 and 1994 (see Fig. 6.4). Still, the position of Dutch philosophy in the academic field has been weak, traditionally. Philosophy is not widely taught at secondary school, and it has struggled to become the formative discipline its proponents hoped it would be, and is in Germany and France. Intellectually, Dutch philosophy has traditionally been dominated by theology. An important current in Dutch philosophy consisted of a Christian-based philosophical work in the confessional universities. The most well-known philosopher in this tradition was Herman Dooyeweerd (1894–1977), who taught at the Free University of Amsterdam (see e.g., Woldring 2012). Another example is the (Belgian) theologian Edward Schillebeeckx (1914–2009), who held a

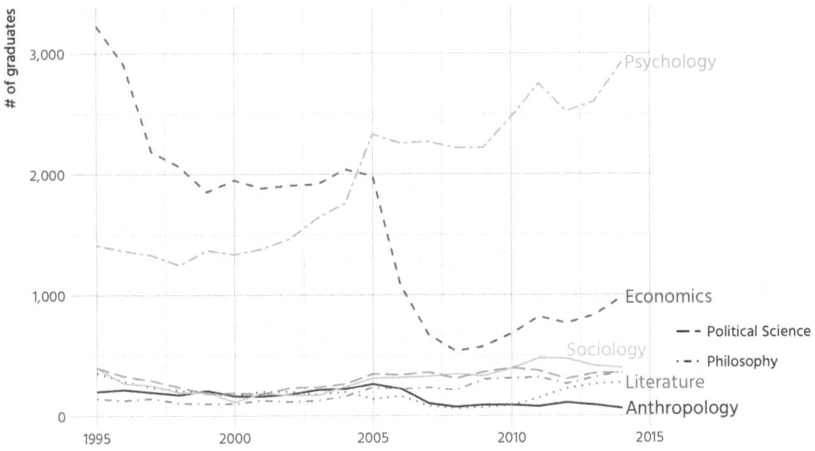

Fig. 6.3 Number of students who graduated 1995–2014 in economics, political sciences, sociology, anthropology (cultural anthropology and non-Western sociology), psychology, sciences of literature and philosophy (*Source* CBS, various statistical publications)

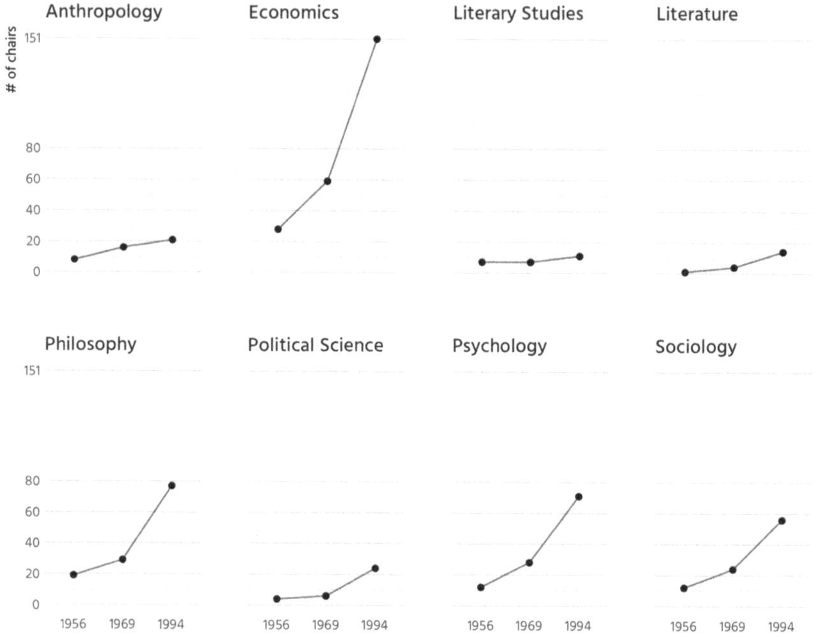

Fig. 6.4 Overview of the number of ordinary chairs in eight SSH disciplines across all faculties and universities (*Source* Adreslijst Universiteiten en Hogescholen, various editions)

chair in dogmatic theology at the University of Nijmegen from 1958 to 1983. The influence of these scholars extended to the social sciences as initially practiced at confessional universities. In general, philosophies such as existentialism and phenomenology, which became popular after the Second World War and replaced neo-Kantism and neo-Hegelianism that were *en vogue* before the war (Kuypers 1952; Sassen 1967), were often given a Christian-philosophical interpretation (Nauta 1990; see also Duyvendak 1997). At the public universities, an opposite pole to theologian philosophy developed in the form of analytical philosophy and logic. In logic, the work of Evert Beth (1908–1964) is especially noteworthy. Beth, who became professor of the philosophy of the sciences at the University of Amsterdam in 1946, continued the line of work of Mannoury and Brouwer.

Economics: Origins in Schools of Commerce

Chairs in political economy (*Staathuishoudkunde* in Dutch) were established at Faculties of Law in the first half of the nineteenth century. The subject was taught to law students, often in combination with other subjects such as statistics (in the traditional sense of state staistics; see Boschloo 1989: 16). The first of these chairs were established at the Law Faculty at the Universities of Leiden (1815; Hendrik Tydeman), Utrecht (1831; Jan Ackersdijck), and Groningen (1858; Otto van Rees).[17] Around that time, other signs of the early institutionalization of economics outside academia were visible. For example, the economics discipline was the first among the social sciences in the Netherlands to establish its own journal (see Table 6.3). In 1852, the journal *De Economist* was founded by Jacob Leonard de Bruyn Kops (1822–1887) and later issued by the Statistical Association/Association for Political Economics (*Vereniging voor de Statistiek/ (Koninklijke) Vereniging voor de Staathuishoudkunde*, KVS). Founded in 1849, it is one of the oldest professional organizations for economists in the world.[18] The *Economist* journal still exists but has been published entirely in English since 2012.

Towards the end of the nineteenth century through to the beginning of the twentieth century, a new generation of Dutch economists further established the discipline. Their principal interest was monetary economics. An important representative was Nicolaas Pierson (1839–1909), who became professor at the University of Amsterdam in 1877 and, later, President of the Dutch Central Bank and Secretary of Finance. He wrote an important two-volume textbook, *Leerboek der Staathuishoudkunde* (1884/1890), which was translated into English, French and Italian. His student Coenraad Verrijn Stuart (1865–1947) was the first director of the Dutch Statistical Agency (CBS) and became professor at the University of Groningen in 1909. Mainly influenced by the Austrian School of economics, his *De Grondslagen der Volkshuishouding* (1920) was an influential textbook and was translated into German in 1923.

---

[17]For a complete list, see Duyverman (1978).

[18]The roots of the association can be traced back even further, to 1849. See Mooij (1994).

The development of economics as an academic discipline gained momentum when the 'Dutch School for Higher Studies in Commerce' (NHH) was established in Rotterdam in 1913.[19] The curriculum offered by the NHH was not, however, named economics. Rather, it was known as 'sciences of commerce' (*handelswetenschappen*). The origins of the School were outside of the traditional university system: it was a private school founded, together with a small group of accountants, by merchants and corporations located in Rotterdam (Dicke et al. 2013; Diemel 2013: 17). They were united in their quest to increase the social standing of their trades by providing a science-based education for prospective commerce professionals.[20] The various parties provided the financial endowment to start the new School. Although it was not recognized as an academic education by the Ministry of Education until 1938 (when economics was adopted in the Academic Statute as an academic discipline and as the official name for commerce studies), the School was a success from the start; in 1917, it had already succeeded in attracting some 250 freshman students (Diemel 2013).[21] Other institutes reacted to the success of the NHH. The University of Amsterdam was granted permission by the Amsterdam municipality to initiate a curriculum in commerce and establish a sixth faculty. In 1922, 35 freshman students enrolled to embark on a four-year study. Six years later, in 1927, a Catholic School of Commerce was founded in Tilburg.

The approaches of the three institutes differed. While in Rotterdam the emphasis was on a scientifically founded, but practical, education (initially taught in a two year, interdisciplinary programme that also included subjects like chemistry and electrotechnics) focusing on applied knowledge (i.e., trade), the curriculum in Amsterdam emphasized the scientific content of business studies as a branch of economics with theoretical aspirations from the start and was taught in a full

[19]In 1863, the subjects of *staathuishoudkunde* and *handelswetenschappen* became part of the newly established HBS secondary education (Gorter 2012).

[20]The role of the private sector in the establishment of the Tilburg School and the Amsterdam economics faculty some years later was much less (Kastelein 1988b: 43).

[21]Besides Rotterdam the trade school of Tilburg was also recognised as academic institutions in 1938.

four-year programme. These differences would come to influence the development of business studies in the Netherlands in the 1960s, with the Amsterdam approach forming the foundation of business economics (first known as *bedrijfshuishoudkunde*, later as *bedrijfseconomie*, institutionalized in economics faculties) and the Rotterdam approach being influential in the institutionalization of separate faculties for business studies (first as *bedrijfsleer*, later as *bedrijfskunde*; Karsten and de Man 1992: 304). The Tilburg approach was inspired by Catholicism and a Catholic social philosophy. A key figure in this was the priest Martinus Cobbenhagen (1893–1954), who stressed the importance of embedding economics in philosophical, psychological and sociological approaches. The two Schools of Commerce each founded an economic journal. In Rotterdam, *Economisch Statistische Berichten* (ESB) was established in 1916. The journal, which is still published in Dutch, has traditionally focused on economic policy issues. The Catholic answer to ESB was the Tilburg-based *Maandschrift Economie* (1935–2010) which was, particularly in the beginning, an important outlet for Tilburg economists.

After the war, the government started to formulate the *dirigiste* framework for the design of economic policies designed to rebuild the country's economy. A consensus emerged around an approach to economic policy based on plans devised by the state and corporatist sector—or industry—organizations. To perform these tasks, these parties were to be supported by a new planning agency. This materialized in the establishment of the Central Planning Bureau (*Centraal Plan Bureau*, CPB) in 1945, which had an important influence on the post-war development of Dutch economics. The Bureau was founded as a planning agency using economic models to design economic policies by the government. Its first chairman was Jan Tinbergen (1903–1994), the first (co-)recipient of the Nobel Prize in economics in 1969 (together with Ragnar Frisch). By 1945, Tinbergen was already an authority on the use of mathematical models of the economy. He rose to fame during the depression of the 1930s when he was able to convincingly demonstrate the use of such models for policy purposes. Under his directorship, the CPB became a hotbed for economic research in the 1950s and 1960s (Van Dalen and Klamer 1996: 64). After working at the statistical agency (the CBS), Tinbergen trained as a physicist; a student of Paul

Ehrenfest. In 1933, he became a professor of statistics at the Rotterdam School of Commerce, where he stayed throughout his academic career. Under Tinbergen, the CPB initially framed itself as the supplier of scientifically grounded plans that could be used by the government to guide economic development and rationalize production in a manner comparable to practices in the business sector (Wilts 1997: 123). After those initial years, the CPB gradually became an evaluator of proposed policies, assessing the economic effects of various policy proposals using its quantitative models (such as the well-known VINTAF model). The CPB became a major centre of econometric research in the Netherlands in the 1950s, with substantial financial and human resources at its disposal. The Bureau developed expertise on the practical use of large-scale macro-economic models and its research was held in high regard among academics (Wilts 1997: 162–163). Its status decreased in later years, but its influence in the 1950s, together with the demand for policy orientated economists from the CPB, CBS and the Ministry of Economic Affairs, can be seen in the fact that Dutch economists still have a relatively practical and policy-orientated disposition (cf. Van de Beek 2010: 136).

In 1948, two further faculties of economics were added to the existing three: the Faculty of Economics and Social Sciences at the Free University of Amsterdam and the Faculty of Economics at the University of Groningen. At both faculties, the teaching of economics initially was placed in a broader context. For the Free University, the teaching of economics was part of the confessional foundation of the university, while in Groningen the aim was to connect to the agricultural economy in the north of the country (Kastelein 1988a: 12; Wilts 1997: 137). As a testament to this wider approach, chairs for sociology were initially part of the new Faculty of Economics. However, this changed very quickly; the sociology chairs migrated to a united interfaculty of Letters, Law and Economics, and the economics faculties became progressively focused on an increasingly narrow interpretation of economics based on neo-classical economic theory (Karsten and de Man 1992: 314). Economists' interest in cooperating with other social scientists disappeared (Rupp 1997: 95), and the difference in emphasis at the different universities receded in favour of a more uniform approach. In Rotterdam, practical subjects like organic chemistry, electro technical engineering and foreign language courses disappeared from the curriculum in 1947 (Klein 1974: 169; Diemel 2013: 153). The widespread adoption of Samuelson's textbook *Economics* (the first edition appeared in 1948) contributed to the virtual disappearance of pluriformity in teaching Dutch economics. A group of socialist economists who were critical of the increasingly orthodox approach formed the critical journal *Tijdschrift voor Politieke Ekonomie* in 1977. The journal is still in print as a Dutch language publication with a focus on unorthodox economics.

In 1957, following the increased emphasis on quantitative and mathematical techniques and the tradition initiated by Tinbergen, a separate quantitative-economic study was sanctioned by the government on the basis of the Econometric Institute (founded by Tinbergen and Henri Theil) at the University in Rotterdam (Wilts 1997: 140). Henri Theil (1924–2000) was Tinbergen's successor in Rotterdam. He played large role in the institutionalization of econometrics as a curriculum separate from economics. This differs from the situation in other countries where econometrics is integrated into the teaching of economics (Van Dalen and Klamer 1996: 38). For a short while in the 1960s, the Econometric Institute had a strong international allure (Plasmeijer and Schoorl 2005: 76).

Compared to other SSH disciplines, economics emerged relatively unscathed from the consecutive rounds of budget cuts and restructuring in the 1970s and 1980s. Student numbers in economics and business studies increased rapidly, especially in the early 1980s, and cutbacks in staff and curricula were avoided. Virtually all other social science disciplines suffered from declining student numbers and corresponding budget cuts at this time. In fact, the reforms of the 1980s were seen as an opportunity to revitalize Dutch economics and increase its international standing by emphasizing international publications and ending the 'provincialism' of Dutch economics (Buyst et al. 2005). As a result of the introduction of the two-phase structure, two graduate schools were established. In 1987, the Tinbergen Institute was created as a cooperation between the Erasmus University Rotterdam and the two Amsterdam Universities. At the University of Tilburg, the Centre for Economic Research (CentER) was founded in 1988.

Despite the influence of US economics and the international orientation, there are still some characteristics of Dutch economics, and economists, which make it different from other national contexts. The long empirical tradition can still be seen; Dutch economists have not contributed much to theoretical economics. By and large, their main interest was (and to some degree, still is) empirical. In fact, the Dutch economists who have gained recognition for their theoretical work—like Henri Theil, Tjalling Koopmans (1910–1985) and Hendrik Houthakker (1924–2008)—have all spent most of their career in the UK or the US. The empirical orientation of a towering figure like Tinbergen, the influence of the CPB and the general policy-orientated bent of Dutch social science have all contributed to this.

Whereas econometrics became a separate subject from economics, the subject of business economics remained institutionally wedded to the overall discipline of economics. This is also a feature of Dutch economics that differs from the situation in other countries, and this can be traced to the initial approach to the sciences of commerce and economics in Amsterdam. The Rotterdam approach of a more practical study of business, together with the influence from developments in the US, culminated in the development of a new subject (de Man 1996; Van Baalen and Karsten 2002). In the 1960s, *Bedrijfskunde*, or business studies, emerged as a separate specialization, first within the technical universities (Eindhoven in 1965 and Twente in 1968). The universities of Rotterdam and Delft cooperated in a joint interfaculty (founded in 1969) to offer a *doctorandus* degree in business (Eindhoven and Twente rewarded engineering—ir.—degrees), modelled after the US MBA degree.[22]

Anthropology: Colonial Influences

The development of anthropology in the Netherlands is intertwined with its colonial past. The links can be traced to the establishment of the Batavian Society for Arts and Science in 1778, before the academic

[22]The University of Groningen followed in 1977. See Daniëls (1985: 7). *Bedrijfskunde* was contained in the Academic Statute in 1972.

institutionalization of anthropology. In 1851, the *Koninklijk Instituut voor Taal-, Land- en Volkenkunde* (KITLV, now the Royal Netherlands Institute of Linguistics and Anthropology) was founded as a learned society, aimed at studying the Dutch colonies from a multidisciplinary perspective. Its primary source of funding was the Ministry of the Colonies (Kuitenbrouwer 2014: 3). The first anthropological journal—*Bijdragen tot de Taal, Land- en Volkenkunde* (Journal of the Humanities and Social Sciences of Southeast Asia)—was established in 1853 and published by the KITLV. It was a broad, multidisciplinary journal with a strong orientation towards the region of Southeast Asia. Colonial roots can also be seen to be the impetus behind the first chair in anthropology, established in Leiden in 1877. It was the result of the government's decision to transfer the course for prospective civil servants in Indonesia from the Royal Academy in Delft to (eventually) the University of Leiden (Kloos 1987: 11). At the time, the field of study was divided between ethnology (*etnologie* or *volkenkunde*) and ethnography. Ethnography (or 'regional anthropology') was the main strand until the early twentieth century, with a particular focus on the colony of Indonesia (or the Dutch East Indies as it was then called; Vermeulen and Kommers 2002: 4). The Leiden chair in anthropology was dedicated to the ethnography of the Dutch East Indies. Internationally, the chair was one of the first of its kind, although its specific focus on Indonesia meant that it could hardly be considered a chair in general anthropology (Vermeulen 2002: 96). Rather, its place must be understood in relation to the Dutch Orientalist tradition (which was particularly strong at the University of Leiden) and to the other subjects, including languages (Javanese, Malaysian), Islamic studies, history and traditional law. Together these comprised a broad regional focus that made Leiden the leading world centre on these subjects at the time (*NRC Handelsblad* 1992).

The first occupant of the chair, the theologist Pieter J. Veth (1814–1895) taught the subject as part of the study of Indology. Indology was an interdisciplinary educational curriculum aimed at training Dutch civil servants—and employees of large companies—who were to be employed in the colony as administrators. Besides indigenous (*adat*) law, Indonesian languages and linguistics, history and economics, ethnology was a key element of the educational programme (Kloos 1976; Kloos 1995: 204). As an academic discipline, indology had a relatively short history, but one that left a distinct mark on the course of the institutionalization and intellectual development of Dutch anthropology. For example, many indology graduates also conducted research during their time in Indonesia and in a number of cases pursued a Ph.D. after returning to the Netherlands. Furthermore, the subject was the immediate predecessor of a new subject, the 'sociology of non-Western people' (see below). It was taught in the Universities of Leiden (from 1902) and Utrecht (from 1925), and was recognized as a full academic subject (the first one that could be classified as a 'social science'; Groen 1989: 106) in the Academic Statute of 1921. Its institutional base became the newly created interfaculty of Law and Letters.[23]

The first chair of which ethnology (as opposed to ethnography) was part of the ordinance was created at the University of Amsterdam in 1877. Its first occupant was Cornelius Kan (1837–1919). In Amsterdam, anthropology developed out of social geography: the Amsterdam chair was based in the Geography department of the united faculty of Natural Sciences and Letters (De Wolf 1998; De Bruijne et al. 2002). After 1907, it was held by Sebald R. Steinmetz (1862–1940). After the chair was split between an ethnology chair and a chair in social geography in 1933 (see below). Johan Fahrenfort (1885–1975), despite being initially employed as lector rather than professor, was to occupy the chair in ethnology.[24]

The first professional association for anthropologists was the Dutch Anthropological Association (*Nederlandsche Anthropologische Vereeniging*, NAV), founded in 1898. In its early days, it was mainly focused on biological anthropology, which in the Netherlands never established

[23]The first anthropological department was formed in Utrecht in 1914 (Vermeulen and Kommers 2002: 8). The first chair in ethnology in Utrecht was established in 1913 and first held by Jacob Kohlbrugge (1865–1941).

[24]Fahrenfort did not become a full professor until 1946. He is described as an armchair anthropologist (Köbben 1998: 53) whose main contribution was theoretical; it was not until 1955, when Fahrenfort's successor A. Köbben was appointed to the chair, that anthropology in Amsterdam became more empirical.

strong connections with cultural anthropology. In 1922, the Amsterdam counterpart to the NAV, the Dutch National Bureau for Anthropology (*Nederlandsch Nationaal Bureau voor Anthropologie*) was established (De Wolf 1998: 18). The Bureau was affiliated with the *Institut International d'Anthropologie* and sponsored the newly founded journal *Mensch en Maatschappij* in 1925 (see below). The two associations merged in 1949, continuing as the Dutch Society for Anthropology (*Nederlands Genootschap voor Anthropologie*).[25]

In Leiden, a supernumerary chair in general anthropology was created in 1922. It was first occupied by J.P.B. (Jan) Josselin de Jong (1886–1964). Josselin de Jong was a linguist. He was a student of Christian Uhlenbeck, with whom he conducted fieldwork on Native American languages in Montana (USA). His linguistic background informed his approach of anthropology. He was influenced by the work of Émile Durkheim and Marcel Mauss, which was brought to his attention by his friend Willem Rassers. Both men worked at the Museum of Ethnology in Leiden at the time and were more influenced by this than the ethnographical tradition that was then strong at academia (Van Bremen and Shimizu 2013: 331). As a university professor, Josselin de Jong and the group around him developed a distinct structural anthropology, about ten years before a similar approach became popular under the influence of French anthropologist Claude Lévi-Strauss.

After the Second World War, additional chairs in anthropology were founded at the Universities of Nijmegen (1948) and the Free University in Amsterdam (1956). In both cases, anthropology was practised according to the confessional foundations of these respective universities. The granting of independence to Indonesia in 1949 meant that a national review of the ethnology study was in order. A committee, which included Fahrenfort, Josselin de Jong and H. Fischer (Utrecht), decided on a new name for the subject of general ethnology in 1953: henceforth, it was to be called cultural anthropology, a name inspired by

[25]In 1965, the NSV (Dutch Sociological Association, see below) established a section for SNW amongst its ranks, which was subsequently converted into a department in 1971, following a merger with the *ethnologenkring*. At this point the NSV changed its name to the Dutch Sociological and Anthropological Association (NSAV).

US-American nomenclature. In the same year, it also became possible to obtain a bachelor's degree (*kandidaats*) in the subject making it, finally, a fully formed curriculum (the master's or *doctoraal* study had already been introduced in 1929 in the Faculty of Letters and Philosophy). The 1950s also saw the introduction of a new subject: the Sociology of Non-Western people (SNW; Vermeulen 2002: 109). SNW was born out of the demise of indology. The reorganization that followed the discontinuation of indology resulted in the establishment of a multidisciplinary curriculum—SNW—that was unique. Chairs were initially created in Leiden, Amsterdam (Free University), Nijmegen and Utrecht, and later in Wageningen and the University of Amsterdam (Kloos 1995: 207).[26] From the start, this new discipline—a status granted by the Academic Statute in 1952 under the name 'non-Western social study'—struggled to differentiate itself from sociology (Breman 2002: 304). In general, it was argued that, apart from a regional specialization, the subject of Non-Western Sociology relied on different methods of data gathering (e.g., ethnography instead of surveys) and was less concerned with issues of urbanization and modernization (see also Köbben 1964; Breman 2006). However, SNW was dominated by sociology in the field of disciplines. The subject was discontinued at the University of Utrecht in 1969. At the other universities, the bachelor's degrees of cultural anthropology and SNW were integrated in 1967, and the separate departments merged a few years later. Leiden was the exception, continuing to teach the subject until the early 1980s. The main reason for this was the first major restructuring operation in an SSH subject. In the 1960s, the Secretary of Education sought to reduce the number of anthropology departments and introduce regional and thematic specialization among the universities. By 1969, the Ministry of Education had concluded that there were too many specializations in anthropology offered by the universities. Subsequent negotiations led to the allocation of regional and thematic specializations among the different faculties. For example, Utrecht would specialize in the anthropology of Latin America and

[26]At the University of Amsterdam, Wim Wertheim had already interpreted his chair in the history and sociology of Indonesia (which he accepted in 1946) in a broader way to encompass the whole Asian continent. See Wertheim (2002: 272).

the Caribbean, while the Free University of Amsterdam would study non-Western religions and political anthropology. Later rounds of budget cuts and further allocation of specializations would lead to the closing of the anthropology department in Groningen in 1983.

The introduction of a new two-phase structure of academic education meant that research had to be organized in research institutes. For anthropology, this resulted in cooperation with sociologists in Amsterdam (in the Amsterdam school of Social Research, ASSR), cooperation with linguists and historians in Leiden (in the Centre for non-Western Studies, CNWS) and organization around the theme of development studies in Nijmegen, Utrecht and Wageningen (in the Centre for Resource Studies, CERES). Around this time, new journals appeared with a more general outlook. The regionally focused *Bijdragen* journal of the KITLV was joined by universal anthropological journals, such as *Focaal* (1985) and *Etnofoor* (1988). *Focaal* was established at the University of Nijmegen, while *Etnofoor* was based at the University of Amsterdam (Table 6.3).

Psychology: A Natural or Social Science?

Compared to other countries, the academic institutionalization of psychology in the Netherlands was late. The autonomization of the discipline took a relatively long time. Before the creation of the first chairs in psychology, the subject of *zielkunde* ('studies of the soul', cf. the German *Seelenkunde*) was part of the ordination of a number of chairs in philosophy. Amongst the first chairs for which psychology formed part was the chair for philosophy and psychology at the University of Groningen (Gerard Heymans, in 1890)[27] and a chair endowed in 1907 at the University of Amsterdam. Its first occupant, Tjitze de Boer, alternated yearly between teaching general and special psychology to philosophy students, but as a scholar, his main contributions remained in philosophy (Pos 1946).

[27]In Groningen, Heymans also founded the first psychological laboratory in the Netherlands in 1892. In general, however, early applied research in psychology took place outside academia and was mostly related to occupational research.

The first chair wholly dedicated to psychology was established at the University of Utrecht in 1922, first held by Franciscus Roels (1885–1962). Shortly before, in 1921, the subject of psychology became a major for philosophy students, making it possible to graduate as a psychologist at the Faculty of Letters and Philosophy. After Groningen and Utrecht, pre-war chairs in psychology were also established in Nijmegen (1923, at first also occupied by Roels), the Free University of Amsterdam (1929, a chair in pedagogy and psychology first occupied by Jan Waterink) and at the University of Amsterdam (1932). The chair in Amsterdam resulted from a split of de Boer's chair between a chair in philosophy and one in psychology. As a testament to the uncertain institutional designation of the discipline, the chair dedicated to psychology, held by the Hungarian Géza Révész (1902–1977), was established at the faculty of Mathematics and Physics and not at the Letters and Philosophy faculty.

Initially, students were only moderately interested in the new subject. By 1940, there were only 47 graduated psychologists in The Netherlands (Busato 2008: 37), and it did not become a full curriculum until 1942. From that year on, it became possible to study psychology at the United Faculty of Mathematics/Physics and Letters/Philosophy and at the Faculty of Letters and Philosophy. This situation of dual curricula reflected the ambiguities surrounding the status of psychology: did it belong to the natural or the social sciences? A temporary solution was found in 1952 by harbouring all psychology curricula in the United Faculty. Eventually, psychology became a part of the new Faculty of Social Sciences, a structure pioneered by the University of Amsterdam some 20 years earlier (see Table 6.2), although the new home was not welcomed by all (e.g. Bak 2013: 42ff.).

Before the Second World War, the main drive for the development of psychology came from outside academia (Eisenga 1978: 130). This is reflected by the oldest surviving association for psychologists, the Dutch Institute for Psychologists (*Nederlands Instituut voor Psychologen*, NIP), founded in 1938 (Table 6.3). The main motivation for founding the NIP was to develop, professionalize and regulate the occupation of psychologist. Older associations, such as the 'Psychological Circle' (*Psychologische Kring*, 1919, founded by Roels and a precursor to the

NIP) and the Dutch Psychological Association (*Nederlandse Vereniging voor Psychologie*, 1927–1955) had a narrower scope.

The first general journal for psychology was the *Nederlands Tijdschrift voor de Psychologie en haar Grensgebieden*, founded in 1933. Also in 1933, the philosophy journal *Algemeen Nederlands Tijdschrift voor Wijsbegeerte* included psychology in its title (remaining until 1970), which signalled its willingness to accept contributions from psychologists, although the journal would continue to be dominated by philosophers.

The content of early Dutch academic psychology was heavily influenced by developments in German psychology. The key figure in those early days was Géza Révész, who founded the internationally orientated journal *Acta Psychologica* in 1935; its early volumes clearly reflect the German influence (Busato 2008: 69). Conceptually, a broad approach was advocated which in Amsterdam was highly marked by the personality and style of Révész. In Utrecht, a distinct approach was founded based on Husserl's phenomenology. This strong philosophical foundation, critical of positivist psychology, inspired a relatively coherent approach to research known as the 'Utrecht School' of phenomenological psychology, although that designation originated more from outsiders than from the Utrecht psychologists themselves (Van Hezewijk et al. 2001; Van Hezewijk 2005). Key figures in this approach were Frederick Buytendijk (1887–1974) and Johannes Linschoten (1925–1964). The phenomenological approach had some domestic influence (mainly psychologists at the University of Nijmegen) and also resonated abroad (through the work of Amedeo Giorgi in the US, for example; see Van Hezewijk et al. 2002), but never developed into a strong, continuous programme. After a short period of bloom in the 1950s and early 1960s, this approach to psychology was increasingly marginalized by the advance of experimental psychology and the further autonomization of psychology from philosophy. A growing orientation towards the US and the influence of the positivistic approach to psychology dominant there contributed to this development. Scholars like Linschoten, who published widely in German, became increasingly rare.

The experimental approach gained ground in Amsterdam and Utrecht. In Amsterdam, Adriaan de Groot (1914–2006) developed a methodological approach to psychology that would become very influential, while in Utrecht Buytendijk's students Johannes Linschoten (most notably in his influential book *Idolen van de Psycholoog*, published posthumously) and Benjamin Kouwer mostly abandoned the phenomenological line of their mentor (Busato 2008: 79). De Groot finished his influential book *Methodologie* (Methodology) while he was a fellow at the *Institute for Advanced Study in the Behavioral Sciences* at Stanford University from 1959 to 1960 (Rupp 1997: 227). In the book, De Groot embraced the Anglo-Saxon approach to psychology, particularly advocating the development of instruments to measure educational performance. By the end of the 1960s, the reorientation of Dutch academic psychology towards the experimental approach was largely complete (Kleber and Schut 2011: 265). The Foundation for Psychonomics (NVP) was founded in 1968 in an effort to unite practitioners of this strand of psychology. Other psychological research outside of academia continued to be practised on a small scale in research institutes affiliated to large companies (such as the Dutch Postal Service) and applied physics research institutes (part of the *Nederlandse organisatie voor Toegepast Natuurkundig Onderzoek*, TNO).

As in other disciplines, publishing internationally (in English) became increasingly important. As a result, a number of journals that relied almost exclusively on contributions in Dutch disappeared or switched to English, such as *Acta Psychologica*, which was initially multi-lingual. The journal *Psychologie & Gezondheid* (founded in 1952) ceased to exist in 2011, while the critical journal *Psychologie en Maatschappij* (1977), which initially aimed to develop an alternative psychology (Abma 2001: 9), folded in 2001. The *Nederlands Tijdschrift voor de Psychologie*, in existence since 1934, was discontinued in 2010 after a brief spell (since 2006) as an English language publication (as the *Netherlands Journal of Psychology*).

The growing influence of Anglo-Saxon psychology can also be seen at the level of textbooks. In the formation years of Dutch psychology, German and, to a lesser degree, French texts were ubiquitous in psychology curricula. After the 1960s, however, these increasingly gave way to texts in English. Student numbers were rising fast in this period, and psychology rapidly became, together with economics, the most popular

SSH study. At some universities, a *numerus fixus*[28] had to be introduced to control the influx of first-year students. Psychology departments were added at the University of Maastricht (1995), the Erasmus University Rotterdam (2001) and the University of Twente (2002). The number of chairs in psychology saw a similar development, more than doubling between 1969 and 1994 (Fig. 6.4).

Sociology: The Influence of Sociography

Although the first chair in sociology was held by criminologist Willem A. Bonger (1876–1940) in 1922 (at the Faculty of Law at the University of Amsterdam), the early history of sociology in the Netherlands was heavily influenced by the subject of sociography. Sociography was a somewhat unusual subject, the history of which, as an academic discipline, can be traced to a chair in geography and ethnology at that same university. Established in 1877 and first occupied by Cornelius Kan (1837–1919), it was the first chair in geography in the Netherlands. Kan's successor in 1907 was Sebald R. Steinmetz (1862–1940).[29] Together with Bonger, Steinmetz can be regarded as the founding father of Dutch sociology (Van Doorn 1964: 34).[30] Trained in law, Steinmetz developed an early interest in ethnology and philosophy as a student in Leiden, where he also wrote his dissertation (in German) on the ethnology of punishment. The programmatic elaboration of sociography was very much the work of Steinmetz and reflected his ethnological approach to the subject of geography. In short, the sociographical approach was based on a positivistic approach to

[28]Restriction on the number of students allowed to commence their studies.

[29]The official designation of the chair was 'political geography, *volkenkunde* and the *land- en volkenkunde* of the Indonesian archipelago' In comparison to the designation of Kan's chair, *Volkenkunde* was added to the ordinate.

[30]In addition, Van Doorn (1964: 16) mentions Jeronimo de Bosch Kemper (1808–1876) as the first to write a sociological work (*Handleiding tot de kennis van de wetenschap der Zamenleving* in 1863) while he was a professor at the Faculty of Law in Amsterdam. In general, there was an opposition between jurists and socialists who debated social issues, but both sides did not systematically draw on sociology as it was developing in France and Germany at the time (Heilbron 1988: 78).

science and was a descriptive, heavily empirical orientation, critical of deductive reasoning and theoretical speculations. Steinmetz proposed a broad approach to ethnographically describe specific, mostly territorially defined social groups (and, most often, in the Netherlands) as accurately and completely as possible. With this approach, Steinmetz drew loosely on geography, early anthropology and sociology, strongly influencing the early development of these disciplines in the Netherlands. For example, the later development of ethnology as practised in Amsterdam was influenced by Steinmetz' approach, making it distinct from the ethnology done in Leiden (Van Doorn 1964: 36), while the sociological ambition differentiated the approach from social geography as practised in Utrecht (Van Doorn 1964: 41; Heslinga 1978: 55). For the most part, the impact of the sociography school and the disputes with others remained a domestic affair. The sociographers had a negligible influence on the social sciences in other countries, nor were they themselves heavily influenced by developments abroad, although there was a general intellectual orientation towards German sociology (Hofstra 1978: 29; Den Hollander 1948: 122). Bonger (who wrote his dissertation in French) was mainly influenced by the scientific socialism based on the works of Marx and Engels (Van Heerikhuizen 1987: 5). His dissertation *Criminalité et conditions économique*, published in 1905, was translated into English in 1916 (Van Heerikhuizen 1987: 63).[31]

Other (local) approaches to sociology also came to fruition, predominantly in the confessional universities. At the Free University of Amsterdam, for example, Roel van Dijk (1906–1962) developed a sociology based on Dooyeweerd's neo-Calvinistic philosophy, while in Tilburg Catholic social scientists founded their own journal *Sociale Wetenschappen* in 1957. These research strands were often supported by Protestant and Catholic research institutes outside of academia. Nevertheless, sociography had the biggest impact on the institutional development of sociology in the Netherlands before and shortly after the war. Steinmetz trained a

[31]One of the reasons it took 17 years for Bonger to be appointed to the chair in sociology was his outspoken socialist ideas. In general, sociology was often equated with socialism in those days, more so because the only other chair in sociology installed before 1940 was a chair in Marxist sociology in Utrecht in 1931 (held by Rudolf Kuyper). See Van Heerikhuizen (1987: 163ff.).

large number of students, and a number of them, unlike Bonger's students, came to occupy the first chairs in sociology. In 1933, Steinmetz's chair in Amsterdam was split into an ethnological component (occupied by Johan J. Fahrenfort, first as a lecturer, later as professor) and a sociographic chair, held by Henri N. Ter Veen (1883–1949). Ter Veen was one of Steinmetz's first students and later initiated a number of sociographical studies often with a strong policy orientation. His study on the population of the newly formed polders in the *Zuiderzee* became well known. The link with policy was further strengthened when Ter Veen founded the research institute *Instituut voor Onderzoek van het Nederlandse Volk* (ISONEVO; Institute for the Research on the Dutch People) in 1940. Institutes like these strengthened the policy orientation of the social sciences in the Netherlands (most notably in sociology and economics) that persists to this day (e.g. Van Heek 1978: 21; Ellemers 1978: 41). They also assured the continuation of sociographic work during the German occupation, and the existence of a network of sociographers was an important factor in their influence in the years immediately after the war (Gastelaars 1985: 78, 87).

Ter Veen also became an important figure in the Dutch Sociological Association, the NSV. This was established in 1936 by a group of social scientists under Bonger's chairmanship. Initially, it was a rather exclusive association for social scientists of various disciplines: until 1964 only social scientists who were employed in academia, and who were invited to join, could become members. After 1964, the association became more orientated towards sociologists (a sociology degree became a prerequisite for membership from 1965 to 1971), but this was reversed when the association explicitly included anthropology from 1971 to 1993.

Although efforts were made to open up the association, the NSV was criticized by a younger generation of social scientists who argued that important developments, both politically and intellectually, were taking place without the NSV's involvement. These youngsters established their own journal (*Sociologische Gids*) as an alternative to the NSV's *Mens en Maatschappij* (founded in 1925; see Table 6.3), and were influenced by Mertonian functionalism, which was popular in the US at the time. They rebelled against Steinmetz's old sociographical approach and instead envisioned a modern sociology as an integrated scientific approach (Gasterlaars 1985: 185, 195).[32] 'Modern sociology' was also the title of an American-style book that appeared in 1959 (Van Doorn and Lammers 1959) that was a landmark for this approach and which would serve as a textbook for many years to come (Goudsblom 1985). Thus the movement for an independent position of sociology among the sciences started as part of a social geography that emphasized social aspects of regionally identified groups under the name of sociography. It continued as a critique of the empirical, theory-less approach under the banner 'modern sociology'. The autonomization of sociology from philosophy, a struggle that marked the development of sociology in other countries, was much less important in the Netherlands.

Dwindling student numbers, combined with budget cuts in the 1980s, also had consequences for sociology. Sociology education in Leiden and at the Free University of Amsterdam was discontinued at the end of the 1980s. The reorganization of curricula into a two-phase structure in the mid-1980s resulted in the founding of two graduate schools: the ASSR (Amsterdam School for Social Science Research, now AISSR) and the ICS (the Interuniversity Centre for Social Science and Methodology, a collaboration between the universities of Utrecht and Groningen, joined later by Nijmegen). The two schools had a different approach towards sociology that in both cases was a continuation of existing local traditions in Amsterdam and Utrecht. The ASSR developed a historically orientated and interdisciplinary approach, while the ICS advanced an empirical-analytical orientation based on methodological individualism. The ASSR approach was influenced by the figurational sociology of the German sociologist Norbert Elias, which had come to fruition in the Netherlands somewhat earlier. Key figures in this Amsterdam school were Johan Goudsblom (1932–) and Abram de Swaan (1942–). An important outlet, especially in the beginning, was the *Amsterdams Sociologisch Tijdschrift* (1974), which merged with merged with the aforementioned *Sociologische Gids* in 2005 to become the journal

[32]One of the forerunners of the new approach, Jacques van Doorn, branded sociography as an anachronistic discipline that was in effect responsible for holding back the development of sociology in the Netherlands.

Sociologie. With the continuing convergence of Dutch sociology towards US practices, the starkest differences between the approaches receded somewhat, although they have not disappeared completely (Engbersen 2005).

Political Sciences: Central Place in a New Faculty

The world's first ever chair in political sciences was arguably established in the Netherlands. In 1612, at the Faculty of Law at the University of Leiden, a chair in Greek languages and Studies of the State was first held by the philologist Daniël Heinsius. The subject of this *scientia politica* was initially taught as an integral part of constitutional law, gyrating in later years increasingly towards *Staatswetenschap* (a multidisciplinary 'science of the state', *cf.* the German *Staatswissenschaft*). Until the end of the Second World War, the study of the state was embedded firmly in Law. The institutionalization of modern political sciences in the Netherlands is closely related to the establishment of the seventh faculty in Amsterdam. Jan Romein—who was one of the initiators of the seventh faculty—was well versed in political science as it was taught at US universities. This duly became an important model for the early implementation and scientific development of political sciences in Amsterdam (and, somewhat later, in Leiden), although existing studies at the London School of Economics and the *Institut d'Études Politiques* (Sciences Po) in Paris also provided inspiration. The first chair in 'the sciences of politics' was established in Amsterdam in 1948, and was held by Jan Barents (1916–1961). Barents—only 31 when he was appointed—had studied law and philosophy. He was actively involved in putting the new discipline 'on the map' (see Barents 1948). From the beginning, political sciences was to be the central discipline of the new faculty. Ambitions were high: an objective study of the political process was not only deemed to be feasible but also necessary to provide a rational assessment of political decision-making. The study of political sciences upon which the first students embarked was very broad. Besides political theory, students had to complete courses in economics, sociology, history and law (Barents 1948: 137; Daudt 1998: 14). This new curriculum also stood out because it did not include other 'political' subjects already taught in universities, such as public administration and international affairs. These subjects are often part of the political sciences in other countries, so this instantiation is typical of Dutch political science (Lieshout and Reinalda 2001).

The new discipline, strongly associated with the left-wing image of the new Amsterdam faculty, was met with scepticism from the established Dutch political elite. These elites, together with top civil servants at the state department, had mostly studied law in Leiden and were hesitant to accept students with a degree specializing in political sciences.[33] A large part of the controversy concerned the supposed lack of objectivity regarding the new science of politics or 'politicology' as it was to be called. None of the initiators of the new faculty had a clear idea what this new science was to entail, and amongst liberals and confessional politicians, there were concerns that the subject would become the domain of leftist professors (Knegtmans 1998: 278).[34] Therefore, chairs in the Catholic-inspired political science at Nijmegen (1950) and in a Reformed-Protestant approach at the Free University of Amsterdam (1953) were established to counterbalance the Amsterdam faculty. Political science curricula were also installed in Leiden (1963) and Rotterdam (1978). In 1950, the *Nederlandse Kring voor Wetenschap der Politiek* (Dutch Political Science Association, NKWP) was founded as the professional association for political scientists in the Netherlands (Table 6.3).

The first graduates in political sciences to become professors in the discipline did not enter the scene until the 1960s: Barents' student Daudt (University of Amsterdam), Hans Daalder (University of Leiden) and Andries Hoogerwerf (University of Nijmegen). Hoogerwerf was the first political science student to graduate from the Free University of Amsterdam. These three men were instrumental in the development of political science after the initial phase. Their differences of opinion regarding the content of the political sciences were to shape the

[33]Leiden had been a stronghold of the teaching of various subjects related to the state ('sciences of the state') since 1815. See Boschloo (1989: 19).

[34]In 1612, there was already a 'professor politices' in Leiden: Daniel Heinsius, who taught Greek language and literature (Tromp 2007: 21).

discipline in the 1960s and 70s. Daalder (1928–2016) later became the first head of the Department of Political and Social Sciences at the European University Institute in Florence. He was also one of the founding fathers of the European Consortium for Political Research (ECPR).

In 1964, the Dutch name for political sciences (*politicologie* or 'politicology') was introduced into the academic statute and was thus officially recognized as a separate curriculum. Another important milestone was the first issue of the journal *Acta Politica* in 1965. This journal, published under the auspices of the NKWP, was originally published in Dutch and English but switched to English in full in 1996. In 2003, the journal ceased to be published by Dutch publisher Boom and migrated to Palgrave (UK). In 1973, another political science journal—which took a broad, multidisciplinary approach to political and policy sciences by also publishing contributions from sociologists and public administration scholars (in Dutch)—saw the light of day: *Beleid en Maatschappij*.

According to Hoogerwerf (1981), the main research areas of Dutch political sciences at the beginning of the 1980s were research on elections and voters—institutionalized in the *Nationaal Kiezersonderzoek*, the result of a cooperation started in 1967 between the professors in political science at the time on the research on elections, held for the first time in 1971 and conducted as a two-wave panel study—political parties (most notably in the *Documentatiecentrum Nederlandse Politieke Partijen* at the University of Groningen) and the Dutch parliament.

Like most other SSH disciplines, political sciences also faced the consequences of the 1980s' budget cuts. These resulted in the discontinuation of the political science department at the Erasmus University of Rotterdam in 1986. An initiative in the early 1990s to establish a national graduate school in political sciences (named PolyBios Graduate School of Political Science) proved to be unsuccessful.

Sciences of Literature: The Newcomer

The study of the Dutch language and literature in academia had long been overshadowed by the classical languages when the first chairs in Dutch linguistics were established at the University of Franeker

(Everardus Wassenbergh) and Leiden (Matthijs Siegenbeek) in 1797. These chairs are seen as the academic genesis of Dutch Language and Letters (*Neerlandistiek*; Van Anrooij et al. 2003: 9). Both professors understood their field of study as including the study of literature, and Siegenbeek's chair was officially expanded by the subject of literature in 1799 (Johannes 2011: 35). In 1815, the Dutch government ruled that the newly formed Faculty of Letters and Philosophy had to include a chair in Dutch language, literature and eloquence. According to Johannes (2011), this initial institutionalization of Dutch studies was driven principally by politics: the state saw the need for a new discipline that could contribute to a strong Dutch identity and provide an education in eloquence in the Dutch language.[35] The subject of '*Hollandse letterkunde en welsprekendheid*' was one of the 14 subjects taught at the faculty of Letters and Philosophy (Groen 1986: 7).

The Society for Dutch Letters (*Maatschappij der Nederlandsche Letterkunde*) founded in Leiden in 1766 was primarily a learned society for the study of the Dutch language, literature and history. It installed a Commission for language and letters in 1847 to differentiate these subjects from history but, despite this, there was a lingering (and sometimes outspoken) criticism of the neglect of the subject of letters by the Society (Kruyskamp 1966: 158). Moreover, contemporary literature was generally not considered to be as important as linguistics or the history of literature (Braat 1966: 73, 80). One of the feats of the Commission was to oversee the publication of the journal *Tijdschrift voor Nederlandse Taal en Letterkunde*, first published in 1880.

Despite the apparent lack of interest for Letters of the Society, and the fact that *Letterkunde* was interpreted mostly as philology in the *Tijdschrift voor Nederlandse Taal en Letterkunde* (Kruyskamp 1966: 165), the nineteenth century witnessed a growing interest in the subject of Dutch letters. This was reflected in its academic institutionalization. In 1876, the study of Dutch Letters (*Nederlandse letterkunde*) was established as a separate curriculum. Examinations in the subject of *Letterkunde*, however, were only conducted as part of the *doctoraal examen* (the *Kandidaats*

[35] Johannes sees this as an example of a 'top down' institutionalization of a discipline.

consisted of Dutch language, Middle-Dutch, Gothic, Dutch history and General history). In the academic statute of 1921, *Letterkunde* finally also became part of the *Kandidaats* examination. At the beginning of the twentieth century, the Society for Dutch Letters—as well as the short-lived Association for the Advancement of Ancient Dutch Letters before that—was successful in publishing and reprinting a number of ancient Dutch texts as part of various book series' (*Vereeniging ter Bevordering der Oude Nederlandsche Letterkunde* 1844; Kruyskamp 1966: 161–162; Braat 1966: 84). After the Second World War, discussions about the establishment of a subject of sciences of literature, separate from philology and the *letterkundes*, gained momentum (Goedgebuure and Heynders 1996). In 1956, the first chair in *Algemene Literatuurwetenschappen* ('General science of Literature'), held by Hubert PH Teesing (1907–1993), was institutionalized at the University of Utrecht. Initially, Teesing taught 'new German literary studies and the theoretical science of literature', but the theoretical literature part of the chair's designation was separated in 1971. In Leiden, a chair was established quickly thereafter, in 1957, held by Jacob G. Bomhoff (1903–1988).

Teesing had argued for a separate discipline of the theory of literature at the 1950 congress of Dutch philologists. The ensuing establishment of his chair constituted a break from the classic disciplines of philology and language-based literary studies by identifying the theory of literature as a separate study domain, applicable to both ancient and modern literature, as well as literature originating from different national contexts. These changes were mainly influenced by similar developments in Germany and the US, but a major domestic influence was Wytze Hellinga (1908–1985). Hellinga became professor of Dutch language at the University of Amsterdam in 1946. In his inaugural lecture, he offered an approach to the study of style differences in literary works from a grammatical perspective. This linguistic interpretation of literature became the basis for the autonomist approach that was central to the early years of literature sciences in the Netherlands. This strand of work resonated in the approach presented in the short-lived periodical *Merlyn* (1962–1966). Partly influenced by the Anglo-Saxon New Criticism movement, the group around *Merlyn* advocated the close reading of literature and rejected the dominant individualistic style of literary criticism.

Criticism of the *Merlyn* approach in the late 1960s and early 70s revealed an early form of contextualism developing in Dutch literature sciences. This strand was reinforced when the first textbook in Dutch 'Science of Literature' was published (*Literatuurwetenschap*; Maatje 1970). The book triggered a debate about the methodology of literature sciences, which continued until the early 1980s. The main critic was Hugo Verdaasdonk (1945–2007) who, in a series of articles published in the literary journal *Revisor* in 1974 and 1975, critiqued all forms of literature studies that fell back on essentialist notions of the text. Verdaasdonk was appointed to a chair in linguistics and the theory of literature at the University of Tilburg in 1981. There he, and a group of scholars around him (the 'Tilburg School'), drew on the work of the French sociologist Pierre Bourdieu. They developed a sociological approach to literature that took the literary field as its research object.

Literature scholars often publish in journals that combine Language Studies, Literature Studies and Dutch Studies. A separate journal for history of literature and literature theory is *Spiegel der Letteren*, founded in 1956. Another example was *Forum der Letteren* (founded in 1960), a continuation of an older philology journal. In 1995, *Forum der Letteren* merged with several other journals to become the *Tijdschrift voor Literatuurwetenschap*, but was discontinued in 1998. As with many other SSH disciplines, Dutch literary scientists have increasingly turned to publishing in English in international journals.

Literature sciences was officially recognized by the government as the separate curriculum of literature (along with other subjects such as modern languages and history of art) for master's students in 1967. By then, chairs had been established in almost all Letters faculties. In contemporary literature sciences, two main currents are still distinguishable: *Algemene Literatuurwetenschap*, or comparative literature sciences (in which literature from different languages, movements etc. is compared), and *Theoretische Literatuurwetenschap* (theoretical science of literature). Chairs in these two ordinations were established in a number of universities. The number of chairs has remained relatively constant over time (see Fig. 6.4; Table 6.3).

Table 6.3 Most important associations and journals per SSH discipline

Discipline	Association	Year of foundation	Major journals
Philosophy	Afdeeling Nederland der Kant-Gesellschaft (1921–1933)[a]; Algemene Nederlandse Vereniging voor Wijsbegeerte (ANVW)	1933	Algemeen Nederlands Tijdschrift voor Wijsbegeerte (1907) Krisis (1980)
Economics	Vereniging voor de Statistiek/ (Koninklijke) Vereniging voor de Staathuishoudkunde (KVS)	1849	De Economist (1852) ESB (1916) Tijdschrift voor Politieke Ekonomie (1977)
Anthropology	Nederlands Genootschap voor Anthropologie/ Vereniging voor Culturele Antropologie en Sociologie van de Niet-Westerse samenlevingen	1949/1970	Bijdragen tot de Taal-, Land- en Volkenkunde (1853) Focaal (1985) Etnofoor (1988)
Psychology	Psychologische Kring / Nederlands Instituut voor (Praktizerende) Psychologen (NIP(P))	1919/1938	Nederlands Tijdschrift voor de Psychologie (1934–2010) Acta Psychologica (1935)
Sociology	Nederlandsche Sociologische Vereniging (NSV)	1936	Mens en Maatschappij (1925) Sociologische Gids (1953)
Political sciences	Nederlandse Kring voor de Wetenschap der Politiek Vereniging voor Bestuurskunde	1950 1973	Acta Politica (1965) Beleid and Maatschappij (1973)

(continued)

Table 6.3 (continued)

Discipline	Association	Year of foundation	Major journals
Science of literature	Maatschappij der Nederlandse Letterkunde	1766	Spiegel der Letteren (1956) Forum der Letteren (1960–1995) Journal of Dutch Literature (2010)

[a]The *Afdeeling Nederland der Kant Gesellschaft* functioned as the general association for Dutch philosophers until 1933. In 1933, the association disbanded, but was immediately re-incorporated as *Landesgruppe Holland van de Kant-Geselleschaft*. See Derckx (1994: 92)

Conclusion

The overview of the institutionalization of seven SSH disciplines over the course of roughly 70 years presented in this chapter remains necessarily broad and general. Nevertheless, some general trends can be identified. First, looking at the relation between the field of higher education and the state, a clear development from a decentralized system to a more centralized arrangement can be observed. Initially, the Dutch approach to higher education was inspired by the German system. Institutes were relatively free to develop research and education according to local preferences. After the Second World War, political discussions about how to contain the increasing costs of the higher education system began in earnest. The growing number of students made the need to constrain the financial resources available for tertiary education even more pressing. From the 1960s onwards, this resulted in a number of laws that drastically changed the institutional landscape. By the 1990s, successive restructuring programmes, carried out under the banner of 'efficiency improvements', resulted in a more centrally organized system and an increased level of government involvement. In many cases, the content of educational curricula had been standardized. Educational trajectories had been restructured into the two-phase structure that bore important elements of the US system of higher education. Loud calls for more democratization at the end of the 1960s and early 1970s resulted in a number of faculty reorganizations, the most notable being the emergence of departments (the so-called *vakgroepen*), which undermined the power of individual chair holders to determine research agendas and the content of studies. However, the wave of democratization that engulfed the universities in the 1970s was later partly scaled back, again often under the banner of 'efficiency requirements'.

Cognitively, the initial autonomy of institutes of higher education resulted in some noticeable differences between them. The most prominent contrasts were those between the technical and the general universities, and between the institutes of (confessional) particular education and public education. The latter contrast was the most significant for the intellectual and institutional development of the SSH in the Netherlands. Local approaches and various 'schools' in philosophy, economics, anthropology and sociology were often, but not always, based on the different confessional vocations of the institutes in which they were embedded. The generally weak, and mostly local, influence of philosophy on the intellectual development of the social sciences in the Netherlands can be understood along these lines. Examples were the Calvinistic philosophy developed at the Free University of Amsterdam, which informed the specific lines of anthropological and sociological work based on this *Weltanschauung*, and Catholic sociology, which was partly inspired by the theology of Edward Schillebeeckx at the University of Nijmegen, for example.

Gradually, however, these local variations gave way to a more uniform approach. In many cases, the model of convergence was that of the US. In the Netherlands, economics and psychology especially became mainstream disciplines relatively early, heavily orientated towards the US and increasingly deprived of their national or local character. The wave of critical science that was associated with the democratization movement of the late 1960s and early 1970s, initiated by students and supported by younger members of the academic staff, was often directed at the one-dimensional orientation towards the US in many disciplines. However, the influence of this critical movement in the SSH proved to be short-lived. Chairs designated to critical approaches were discontinued and in the 1980s most critical SSH journals were confronted with declining subscriptions and diminishing submissions. Somewhat later, this became a general trend for many journals that were published in Dutch; they too were faced with a withering interest of readers and contributors. The pressure to publish work in Anglo-American journals meant that writing articles in Dutch became less attractive, and most periodicals in Dutch faced increasing difficulties to attract copy and find editors. Quite a few of them eventually disappeared or made the transition to English. The same trend can be observed when looking at textbooks used in introductory courses: Dutch, German and occasionally French texts were gradually replaced by English books, with economics and psychology being the forerunners. Part of this shift towards English is rooted in changes in secondary education. With a new law on secondary education, the 'Mammoetwet' (1968), only Dutch and English remained compulsory languages in the higher levels of secondary school, German and French were henceforth optional.

From an international perspective, contemporary Dutch SSH work can be characterized as relatively solid, mainstream research, but not very original. There are very few Dutch scholars who have made a significant theoretical contribution to their discipline. Sociology and economics have retained their policy orientated disposition, while their intellectual focus on the US remains strong. By and large, this observation also holds for psychology and political science in the Netherlands. The picture for anthropology and sciences of literature is a bit more nuanced. Dutch anthropologists have had some success through their regional specialization in some areas, while the science of literature remains a small, but vibrant, field open to influences from developments in various countries and disciplines. However, in general, the intellectual resources for innovative work seem to have become ever more reduced. The access to German and French work has become very rare, while in the Netherlands a more general discipline, such as philosophy, never had the role it has played in France or Germany. Dutch philosophy has often played an institutionally subordinate role in the faculties in which it was embedded. It was finally given its own Interfaculty in order to strengthen its position as a 'central discipline', only to see that status removed once again some twenty years later as a result of budget cuts. Intellectually, a significant part of Dutch philosophy long remained confined in the confessional framework of the private universities. Its international impact in most cases is restricted to academic niches, or limited to solid incremental work within other (often Anglo-Saxon) traditions.

References

Abma, R. 2001. Afscheid van Psychologie en Maatschappij. *Academische Boekengids* 29: 9–10.

Bak, P. 2013. *Gedonder in de Sociale: Vijftig Jaar Sociaal-Culturele wetenschappen aan de Vrije Universiteit, 1963–2013*. Zoetermeer: Meinema.

Barents, J. 1948. *De Wetenschap der Politiek: Een Terreinverkenning*. The Hague: Stols.

Bijleveld, G. 1996. *Geleerd Genootschap of Beroepsvereniging? Nederlandse Sociologen en Antropologen 60 jaar Verenigd*. Amsterdam: NVMC.

Boschloo, H. 1989. *De Productiemaatschappij: Liberalisme, Economische Wetenschap en het Vraagstuk der Armoede in Nederland, 1800–1875*. Hilversum: Verloren.

Braat, W. 1966. De Maatschappij der Nederlandse Letterkunde van 1803 tot 1966. In *Gedenkboek bij het 200-jarig Bestaan van de Maatschappij der Nederlandse Letterkunde te Leiden*. Leiden: Brill.

Brandt, E. 1998. De rode mythe. *De Groene Amsterdammer*, januari 7.

Breman, J. 2002. The sociology of non-western societies at the University of Amsterdam. In *Tales from Academia: History of anthropology in the Netherlands, Part I*, ed. H. Vermeulen and J. Kommers. Saarbrücken: Verlag für Entwicklungspolitik.

Breman, J. 2006. De sociologie van het polderland. *Sociologie* 29 (3): 329–354.

Busato, V. 2008. *Leve de Psychologie! Honderd Jaar Psychologische Wetenschap aan de UvA*. Amsterdam: Bert Bakker.

Buyst, E., I. Maes, H. Plasmeijer, and E. Schoorl. 2005. Comparing the development of economics during the twentieth century in Belgium and the Netherlands. *History of Political Economy* 37 (1): 61–78.

CBS. 1956. *Berekeningen Omtrent de Toekomstige Omvang van het Nederlands Hoger Onderwijs en de Aantallen Afstuderenden 1956–1960*. The Hague: CBS.

Coumou, H. 2013. *De dubbelzinnige 'codificatie' van de sociale pedagogiek in het Academisch Statuut*. Amsterdam: Jeugdzaken.

Daalder, H. 1995. Over vasthoudendheid en lafhartigheid in de akademie: Een geschiedenis van de zaak Daudt. In *Over Echte Politicologie: Opstellen over politicologie, Democratie en de Nederlandse Politiek*, ed. H. Daudt. Amsterdam: Bert Bakker.

Daniëls, M. 1985. *Bedrijfskunde - Uit Heden en Verleden*. Eindhoven: Technische Hogeschool Eindhoven.

Daudt, H. 1998. Jan Barents: Grondlegger van de Nederlandse politicologie. In *In de Zevende: De Eerste Lichting Hoogleraren aan de Politiek-sociale Faculteit in Amsterdam*, ed. J. Goudsblom, P. de Rooy, and J. Wieten. Amsterdam: Het Spinhuis.

De Bruijne, A., T. Dietz, and W. van Spengen. 2002. De ontwikkelingsgeografische traditie: Van nuttige exotica, via ontwikkelingsvraagstukken naar glocalisering. In *Alles heeft zijn Plaats: 125 Jaar Geografie en Planologie aan de Universiteit van Amsterdam*, ed. H. Knippenberg and M. van Schendelen. Amsterdam: Aksant.

Den Hollander, A. 1948. Sociografie en sociologie. In *Geestelijk Nederland 1920–1940 Deel 2: De Wetenschappen van Natuur, Mens en Maatschappij*, ed. K. Proost and J. Romein. Amsterdam and Antwerp: Kosmos.

De Man, H. 1996. Continuities in Dutch business education: Engineering, economics and the business school. In *Management, education and competitiveness: Europe, Japan and the United States*, ed. R.P. Amdam. London and New York: Routledge.

De Wolf, J. 1998. *Eigenheid en Samenwerking: 100 jaar Antropologisch Verenigingsleven in Nederland*. Leiden: KITLV Uitgeverij.

Dicke, M., P. van de Laar, and J. Visser. 2013. *Ambitie en identiteit: Van Nederlandsche Handels-Hoogeschool tot Erasmus Universiteit Rotterdam 1913–2013*. Rotterdam: Erasmus University Rotterdam.

Derckx, P. 1994. *H.J. Pos, 1898–1955, Objectief en Partijdig: Biografie van een Filosoof en Humanist*. Hilversum: Verloren.

Diemel, B. 2013. *Erasmus School of Economics 100 Jaar*. Rotterdam: Stad en Bedrijf.

Duyvendak, J. 1997. Publieke filosofen in Groningen en de rest van Nederland 1975–1995. In *Zeer Kundige Professoren: Beoefening van de Filosofie in Groningen van 1614 tot 1996*, ed. H. Krop, H. van Ruler, and A. Vanderjagt. Hilversum: Verloren.

Duyverman, J. 1978. Feiten en feitjes betreffend de groei van de economische wetenschap in Nederland. *De Economist* 126 (1): 1–36.

Eisenga, L. 1978. *Geschiedenis van de Nederlandse Psychologie*. Deventer: Van Loghum Slaterus.

Ellemers, J. 1978. De Nederlandse sociologie en de Amsterdamse sociografie in de jaren dertig: een voorbeeld van onvolledige institutionalisering. In *Toen & Thans: De sociale Wetenschappen in de Jaren Dertig en Nu*, ed. F. Bovenkerk, H. Claessen, B. van Heerikhuizen, A. Köbben, and N. Wilterdink. Baarn: Ambo.

Engbersen, G. 2005. Een eeuwige jeugd: constanten en vernieuwingen binnen de Nederlandse sociologie. *Sociologie* 1 (1): 90–105.

Foppen, J. 1989. Gistend Beleid: Veertig Jaar Universitaire Onderwijspolitiek. PhD dissertation, Rotterdam.

Gastelaars, M. 1985. *Een Geregeld Leven. Sociologie en Sociale Politiek in Nederland 1925–1968*. Amsterdam: SUA.

Goedgebuure, J., and O. Heynders. 1996. *Literatuurwetenschap in Nederland: Een Vakgeschiedenis*. Amsterdam: Amsterdam University Press.

Gorter, 2012. *Anderhalve Eeuw Economieonderwijs in Nederland: Biografie van een Schoolvak*. Delft: Eburon.

Goudsblom, 1985. De Nederlandse en de Amerikaanse sociologie in de jaren vijftig: een blik van achter de doorkijkspiegel. *Tirade* 29 (296–300): 631–639.

Groen, M. 1983. *Het Wetenschappelijk Onderwijs in Nederland van 1815 tot 1980 I: Een Onderwijskundig Overzicht*. Eindhoven: Technische Universiteit Eindhoven.

Groen, M. 1986. *Het Wetenschappelijk Onderwijs in Nederland van 1815 tot 1980 VIII: Letteren*. Eindhoven: Technische Universiteit Eindhoven.

Groen, M. 1989. *Het Wetenschappelijk Onderwijs in Nederland van 1815 tot 1980 III*. Eindhoven: Technische Universiteit Eindhoven.

Haas, E. 1975. Geesteswetenschappelijk onderzoeksbeleid. In *Geesteswetenschappelijk Onderzoeksbeleid*, ed. W. De Vroomen. The Hague: Smits.

Heilbron, J. 1988. Particularités et particularismes de la sociologie aux Pays-Bas. *Actes de la Recherche en Sciences Sociales* 74 (1): 76–81.

Heslinga, M. 1978. Sociografie versus sociale geografie. In *Toen & Thans: De sociale Wetenschappen in de Jaren Dertig en Nu*, ed. F. Bovenkerk, H. Claessen, B. van Heerikhuizen, A. Köbben, and N. Wilterdink. Baarn: Ambo.

Hofstra, S. 1978. Sociologie in de jaren dertig en later. In *Toen & Thans: De sociale Wetenschappen in de Jaren Dertig en Nu*, ed. F. Bovenkerk, H. Claessen, B. van Heerikhuizen, A. Köbben, and N. Wilterdink. Baarn: Ambo.

Hoogerwerf, A. 1981. Schets van de Nederlandse politicologie. *Ons Erfdeel* 24 (1): 39–46.

Jensma, G., and H. De Vries. 1997. *Veranderingen in het Hoger Onderwijs in Nederland tussen 1815 en 1940*. Hilversum: Verloren.

Johannes, G. 2011. 'Nationale filologieën' en het historisch onderzoek naar disciplinevorming in de geesteswetenschappen. *Studium* 1 (2011): 31–46.

Karsten, L., and H. de Man. 1992. Bedrijfskunde: een nieuwe loot aan de academische stam? *Jaarboek voor de Geschiedenis van Bedrijf en Techniek* 9: 292–329.

Kastelein, T. 1988a. Het ontstaan van de Faculteit der Economische Wetenschappen. In *Een Eigenzinnige Koers – 40 jaar Faculteit der Economische Wetenschappen in Groningen*, ed. T. Kastelein, K. Bijsterveld, and H. van der Meulen. Groningen: Rijksuniversiteit Groningen.

Kastelein, T. 1988b. Het verleden zet zijn stempels. In *Een Eigenzinnige Koers – 40 jaar Faculteit der Economische Wetenschappen in Groningen*, ed. T. Kastelein, K. Bijsterveld, and H. van der Meulen. Groningen: Rijksuniversiteit Groningen.

Kleber, R., and H. Schut. 2011. De periode'Gedrag & Gezondheid': een selectieve en ongetwijfeld gekleurde impressie van twee oud-(hoofd)redacteurs. *Psychologie & Gezondheid* 39 (5): 265–268.

Klein, P. 1974. *De Nederlandse Economische Hogeschool 1963–1973*. Rotterdam: Universitaire Pers Rotterdam.

Kloos, P. 1976. Culturele antropologie in Nederland. In *Culturele Antropologie: Portret van een Wetenschap*, ed. P. Kloos. Utrecht: Boom.

Kloos, P. 1987. L'Anthropologie des années 80 aux Pays Bas. *Anthropologie et Sociétés* 11 (3): 11–34.

Kloos, P. 1995. *Culturele Antropologie*. Assen: Van Gorcum.

Knegtmans, P. 1998. *Een Kwetsbaar Centrum van de Geest. De Universiteit van Amsterdam tussen 1935 en 1950*. Amsterdam: Amsterdam University Press.

Knegtmans, P., and H. Keman. 1998. De wording van 'de zevende' en de politicologie in Nederland. In *Uit de Zevende: Vijftig Jaar Politieke en Sociaal-culturele Wetenschappen aan de Universiteit van Amsterdam*, ed. A. Gevers. Amsterdam: Spinhuis.

Köbben, A. 1964. *Van Primitieven tot Medeburgers*. Assen: Van Gorcum.

Köbben, A. 1998. J.J. Fahrenfort: Tegen pompeuze en duistere taal. In *In de Zevende: De Eerste Lichting Hoogleraren aan de Politiek-sociale Faculteit in Amsterdam*, ed. J. Goudsblom, P. de Rooy, and J. Wieten. Amsterdam: Het Spinhuis.

Krop, H. 2006. Laat de universiteit meer dan een vakschool zijn – Het pleidooi voor de Centrale Interfaculteit. In *Universitaire Vormingsidealen: De Nederlandse Universiteiten sedert 1876*, ed. L. Dorsman and P. Knegtmans. Hilversum: Verloren.

Kruyskamp, C. 1966. De Commissie voor Taal- en Letterkunde. In *Gedenkboek bij het 200-jarig Bestaan van de Maatschappij der Nederlandse Letterkunde te Leiden*. Leiden: Brill.

Kuitenbrouwer, M. 2014. *Dutch scholarship in the age of empire and beyond*. Leiden: Brill.

Kuypers, K. 1952. De filosofie in Nederland. *De Gids*, 115.

Langendorff, T. 1990. *De Geesteswetenschappen in Nederland: een Overzicht*. The Hague: RAWB.

Lieshout, R., and B. Reinalda. 2001. The Dutch Political Science Association 1950–2000. *European Political Science* 1 (1): 60–65.

McAllister, J. 1997. Philosophy of science in the Netherlands. *International studies in the Philosophy of Science* 11 (2): 191–204.

Mooij, J. 1994. *Denken over welvaart: Koninklijke Vereniging voor de Staathuishoudkunde 1849–1994*. Utrecht: Lemma.

Nauta, L. 1990. De subcultuur van de wijsbegeerte. *Krisis* 38: 5–19.

NRC Handelsblad. 1992. Handel en wandel, *NRC Handelsblad*, September 12, zaterdag bijvoegsel, pp. 5–7.

NRC Handelsblad. 1993. Stroomlijnen universiteiten is winst, *NRC Handelsblad*, December 16, p. 7.

Overbeek, A. 2010. De afschaffing van de studie klassieke talen aan de RU Utrecht 1982/86, Utrecht University repository. Accesed via http://dspace.library.uu.nl/handle/1874/40343.

Peijnenburg, J. 2009. De Faculteit Wijsbegeerte in Groningen. *Tijdschrift voor Filosofie* 71 (3): 469–474.

Pierson, N. 1884/1890. *Leerboek der Staathuishoudkunde*, 2 vols., Haarlem: Erven F. Bohn.

Plasmeijer, H., and E. Schoorl. 2005. Postwar Dutch economics: Internationalization and homogenization. In *The development of economics in Western Europe since 1945*, ed. A. Coats. London: Routledge.

Pos, H. 1946. Levensbericht Tj. De Boer. In *Jaarboek van het Huygens Instituut 1945–1946*. Amsterdam: KNAW.

Roelink, J. 1955. *Vijfenzeventig Jaar Vrije Universiteit 1880/1955*. Kampen: JH Kok NV.

Rupp, J. 1997. *Van Oude en Nieuwe Universiteiten: De Verdringing van Duitse door Amerikaanse Invloeden op de Wetenschapsbeoefening en het Hoger Onderwijs in Nederland 1945–1995*. The Hague: SDU Uitgevers.

Sassen, F. 1959. *Geschiedenis van de Wijsbegeerte in Nederland tot het Einde der Negentiende Eeuw*. Amsterdam and Brussel: Elsevier.

Sassen, F. 1967. Dutch philosophy. In *The encyclopedia of philosophy*, ed. P. Edwards. New York: Macmillan.

Selten, P. 2005. De geschiedenis van de Algemene Sociale Wetenschappen. In *De Sociale Wetenschappen in Utrecht: een Geschiedenis*, ed. W. Koops, H. van Rinsum, and J. van Teunenbroek. Hilversum: Verloren.

Tromp, B. 2007. *De Wetenschap der Politiek: Verkenningen*. Amsterdam: Amsterdam University Press.

Uhlenbeck, E. 1975. Wetenschapsbeleid en geesteswetenschappelijk onderzoek. In *Geesteswetenschappelijk Onderzoeksbeleid*, ed. W. De Vroomen. The Hague: Smits.

Van Anrooij, W., D. Hogenelst, and G. Warnar. 2003. *Der Vaderen Boek: Beoefenaren van de Studie der Middelnederlandse Letterkunde-studies voor Frits van Oostrom ter Gelegenheid van diens Vijftigste Verjaardag*. Amsterdam: Amsterdam University Press.

Van Baalen, P., and L. Karsten. 2002. Interdisciplinariteit, professies en Amerikanisering. Een geschiedenis van het ontstaan van de Nederlandse bedrijfskunde. *NEHA-Jaarboek voor economische-, bedrijfs- en techniek-geschiedenis* 65: 256–304.

Van Bremen, J., and A. Shimizu. 2013. *Anthropology and colonialism in Asia: Comparative and historical colonialism*. London: Routledge.

Van Dalen, H., and A. Klamer. 1996. *Telgen van Tinbergen: Het Verhaal van de Nederlandse Economen*. Amsterdam: Balans.

Van de Beek, J. 2010. *Kennis, Macht en Moraal: De Productie van Wetenschappelijke Kennis over de Economische Effecten van Migratie naar Nederland, 1960–2005*. Amsterdam: Amsterdam University Press.

Van den Hoven, M. 1986. Internationalisme, volwasseneducatie en syntheticisme – drie determinanten in de onstaansgeschiedenis van de Internationale School voor Wijsbegeerte. In *Filosofie in Nederland: De Internationale School voor Wijsbegeerte als Ontmoetingsplaats*, ed. A. Heijerman and M. van den Hoven. Meppel and Amsterdam: Boom.

Van Doorn, J. 1964. *Beeld en Betekenis van de Nederlandse Sociologie*. Utrecht: Bijleveld.

Van Doorn, J., and Lammers. 1959. *Moderne Sociologie*. Utrecht/Antwerp: Het Spectrum.

Van El, C. 2002. *Figuraties en Verklaringen: Stijlgebonden Schoolvorming in de Nederlandse Sociologie na 1968*. Amsterdam: Aksant.

Van Heek, F. 1978. Omzien naar het sociografisch intermezzo (1925–1953). In *Toen & Thans: De sociale Wetenschappen in de Jaren Dertig en Nu*, ed. F. Bovenkerk, H. Claessen, B. van Heerikhuizen, A. Köbben, and N. Wilterdink. Baarn: Ambo.

Van Heerikhuizen, B. 1987. *W.A. Bonger: Socioloog en Socialist*. Groningen: Wolters-Noordhoff.

Van Hezewijk, R. 2005. De geschiedenis van de psychologie. In *De Sociale Wetenschappen in Utrecht: een Geschiedenis*, ed. W. Koops, H. van Rinsum, and J. van Teunenbroek. Hilversum: Verloren.

Van Hezewijk, R., H. Stam, and G. Panhuysen. 2001. Existential questions: No, one or two Utrecht schools? Paper presented at the 20th conference of the European Society for the History of the Human Sciences, August 14–18.

Van Hezewijk, R., H. Stam, and G. Panhuysen. 2002. The double dissociation of phenomenological and experimental methods in psychology. Paper presented at the 21th conference of the European Society for the History of the Human Sciences, August 27–31.

Van Kemenade, J. 1982. Uit de rede van prof. Dr. J.A. van Kemenade. *Universiteit en Hogeschool* 29: 62–64.

Vereeniging ter Bevordering der Oude Nederlandsche Letterkunde. 1844. *Verslagen en Berigten Uitgegeven door de Vereeniging ter Bevordering der Oude Nederlandsche Letterkunde 1*. Leiden: D. du Mortier en Zoon.

Vermeulen, H. 2002. Contingency and Continuity: Anthropology and other non-western studies in Leiden, 1922–2002. In *Tales from academia: History of anthropology in the Netherlands, part I*, ed. H. Vermeulen and J. Kommers. Saarbrücken: Verlag für Entwicklungspolitik.

Vermeulen, H., and J. Kommers (eds.). 2002. *Tales from academia: History of anthropology in the Netherlands, part I*. Saarbrücken: Verlag für Entwicklungspolitik.

Weijers, I. 1995. Filosofie en wederopbouw 2 – Het streven naar een Centrale Interfaculteit. *Krisis* 58: 74–85.

Wertheim, W. 2002. Globalisation of the social sciences: Non-Western sociology as a temporary panacea. In *Tales from academia: History of anthropology*

in the Netherlands, part I, ed. H. Vermeulen and J. Kommers. Saarbrücken: Verlag für Entwicklungspolitik.
Wilts, A. 1997. Economie als Maatschappijwetenschap: Een Sociologische Geschiedenis van de Economische Wetenschap in Nederland (ca. 1930–1960). Dissertation, Amsterdam.
Woldring, H. 2012. *Een Handvol Filosofen: Geschiedenis van de Filosofiebeoefening aan de Vrije Universiteit*. Hilversum: Verloren.

7

A Reversed Order: Expansion and Differentiation of Social Sciences and Humanities in Sweden 1945–2015

Tobias Dalberg, Mikael Börjesson and Donald Broady

Introduction

Swedish social sciences and humanities have expanded dramatically since 1945. The augmentation has been especially strong in the 1950s, 1960s and 1990s. This is part of a general enlargement of the higher education sector, which has transformed from an elite system, only recruiting a few percent of the youth cohort, to admitting almost half. The expansion has been very uneven, however. The social sciences have surpassed the humanities in a number of aspects such as student enrolment, research financing and demand for their expert knowledge. Thus, a long-established order has been reversed. Further, the expansion has

T. Dalberg (✉) · M. Börjesson · D. Broady
Uppsala University, Uppsala, Sweden
e-mail: tobias.dalberg@edu.uu.se

M. Börjesson
e-mail: mikael.borjesson@edu.uu.se

D. Broady
e-mail: Donald.broady@edu.uu.se

C. Fleck et al. (eds.), *Shaping Human Science Disciplines*, Socio-Historical Studies of the Social and Human Sciences, https://doi.org/10.1007/978-3-319-92780-0_7

facilitated increased differentiation, both in terms of creation of new disciplinary chairs and divisions within disciplines. In this chapter, we study this development on the basis of an analysis of seven disciplines.

The Particularities of the Swedish Case

Nineteenth-century Sweden was a poor rural country in Europe's Northern periphery. The change during the latter half of the twentieth century, from a largely agriculture-based economy to a close integration into the global economy and very high general domestic product per capita was dramatic. This transformation was accompanied by a long Social-Democratic political dominance covering the major part of the twentieth century, and the development of a large and omnipresent welfare state. Although since the 1990s, Sweden has, in some areas, taken the lead in the rapid neoliberal turn of the economy—as in the deregulation (or reregulation) and privatization of the public sector—the starting level was, by international comparison, a rather egalitarian distribution of resources and public welfare institutions, that has still kept much of its legitimacy among Swedish citizens (although less so within the political field).

Swedish higher education and research played an important role in this transformation. Research in technology, science and medicine was crucial for industrial development and the social sciences were equally decisive for the formation of the welfare state (Lundin and Stenlås 2015).

Many of the changes in the Swedish system of higher education are similar to those in other developed countries. These similarities concern expansion periods resulting in a more complex landscape of educational institutions and programs, increased internationalization, and (in the Swedish case since the 1990s) marketization, reregulation and the introduction of new public management models.

Despite these recent turns toward what has been called academic capitalism (Slaughter and Leslie 1997), peculiar to Swedish is the lasting, comparably strong, legitimacy enjoyed by the conception of higher education as a public good and as a democratization project with an emphasis on widened social access (Börjesson and Broady 2016). This implies state-owned educational institutions (there are almost no private ones), no tuition fees anywhere (with the exception of students from outside the EU/EES since 2011), comparably generous financial aids and loans to all undergraduates since the mid-1960s, and currently even state funding in the form of a good four year, full-time salary for all accepted Ph.D. candidates. There is a general political consensus on the importance of higher education and research. Even in times of economic crises, the sector has—regardless of which political party or coalition is in power—been subject to smaller cuts than other sectors, sometimes even receiving increased funding. In addition, given Sweden's relatively small size and dependence on its export industry, international relations and influences have preconditioned the development of both the industry and the higher education sector. Finally, we should also stress the tight connection between higher education and research. Publicly-funded research is carried out at higher education institutions, very seldom at specific research institutes and every higher education institution is supposed to harbor research.

Analytical Dimensions

In this section, we will present the theoretical framework and the analytical dimensions employed for the study of social science and humanities in Sweden from 1945 onwards. Our ambition is to combine a structural and temporal approach. We take as our starting point the institutionalization of the disciplines and go on to link this to their social transformation.

There are good reasons to place the development of disciplines within the wider framework of higher education and its social transformation. First, the student body forms the material basis of the academic life of the discipline. It provides teaching opportunities and positions. Increasing the student body increases the number of positions and also the audience—within and outside of academia—for the knowledge produced. In fact, for the major part of the twentieth century, upper secondary and higher education were the main suppliers of positions for trained Ph.D.s in the humanities and social sciences. In addition, the student body or, more precisely, the undergraduates and Ph.D.s, are the recruitment pool for the group of senior researchers in the disciplines.

Disciplinary Variations

Following Heilbron (2004: 35), we compare the disciplines with regard to their genesis and type while relating them to the specific national context. The genesis can typically be described as either differentiation of an existing discipline, upgrading of some kind of amorphous professional activity to an academic discipline, a synthesis of previously divided disciplines, or a combination of these factors.

A crucial aspect of modern disciplines, taking shape from 1750 to 1850, again drawing on Heilbron (ibid.: 27, 30), is that they primarily function as nexuses for assembling three historically separated activities: research; teaching or professional organization. This implies, as Heilbron advocates, that disciplines can be analyzed from the perspective of the weight of these three different activities (ibid.: 37). From a "bird's eye view," these three poles of activity correspond roughly to the structure of university faculties. In the late nineteenth century, the faculties of law, medicine and theology were primarily sites for the professional training of lawyers, medical doctors and priests, while the faculty of philosophy was divided between the teaching-oriented humanistic section and the research-oriented science section.

Transformations of Social Characteristics of Disciplines

One way to determine the weight of different activities is to simply compare the number of students, research students, academic positions and other institutional developments, such as professional organizations, journals and extra-academic institutions within and between disciplines. A discipline with a lot of students, but few professors, might be thought of as oriented toward teaching, but the ratio between these two categories may still indicate a larger weight of research activities than another discipline. We are thus working in a spirit close to the branch of

Durkheimian sociology that deals with the study of groups, their number and their distribution—the so called *morphologie sociale* (Mauss and Fauconnet 1901; cf. Bourdieu 1984: Chapter 4).

In order to account for the social morphologies of the disciplines, we will draw on a diverse material. In order to count the number of students, we have relied on individual level register data from 1977 onwards (corresponding to the organizational changes described below) and statistical yearbooks and memos for earlier periods. Doctoral students have been counted in a similar way, combining individual-level register data for some disciplines from 1971 onwards with dissertation bibliographies for the earlier periods. The numbers of professors and other teaching categories have been collected from listings in state calendars.

Educational and Disciplinary Fields

While social morphologies trace the development of transformations over time, we use the Bourdieusian concept of field to capture the structure of higher education and research at a given moment. Further, the notion of field implies adding a dimension of power and accounting for relations within and between disciplines. It is thus important to distinguish between a field of higher education (Bourdieu 1989; Börjesson and Broady 2016) and specific academic disciplinary subfields, such as the fields of history, literature or economics (Lebaron 1997). These are social microcosms, with their own specific stakes and rewards, whose development over time depends on internal struggles as well as their position in the hierarchy of disciplines. Disciplinary expansion and differentiation may be conceived as the increase or decrease in relative autonomy of certain subfields toward other disciplines and faculties within universities, as well as toward extramural institutions and fields.

Thus, there are fields at different levels with different prime agents. While the field of higher education is constructed primarily on bases of the students enrolled in higher education programs and courses at different institutions, in the fields of disciplines it is the professors, lecturers and research students that are the principal analytical entities. The former is closer to a consumption field, while the latter is more of a production field, producing scientific works, ideas, theories, concepts, methods and people; namely specialists. Relating fields to Heilbron's three activities of disciplines, the field of higher education accounts for the educational dimension of a discipline, and lays open the structure of relations *between* disciplines, while the subfields of disciplines focus on the second activity, research, especially its *internal* differentiation. (This distinction is of course not absolute. In Bourdieu's analysis of Parisian professors in the four university faculties of humanities, science, law and medicine, differences between the faculties and disciplines are of central interest (1988).)

In this context, we will use the notion of the field of higher education and of specific disciplines in a heuristic sense. We will not perform any proper field analysis but, when relevant, refer to existing full-scale field analyses.

Disposition of This Chapter

First, in order to provide a context for the analysis of the individual disciplines, we outline the overall development of the higher education sector, its organization, funding, expansion and employment structure. Then we focus on, and compare, the institutional histories of the disciplines in terms of chairs, departments, external research funding, scientific journals and associations. The following considerations are based on studies of the social morphology of these disciplines including students, Ph.D.s, teachers and professors. We conclude by discussing the phases of establishment and consolidation of the disciplines and analyze the transformation of their profiles and internal order.

Background: The Establishment of a System for Higher Education and Research

In this section, we discuss the general conditions for Swedish higher education and research up to the mid-twentieth century which, by and large, comprises the history of a state-driven forming of a national *system* for higher education and research.

The Forming of a System of Higher Education

There have been higher education institutions in Sweden since the fifteenth century (the first university, at Uppsala, was founded in 1477). However, the growth was slow. If we exclude universities outside the current borders of Sweden, the second university was established in Lund in 1666 as part of the policy of "Swedishizing" the formerly Danish, and recently conquered, southern regions. Not until after the mid-twentieth century did Sweden obtain its third and fourth universities, when the private university colleges in Gothenburg and Stockholm were taken over by the state and given university status. In the eighteenth century there were also professional schools, mostly in the arts, which were joined by schools of medicine, engineering and agriculture.

A national public *system* for higher education took form from the latter half of the nineteenth century, at which time a set of conditions were put in place (cf. Agevall and Olofsson 2013: 5–7). The number of professional schools and university colleges that existed alongside the two universities was growing. The state became more active, taking over the financing and the control of the universities, including a regulation of the universities introduced in 1852 and a national office to coordinate their activities. During the twentieth century, new institutions were created and some existing private institutions were taken over by the state, increasing the size and scope of the national system. Currently, six of the former professional schools have obtained the status of universities (however, in contrast to the general universities, they have only one scientific domain each). This group includes high prestige institutions such as the Karolinska Institute, the Royal Institute of Technology and the Stockholm School of Economics.

The higher education system has historically been divided by types of institutions and fields of studies and still is. These two principles of division are intersected. Applied fields, such as engineering, agricultural sciences and fine arts, are most strongly represented at the professional schools, especially in the Stockholm region. Social sciences and humanities, which are the focus of this chapter, are predominantly associated with the universities and have grown out of the faculty of philosophy: in other words, the lower faculty in the medieval university structure.

The Division of the Faculty of Philosophy and Increased Specialization

At the end of the nineteenth century, increasing specialization shaped the state universities and their organization. At that time, there were still only two universities; Uppsala and Lund. An important step, taken in 1876, was the splitting of the faculty of philosophy into two sections: humanistic and science. At its inception, the section of humanities consisted of chairs in practical and theoretical philosophy, Semitic languages, political science, history, classical philology, modern European linguistics, Scandinavian languages and combined chairs of aesthetics, art and history of literature. Since both Uppsala and Lund were state universities, both they and their humanities sections had strong ties to the evolving educational system through the provision of upper secondary school teachers. To teach at upper secondary schools, one needed a degree in the subject one was to teach, and a subject that led to a degree needed an examiner; the professor. These strong ties between upper secondary school and the humanistic section at university level meant that changes in school subjects being taught often resulted in changes in the disciplinary structure at this section.

The privately founded and funded university colleges in Stockholm (1878) and Gothenburg (1891) were not tied to upper secondary education in the same way and could thus establish chairs in whichever field for which a private donor could be found. They did, however, stick to much the same disciplines as their university counterparts in Uppsala and Lund. Because of this, the rise in number of chairs within humanistic disciplines in the late 1800s was not so much a differentiation as an expansion. The first differentiation would take place during the first decade of the twentieth century, which saw chairs in geography established at the universities and in economics at the university colleges. European linguistics became differentiated into specific language chairs, such as roman, German or English. In 1908, it was decided that the previous *extraordinary* professors in history were to become ordinary, and from that moment history became numerically dominant for a large part of the twentieth century.

Shifting Balance Between Research and Teaching

The idea of the Humboldtian research university—that the professor should not only procure the teaching but also contribute to the production of knowledge—gradually became the new regime. This manifested itself in several ways. The hiring of professors came to put increasing weight on demonstrable scientific skill and, consequently, less weight on teaching ability or seniority. The 1876 reform that cleaved apart the faculty of philosophy was also a sign of this new regime since not only should the applicant for the chair demonstrate scientific skill but those who were to judge the skill level should be his disciplinary peers. This institution of experts, or hiring panels, was to be developed in the decades following the 1876 reform.[1]

At the universities, professors were responsible for teaching a certain amount each week. The rest of the time could be used for meetings with the faculty and university boards as well as various other commitments such as being an expert in appointments of new professors, taking places in government commissions, or engaging in research. These commitments usually took a substantial amount of time to fulfill, and it was common—during times of intense commitments—to put one of the docents as acting professor to carry the teaching load. University departments were not common until the mid-twentieth century. University-based research was carried out by the professor, a couple of docents on scholarship and an assistant or amanuensis.

By the 1940s, a labor market for academics had evolved in addition to university and upper secondary school teaching positions. Despite almost no independent research institutes, there was a demand for scientifically trained personnel. Insurance companies demanded statisticians, often with a licentiate degree. The commercial sector sought degree-holders in languages, economics and statistics. In addition to the libraries and the archives, museums in Stockholm and Gothenburg now offered further positions, both as a result of new museums and new departments at older museums (Kock 1941).

[1]See Blomqvist (1993) for a detailed account of these late nineteenth century transformations.

In Sweden, where dedicated research institutes were—and still are—few in number, the establishment of research council funding became crucial to promote research versus teaching. Organizing research financing in the form of research councils was the main current of Swedish research policy in the 1940s. All major fields were provided with a research council. Medicine was first in 1945, followed by natural science in 1946 and social science in 1947. Although the humanities would not formally get their own research council until 1959, from 1927 onwards they had been able to apply for funding from the Humanistic Fund (*Humanistiska fonden*).

In addition to the state-financed research councils, private foundations have played an important role research funding. Since 1917, the Wallenberg finance dynasty has placed large amounts in foundations that have invested significantly in Swedish research. Since the early 1960s the competing financial group—centered round the commercial bank Handelsbanken—has supported social sciences, and economical sciences, in particular. International private sources have also made significant contributions to research in social science and humanities. The Rockefeller Foundation, for example, invested large amounts in the establishment of social science and economics at the Stockholm University College.

Organization and Expansion of Higher Education and the Social Sciences and Humanities

A Swedish national system for higher education had developed by the mid-twentieth century, including two state universities, two private university colleges and a set of professional schools. This comprised almost 400 professors educating about 12,000 students. This system was about to undergo a far-reaching transformation, including a series of reforms, a vast expansion and disciplinary differentiation. This is also the period when the social sciences "took off," surpassing the humanities in terms of research funding and in the number of students, becoming the largest of all educational domains.

Recent Reforms: Unification, Marketization, Internationalization

The Swedish system of higher education has been substantially reformed three times during the last 60 years: in 1977; 1993 and 2007.[2] These three reforms can be put under three very different headings: *unification*; *marketization* and *internationalization*. All of them created distinctively different conditions for the educational offer to take shape. In addition, a number of commissions have been vital to the development of the sector, the most important was appointed in 1955, which prepared the ground for the extensive expansion of the system in the 1960s.

The commission appointed in 1955 was given the task of meeting the rising demand for higher education that was expected to coincide with an increasing number of people reaching university age. The commission paved the way in two major ways for the extraordinary expansion of higher education enrolment that occurred during the 1960s (see below). First, the financing of the faculty of philosophy—open to all those who passed the upper secondary school exam (*studentexamen*)—was automatically tied to the numbers of enrolled students. This principle, *universitetsautomatiken*, brought about the financial conditions for universities to grow in a way that was not theirs to control; they had to adapt to demand. Second, the new position of university lecturer (*universitetslektorer*) was introduced, with the main task of teaching. This increased opportunities for the higher education institutions to expand their permanent teaching staff beyond the professoriate. In addition to the 1955 commission, further initiatives were taken in the 1960s to facilitate expansion. The most important of these were: the introduction of general study loans; the geographical expansion of the universities; the establishment of outsourced regional branches of the old universities and the addition of two new universities, in Umeå 1965 and Linköping 1975 (Askling 2012: 43–49).

[2] The following section is based on Börjesson et al. (2014).

A further reform in 1977 was a result of a long process undertaken by the commissions largely appointed in the 1960s (UKAS, PUKAS and U68). The *universitetsautomatiken* had resulted in higher education becoming very costly. The watchwords for the commission of the Swedish Higher Education Authority (UKAS) in 1966 were "efficiency" and "throughput." The answer to the problems was fixed study programs, though with some openings for local variation (HSV 2006: 10–11), and the introduction of a *numerus clausus* for all faculties. One of the most important and lasting contribution of U68 and subsequent government bills was the new definition of higher education. With the 1977 reform, higher education was to include a number of programs and institutions that had not been previously regarded as part of the sector. Overnight virtually all public postsecondary education—including previously nonacademic training of nurses or elementary school teachers—was administratively incorporated into *Högskolan*. This almost doubled the number of students in what was now officially categorized as higher education. A new type of institution, the university college, was introduced. These were mainly teaching institutions in regions where higher education had previously had no presence. Over time, their lack of resources for research has become increasingly problematic and certain initiatives to diffuse research resources have been implemented.

In 1989, a new higher education commission was appointed, which resulted in the 1993 reform. The government bill proposing this reform was named "Freedom for quality." The bill was based on a combination of previously observed decline in quality of higher education and fear of further decline in the future. Enhanced quality was to be achieved by increased efficiency in resource management, as well as increased mobility and innovation. To reach these targets, a double strategy was utilized: higher education institutions were to be given enhanced autonomy with regard to organization of studies, the educational offer, student admissions, professor appointments, etc. This increased autonomy—understood as more power to the local university management and board—and institutional diversity were to be combined with economic incentives to be more productive. The shape and orientation of the educational offer and resource allocation would be determined by student choice (Prop. 1992/93:1: 21–22). The reform implied the creation of market-like conditions in higher education (Bauer et al. 1999: 85–88).

The most recent higher education reform, implemented in 2007, might be described as the incorporation of the Bologna process degree structure (3 + 2 + 3, or 3 – 5 – 8 years) into the Swedish system. It began to take shape in 2002. In the Government bill *Ny värld—ny högskola* [New world—new higher education] (2004/05: 162) the most important watchword was "internationalization." All other proposals in the bill can be interpreted as ancillary to this general intention. For example, the enhancement of quality is not only related to the benefit of those who get a better education, but also to its presumed effect on the attractiveness and position of Swedish higher education in international rankings.

A Transformation from Elite to Mass to Universal Access

The Swedish system of higher education has followed the global trend of expansion of the number of students (Verger and Charle 2012). The number of registered students has increased from 17,000 in 1950 to 431,000 in 2011—a 2500% increase. Relating the expansion to the schemes of Martin Trow (1974), this implied a transformation of an elite system into one of mass admission before then becoming a system of universal access. The expansion has not been linear and it is possible to identify six phases, all but two characterized by growth of different magnitude. The 1950s form a first phase with a steady increase from 17,000 to 37,000, implying more than a doubling in a decade and a yearly average growth rate of 8%. A second phase, in the 1960s, was marked by an exceptional expansion, with numbers rising from 37,000 to 120,000, tripling the number of students and an annual average increase of 12%. A number of factors explain this extraordinary expansion (Börjesson 2011). Until 1969, anyone with an upper secondary degree (*studentexamen*) was entitled to enroll in any of the liberal arts faculties (humanities, social science, theology and law). Limitations on the number of places were to be found only in some areas in the faculty of science, and at medical, technological and social institutes; i.e. more professional oriented institutions outside of the universities.

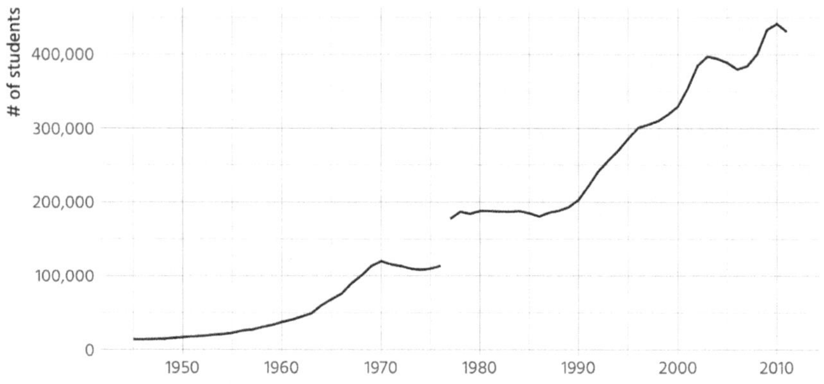

Fig. 7.1 Number of enrolled students 1945–2011

The rise in student numbers in the 1960s could, to some extent, be attributed to the extension in the number of places at the gymnasiums from which one could receive the *studentexamen*. Another important factor, as noted above in the policy context, was the introduction of the so-called *universitetsautomatik* in 1958, which meant that liberal arts faculties would be given teaching resources in proportion to the number of students enrolled. A third important factor was demographic: the generation born immediately after the Second World War, known as the baby boomers, very strong in numbers and now graduating from upper secondary schools (Fig. 7.1).

The subsequent period, phase three, covers the years from 1970 to 1976 in which expansion ruptured and numbers decreased. As we have said, the jump in the figures from 1976 to 1977, from 114,000 to 177,000, was a result of the changing definition of higher education, one of the most important traits of the 1977 higher education reform (see above). Simultaneously, measures were taken to establish a framework for a dimensioning of higher education that would be more in accordance with anticipated labor market demands. In 1979, the entire system became subject to *numerus clausus*, with the government deciding the number of places in each study program. The trend of non-expansion from 1970 to 1976 was carried over to the period 1977–1988 (phase four), when the numbers stayed at around 185,000. Not until in 1989 and into the 1990s did student numbers start climbing again.

The growth lasted until 2003 (the fifth phase). This expansion was not as extraordinary as that seen in the 1960s, but is still important. Student numbers grew from 188,000 in 1988 to 398,000 in 2003, more than doubling in 15 years and an annual growth rate of 5%. The demographic conditions were quite different in this third wave of expansion, occurring in the 1990s, since it coincided with a diminishing young population; the number of 20-year olds fell almost steadily from 125,000 in 1986 to 101,000 in 2003. This implies that, in order to expand the system, it was necessary to attract larger shares of students from homes without study traditions. The proportion of this age cohort in higher education expanded rapidly during the 1990s, from approximately 25% to well over 40%. The subsequent growth in the young population, from 2003 onwards, translates into an oscillating development of the number of students during the sixth period from 2004 to 2009.

Expansion of Social Sciences at Undergraduate Level

Up until 1977, the official data on registered students were organized according to faculty and specialized professional institutions. After 1977, individual-based data on educational programs and courses were collected and digitalized. Therefore, it is difficult to follow the expansion of certain disciplines over time before 1977. However, a general picture of the development of the faculties can be given on basis of the aggregated data available in statistical yearbooks.

For a long time, the faculty of philosophy had been the largest in terms of students, representing approximately 40%. In 1949, its 5500 students accounted for 42%. Expansion in the 1950s and 1960s made it even more dominant in terms of student numbers. By 1960, the faculties of humanities and natural sciences, which combined correspond to the former faculty of philosophy, made up 62% of all students, with 46% in the faculty of humanities. By 1970, after the further division of the faculty of humanities and the forming of the faculty of social sciences, the former faculty of philosophy made up 70% of all students. Out of these, the faculty of social sciences had become the most important, comprising 34% of all students, followed by 24% at the faculty of humanities and 13% in natural sciences. This basic pattern was also apparent in 1976, the year immediately before the substantial reform of the sector. The numbers were slightly smaller, but the relations almost the same, with a small further increase of the relative importance of the social sciences (37, 22 and 10%, respectively).

The close relation in the 1960s between the vast expansion of the overall number of students and the growth of the social sciences and humanities has to be understood in relation to a set of factors. There was clearly a growing demand for higher studies relating to the demographic conditions with rising numbers of young people. A larger share of these had obtained the necessary degrees in upper secondary school and thus was eligible for higher education, further increasing the pool of applicants. This coincided with a political will to expand the higher education system and the implementation of the automatic system of university funding based on numbers in the faculty of philosophy. The latter largely explains the vast expansion of students in the humanities and social sciences, where the absence of *numerus clausus* enabled the growth in student numbers. Further, expansion in these areas, and especially in the social sciences, also corresponded to an important societal transformation: the consolidation of the welfare state and its institutions, based on the belief that more services should be carried out by society and that development should be directed by rational planning. This gave rise to a new set of professions and an increased need for higher administrators and experts—labor that the social sciences were able to produce (Fig. 7.2).

With the reform of higher education in 1977, and the expansion of the definition of higher education to include a large set of educational programs previously lacking the status of higher education, the social sciences held onto its dominant position, although at a somewhat lower level. During the previous three decades, social sciences had been the largest scientific domain with just under 30% of enrolled students. At the end of the 1970s, humanities was the second largest scientific domain with circa 18% of all students, followed by education, health and medicine with 16%. Over time, the humanities decreased in relative terms, stabilizing at 14–16% during the last decade. Other fields have gained shares at the expense of the humanities. Engineering has expanded faster, for example, peaking at 19% around 2000.

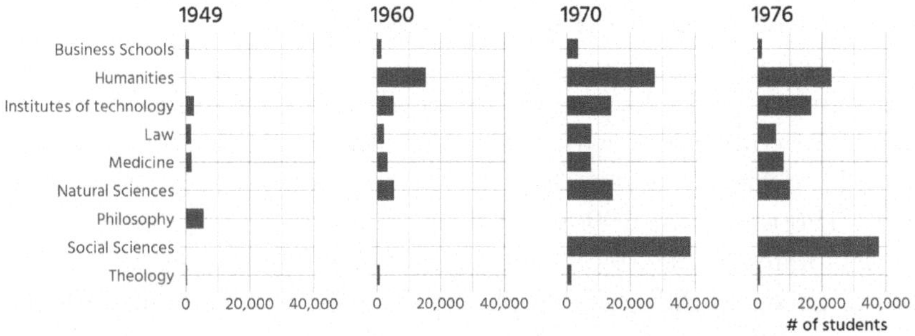

Fig. 7.2 Students by fields of study, 1949–1976 (*Source* Statistical Yearbook of Sweden)

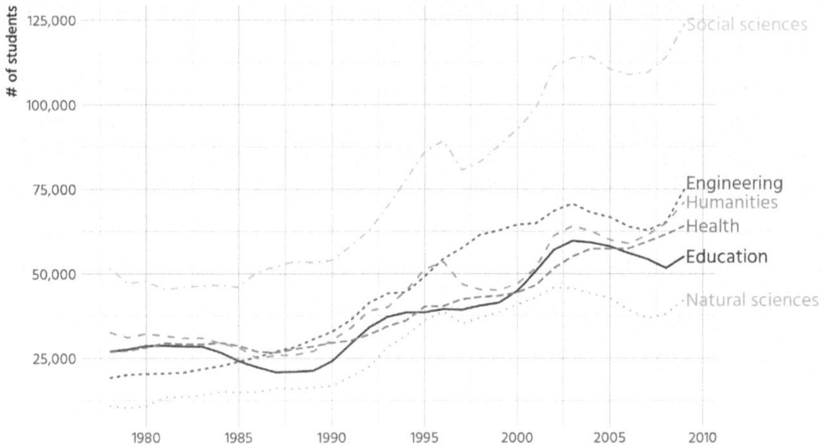

Fig. 7.3 Students by fields of study, 1978–2009 (*Source* Individual based data from statistics Sweden)

The general expansion of higher education translates into a very pronounced expansion of student numbers in the social sciences, from below 50,000 in the early 1980s to over 120,000 in 2009 (a 140% increase). Over the same period, the humanities expanded from 30,000 to over 70,000 (120%). An even more impressive expansion occurred in engineering, rising from 20,000 to 75,000 (almost 300%) and natural sciences going from 10,000 to over 40,000 (also almost 300%). These rates of expansion were, of course, crucial conditions for expansion of the teaching staff and for differentiation of the disciplines (Fig. 7.3).

Steady Increase of Ph.D. Degrees Awarded

The number of Ph.D. degrees awarded also grew steadily over the period. In comparison with the overall student expansion the growth of Ph.D. degrees was time lagged. The large student expansion phases of the 1960s and the 1990s had their counterpart in expansion phases for doctorates in the 1970s and the 2000s. Further, there are different growth curves in different disciplinary domains for doctorates. Whereas medicine had a steady growth rate over the whole period, the human sciences and the sciences start lagging behind medicine from the 1980s onwards. The human sciences saw a rapid increase from the latter part of the 1990s to the early 2000s, an expansion that took place later in the sciences (Fig. 7.4).

Yet another conclusion is that there were different balances between the undergraduate and the graduate levels. The human sciences were the largest fields in terms of the number of undergraduate students, but had inferior shares in relation to medicine at the graduate level, while medicine had a relatively small area at undergraduate levels. The high level of medicine can be explained by the organization of the medical profession, where doctoral degrees are crucial for advancement in a professional career.

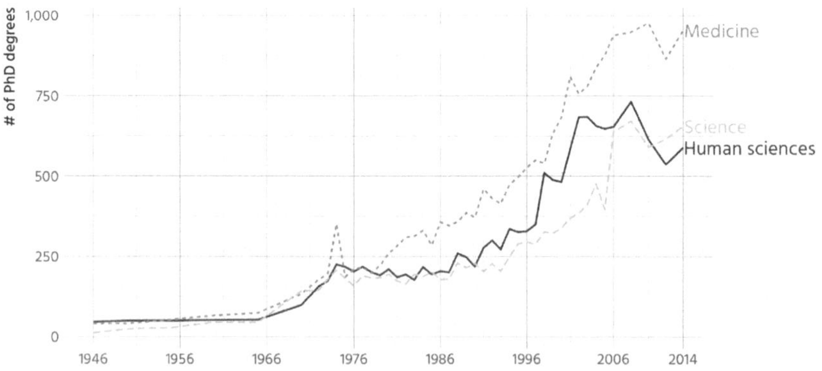

Fig. 7.4 Number of Ph.D. degrees awarded 1946–2014 (*Sources* Statistical Yearbook of Sweden (1946–): Universitet & högskolor: Högskoleverkets årsrapport (2015))

Institutional Patterns of Seven Disciplines

We will now shift our focus from the larger scientific domains to seven disciplines: two in the humanities, literature and philosophy; four in the social sciences, economics, political science, psychology and sociology; and the conglomerate of ethnology and anthropology that is administratively categorized as both humanities (ethnology) and social sciences (anthropology). We will focus on the institutionalization processes, both at the national level and at individual seats of learning.

Chairs and Department Structures

In the modern conception of scientific disciplines, research and education were brought together. Higher education institutions became crucial sites for discipline formation since they often monopolized the function of higher learning, whereas research was more dispersed institutionally. Institutionalization, in the sense of people doing anthropological, economic, literary, psychological, political and sociological research, while having both permanent employment at an institution of higher education for doing just that, and a group of students to whom to teach this subject matter was, for most of the relevant disciplines, already accomplished by the mid-1940s. In fact, during the 1940s and 1950s, the institutionalization processes of the preceding 50 years or so were to be crystalized in the form of signs above doors displaying the name of the discipline (cf. Lundgren 1993: 100–101) at all four universities and colleges. In other words, each discipline got one or more university departments. In general, the process includes starting to teach, establishing more specific chairs and the proper departments centered round the chairs.

On a more precise level, the institutionalization process shows varied origins and timing. Philosophy and literature, the two distinct humanistic disciplines, have ancient roots. Studies in philosophy were central at Uppsala University from the seventeenth century onwards and, in Lund, chairs in philosophy existed from the establishment of the university in 1668, and have done ever since. Literature and rhetoric have been taught at the two universities since their early days and chairs focusing on aesthetics, including literature, have existed since 1742. Independent chairs in "history of literature" were established through a differentiation of the chairs in "history of art and literature" that began at Stockholm University College in 1899 and continued with Uppsala in 1917, Lund in 1920 and Gothenburg in 1932.

The picture is more diverse when it comes to the social sciences. Political science received a chair in 1622 in Uppsala (argued to be the oldest in the world), and economics obtained its first chair in 1741 at the Faculty of Law in Uppsala University. In economics, the development at the turn of the nineteenth century may be described as something of an upgrading, or institutionalization, of a new research practice imported from abroad by Knut Wicksell during his visit to the Fabian Society. Back in Sweden, during the last years of the nineteenth century, Wicksell's works were more in line with neoclassical economics than anything had been before in Sweden, and he, as well as his friends Gustav Cassel and Gustaf Steffen, worked closer to the margin approach than the historical approach. All of the work of economists holding chairs around 1900 had been funded by the Loren Foundation, who had paid for their travels abroad as well as lectures at Stockholm University College.[3]

The political science chair at Uppsala University, established in 1622 in *eloquentia et politice* (lat.), had no equivalent at Lund University where political science was institutionalized as one of the degree subjects at the faculty of philosophy in 1870. The professor of history became responsible for the examinations and the debates surrounding the establishment of political science at the turn of the century dealt with the need to separate political science from history. Political science thus came about as a discipline differentiated from history (Odén 1991: 65–67).

This can be contrasted with sociology and psychology, which obtained their first proper chairs in 1947 and 1948. Ethnographic

---

[3]For a detailed account of the dawn of social sciences in Sweden and the part played by the Loren Foundation, see Wisselgren (2015).

research was carried out at the departments of ethnography at museums in Stockholm and Gothenburg before it got its first chair at Gothenburg University College in 1923. It was a personal chair donated to Erland Nordenskiöld by Naval Minister Dan Broström, meaning that when the professor left in 1932 there would be no successor, and it was not until 1955 that Gothenburg University (which became a university in 1954) got a new chair in Karl Gustav Izikowitz (1903–1984). Stockholm got its first chair in ethnography in 1928, but it was not until 1962 that the first chair in ethnography independent from the museums was established in Uppsala.

For all social science subjects, except anthropology, the 1940s and 1950s were a phase of distinct expansion. In sociology, four chairs were established between 1947 and 1957; in psychology, the same number was reached in an even shorter period, from 1948 to 1956. In political science, the second chair was established in 1926 and, by 1952, chairs existed at the four major universities and university colleges. The same occurred for economics in 1953, where the Stockholm School of Economics had a professor in economics from 1909. Department formation for the social sciences came later than the establishment of the chairs, stretching from the 1940s to the 1960s. In sociology, however, the departments in Lund and Gothenburg preceded the chairs.

By the mid-twentieth century, the university department had become the model for the institutionalized organization of research and teaching within a single discipline. There really were no university departments in 1900, and the most recently established departments are seldom restricted to one clearly defined discipline. From this perspective, the disciplinary department, as a central organizational feature of a discipline, is in decline. This is most apparent for literature in Sweden with just one single-disciplinary department remaining in Uppsala. It is, however, not the case for economics or psychology who both find themselves with 11 distinct, self-contained departments. Important steps toward disciplinary institutionalization also arose outside of the seats of higher learning, to which we will now turn.

Scientific Journals and Organizations

Scientific journals and organizations have been crucial elements in institutionalization on a national, or even Nordic, level. The presence of scientific disciplinary journals is further evidence of how established these disciplines were in the mid-twentieth century: all had a specialized journal in the interwar period, with the exception of sociology. Between 1880 and 1899, no less than four longstanding journals were founded. In 1880, both the Swedish Literature Society and its journal *Samlaren* were founded, with the aim of promoting Swedish research in literature; the journal has been published annually since. The journal *Ymer* (1881) was published by The Swedish Society for Anthropology and Geography (SSAG), established by that name in 1878. A second journal for anthropology and its neighboring disciplines was published in 1936 under the name of *Ethnos* by the Ethnographical Museum of Sweden.

These examples also point to a close link between the establishment of scientific journals and the creation of scientific societies and organizations, which often functioned as the founders and financiers of the journals. We will, therefore, treat journals and organizations in parallel, though, as will be shown, it is not a necessary condition to first establish an organization and subsequently a journal.

At the end of the nineteenth century two social science journals were published, one in political science (*Statsvetenskaplig tidskrift* 1897) and one in economics (*Ekonomisk tidskrift* 1899). *Statsvetenskaplig tidskrift*, still published in Swedish, published works on state law and political science of interest to both scholars and other citizens. When *Ekonomisk tidskrift* launched its first issue in 1899, there was no declaration of principles or any other statements on the direction of or motivation behind the launching of the journal. Instead, the reader would be met by "Some economic data from the past years," as the title of the first article read. Unlike its contemporary *Statsvetenskaplig tidskrift*, the journal would not remain the same over the following century. It changed title and language in 1965 to *The Swedish Journal of Economics* and then changed its name and scope in 1976, becoming *The Scandinavian Journal of Economics*. The political science journal *Scandinavian Political Studies*, on the other hand, did not take over the role of *Statsvetenskaplig tidskrift* when it was launched in 1966, but was, by contrast to *Ekonomisk tidskrift*, an independent journal, adding a Nordic level to the national one. The link to disciplinary organizations is weak for journals in political science and economics. *Statsvetenskaplig tidskrift* was not initially associated with any organization, but from 1919 it obtained financial support from the *Fahlbeckska stiftelsen* and became associated with *Statsvetenskapliga förbundet* (Swedish Political Science Association), founded in 1970. The organization for economists, *Nationalekonomiska Föreningen*, was already founded in 1877, but did not launch a journal until 1973, when *Ekonomisk debatt* was established.

Philosophy's first modern scientific journal was *Theoria*, established in 1935. Its first editor, Åke Petzäll, wanted the journal to be a medium of communication not only within the Nordic countries but also with the rest of the philosophical community. This was manifested in the languages of the published articles and entries. Apart from the first issue, in which all the articles were in Swedish, foreign languages would make up the major part of the journal and, from 1939 onwards, all articles and entries were in a foreign language. German was the most common until English took over in 1947 (Strang 2010: 71–72).

During the first half of the twentieth century, there were a couple of journals aimed at psychology, often combined with pedagogy such as *Psyke* (1906–1920) and *Tidskrift för psykologi och pedagogik* (1942–1950). This corresponded with the organizational principle at the time, when psychology partnered with pedagogics. None of these journals would survive to the present day. It was *The Scandinavian Journal of Psychology*, published since 1960 by the Nordic Psychological Association, that would last. As a reaction against what was perceived as mainstream psychology, meaning "positivistic, removed from socially relevant problems, restricted to laboratory work and biologistic" (Rigné 2002: 272), associations and journals were founded. One of the more lasting journals was *Kritisk Psykologi* (1981–1994).

If the social sciences typically began with a Swedish language journal aimed at a Swedish audience, for sociology it was the other way around. *Acta Sociologica*, the Nordic journal was launched in 1955, a decade before the first exclusively Swedish journal *Sociologisk forskning* in 1964. This also corresponded to the order in which the sociological associations were founded with the Nordic one preceding the Swedish, although only by two years, 1960 versus 1962.

Most of these journals play only a secondary role today, at best, since most publishing activity is aimed at international journals. *Theoria* and *Ethnos* are different in that they are considered international journals.

Research Funding

Because we lack complete historical data on research funding that allows state grants to be differentiated from external funding from private and public sources, we have used the external funding by the research councils as an indicator of the level of research funding for the disciplines in question. Since the level of funding increases drastically over the period (in the prices of 1947 the total sum for the seven disciplines rises from 240,000 to 11,400,000 Swedish kronor, SEK, over the period), we have chosen to present the development in three periods indicating different funding regimes and sizes of the system.

In the first period, covering the expansion of higher education system during the 1950s, it is clear that it was most invested in the newly founded discipline of sociology, especially in the period from the late 1940s to 1953/1954. After this political science, another social science discipline, reached the same level of funding as sociology. The third most important discipline in funding terms changes between economics and literature. It is clear that, from the start, the social sciences came to dominate the external research funding channeled through the research councils.

In the second period, covering the expansive phase of the 1960s and parts of the 1970s, the picture differs somewhat. Psychology rises to a giant in external funding and, by 1970, it attracted five times the sum of the second most funded discipline. Once again, a social science discipline dominates, but the differences are quite small.

Our last period covers the 2000s when the social sciences also gain the most funding. Three disciplines shift in the leading position, sociology, psychology and political science. Among the other disciplines, philosophy rises over the period, reaching the same level as sociology and psychology by 2012 (Fig. 7.5).

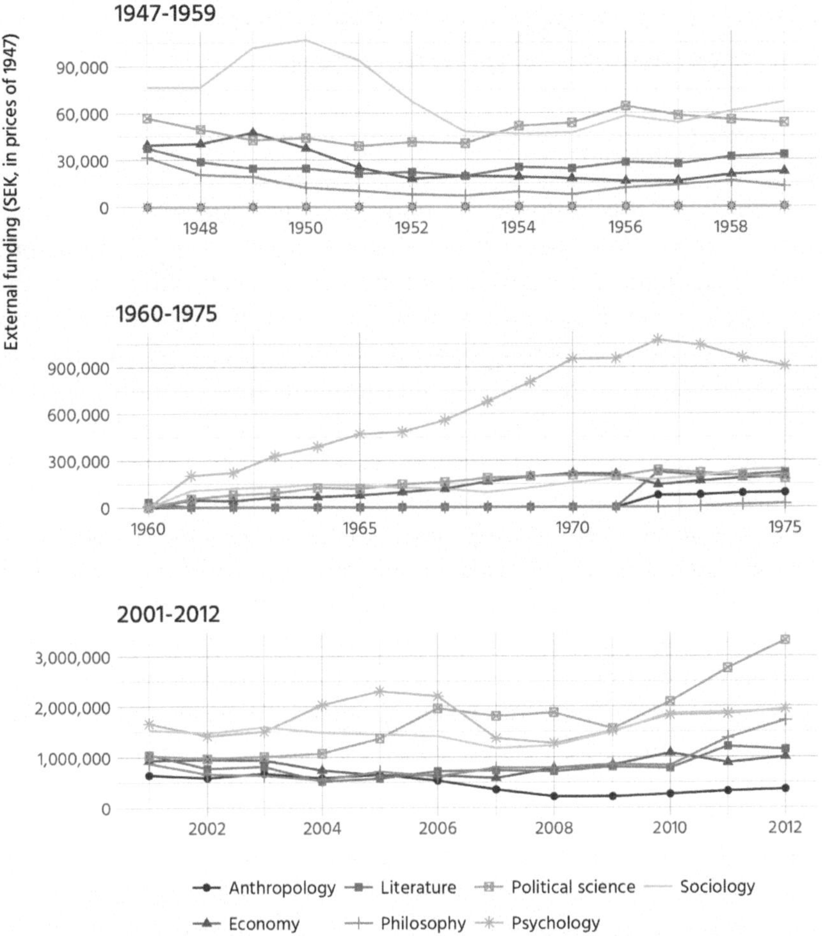

Fig. 7.5 External funding by the research councils in prices of 1947 (*Sources* Annual reports of the Social Science Research Council (1947–1976), Humanities Research Council (1947–1976), and the Swedish Research Council project database (2001–2012))

Humanities Struggling to Keep Pace: Morphological Transformations

Professors

To begin with, the jump in the number of professors for most disciplines between 1985 and 1995—and between 1995 and 2005 for some others—highlights the changing conditions for establishing chairs that resulted from the higher education reform in 1993. As we mentioned earlier, budget decisions were handed over to each institution, in effect allowing them to create chairs by allocating the funds required for such a position. An additional reform in 1999 created the possibility to be promoted from a permanent position as a lecturer to a position as professor without the institution having to advertise the position. Thus, if one was deemed sufficiently qualified by an appointed expert, one could become a professor without the competition of an open hiring process. While there are no formal regulations that distinguish a more traditional chair from this new type of promoted professor, praxis is that being appointed to a chair resembles the traditional full professor, with a guaranteed level of research funding, whereas the promoted professor only gets the title and possibly an often relatively small rise in salary (Fig. 7.6).

For almost 50 years following the Second World War, the only way of increasing the number of professors was to either increase the number of higher education institutions where the discipline is taught, or to create specialized chairs ("sociology of...," "cognitive psychology," "international economics" and so on). These two strategies are what sustained the relatively modest increase in chairs between 1945 and 1985. Economics clearly benefitted from being represented at the Stockholm School of Economics, which added one or two chairs to their national total, something other disciplines could not do.

When budgetary decisions were moved from central government to local institutions with the higher education reform in 1993, disciplines rich in student numbers seem to be the ones that were granted more professors by their local institutions, with economics more than

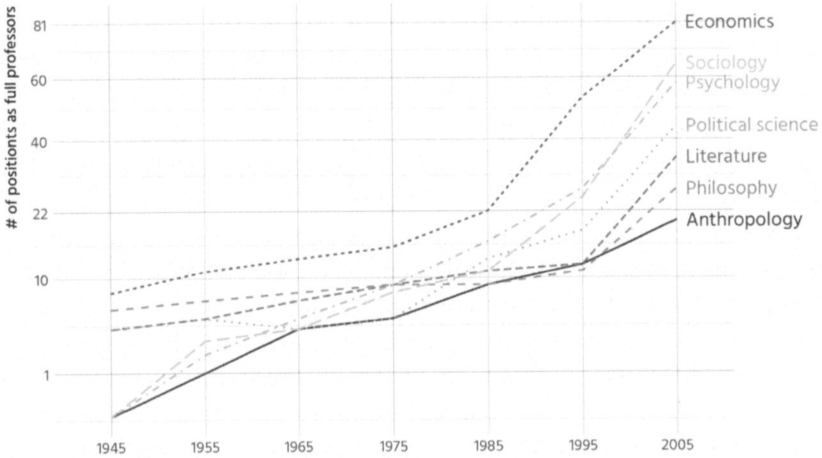

Fig. 7.6 Number of positions as full professors 1945–2005 (*Sources* Sveriges statskalender, editions 1945, 1955, 1965, 1975, 1985, 1995 and 2005)

doubling the number of professors between 1985 and 1995. After the promotion reform in 1999, all of the disciplines rapidly increased their number of professors. The discipline with the lowest increase between 1995 and 2005—economics—still had more than a 50% increase while literature, as the most rapidly growing discipline, almost tripled its number of professors. This indicates that the logic was different and not primarily driven by the number of students and general resources, but rather by the age composition of the staff; older staff had greater periods—and thus greater opportunities—over which to acquire the necessary merit to obtaining a promotion to professor.

Studying the relations between the different disciplines, it becomes clear that the social sciences, in general, have gained momentum over the period. Philosophy increases modestly between 1945 and 1985, going from 6 to 9, but is surpassed by sociology and psychology, which both, having had no chairs in 1945, reached 11 and 16 respectively by 1985. By 2005, a distinct order was established with the social sciences at the top, ranging from 81 professors in economics to 44 in political science, followed by 35 professors in literature and 27 in philosophy. Only anthropology/ethnology stands out with the lowest number (20).

Women are almost completely absent among the professors of the INTERCO disciplines until the 1980s. At the beginning of the twentieth century, women were not legally allowed to become professors at state universities; the geographer Gerd Enequist became the first female professor at a state university in 1949.[4] Having been legally excluded for a large part of the century, not to mention the cultural obstacles faced by a woman entering a completely male-dominated world, women pursuing a career in academia had to make up a lot of ground in order to be as prevalent in the recruitment pool as they are in the general population. This is one of the demographic properties that helps to explain the continued absence of women in the professoriate, even after Gerd Enequist cracked the "glass ceiling." Another demographic property is related to the policy regarding the number of professors per discipline and institution. If the professor installed was young, he would block any pretender for a couple of decades, regardless of the gender of the pretender or the gender proportions among the faculty. This is why the loosening of the "one professor per discipline and institution" in 1993 probably meant a diminishing effect of the "demographic blocking property." Evidence of this can be seen in that disciplines (economics, political science and psychology) that had no female professors in 1985 would—with the exception of philosophy—have at least some today. The increase of women among professors was not fast enough for the government which, in 1995, created 31 new professor positions in 1995 specifically designed to recruit the underrepresented gender in each specific discipline. For the INTERCO disciplines, positions were created in "political science with gender research," "history of philosophy," "psychology, especially cognitive" and all three were filled with women.[5] The share of women continued to rise between 1995 and 2005. Part of this can probably be attributed to the promotion reform in 1999, but at this point, the supply of doctors had become fairly gender-balanced—at least in the total population (Fig. 7.7).

---

[4]For a thorough account of the troubles facing women in academia from the first female student in 1873 to the first professor in 1949, see Markusson Winkvist (2003).

[5]For an evaluation of this reform, see Jordansson (1999).

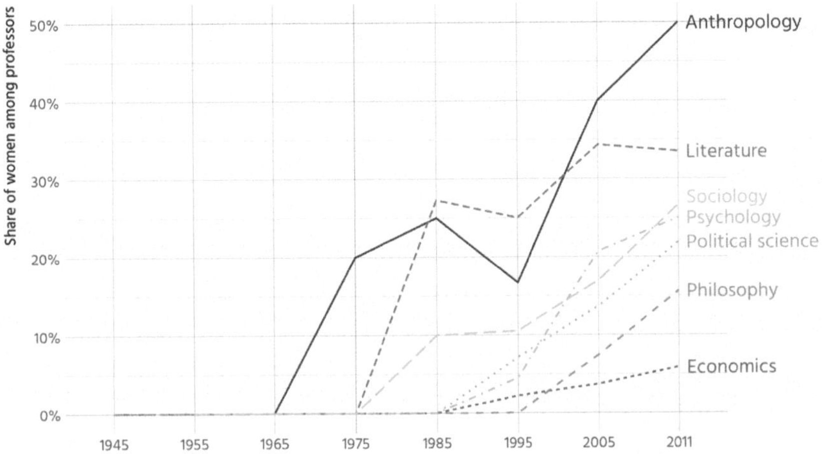

Fig. 7.7 Share of women among professors 1945–2011 (*Sources* Sveriges statskalender, editions 1945, 1955, 1965, 1975, 1985, 1995 and 2005; register data from SCB for 2011)

Ph.D.s Awarded

The first two decades following the Second World War was a period when the annual production of doctors was generally low, with seldom more than a handful (Fig. 7.8). Among the more consistent producers, output from literature and philosophy was fairly balanced, as well as from economics and political science. Partly due to a large project launched in 1965—following the extensive municipality reform—political science managed to produce an extraordinary large number of doctors just before the huge wave of student enrolment of the 1960s would make itself felt at the doctoral level in the early 1970s.

In the 20 years between 1970 and 1990, philosophy and anthropology were left behind as the other four disciplines expanded, only literature managed to keep pace with the social sciences during this period. After the financial crisis of the early 1990s, the divide between the humanities and the social sciences grew deeper. Indeed, there has been some expansion even among the humanities, but nowhere near as steep as in the social sciences, with economics becoming as dominant in the production of doctors as it has been among the holders of chairs.

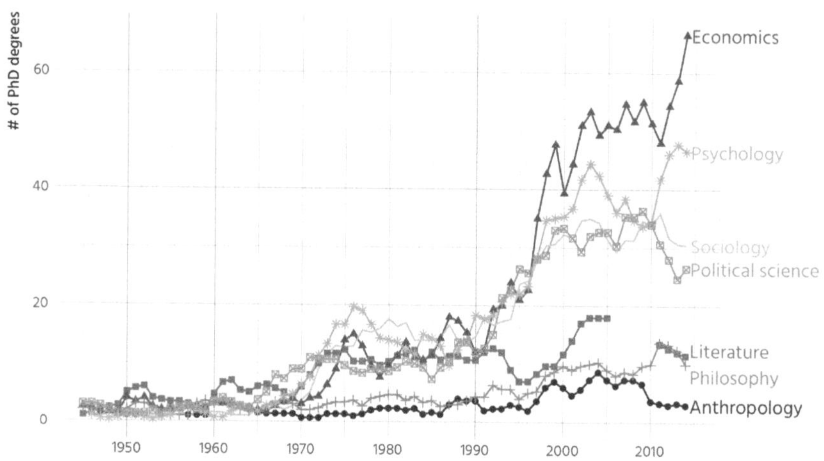

Fig. 7.8 Number of Ph.D. degrees 1945–2014 (EWMA of the year before and after) (*Sources* Register data from SCB 1971–2014; Libris library catalogue for economics, philosophy, literature, psychology 1945–1970; Allardt et al. (1988) for sociology 1951–1974; Johansson (2009) for political science 1945–2009)

The INTERCO disciplines have generally had a larger share of women doctors than the Ph.D. population in general. Economics and philosophy have, however, consistently produced fewer women doctors than the other INTERCO disciplines and the Ph.D. population in general. Instead, women have populated the disciplines that grew out of philosophy, i.e. psychology and sociology, leaving philosophy to the men.

Students

A detailed nomenclature distinguishing between different social science disciplines was not in place until 1986. From the government's official report, produced by the commission on social sciences in 1946, we know that in 1945 there were an estimated 650 students in economics, 278 students in political science and 118 students in sociology nationwide (SOU 1946: 74).

Looking at the number of registered students each fall semester from 1978 onwards, psychology is clearly the most numerous throughout the period. It was the only discipline among the social sciences where *numerus clausus* was introduced to restrict the student intake. However, it was also the discipline with a monopoly on the longer professional program training future psychologists. Psychology, then, would have relatively more students continuing to more advanced levels in the course structure than would the other disciplines, despite its restricted intake at the entry level. Economics, of course, would be a major component in the training of economists. But it also had to compete with business studies in the program structure for economists, since the students could choose an orientation between the two disciplines. Business studies was the discipline with the largest number of registered students over the past 40 years.[6]

Differences between disciplines other than psychology were not that evident from the late 1980s to the early 1990s. Beginning in the midst of the Swedish banking crisis in the early 1990s, the number of economics students began to increase more significantly than the otherwise expanding disciplines of sociology and political science in 1995 and 1996. Since 1997, the seven disciplines can be differentiated into three groups, one of which would contain only the largest discipline of psychology, completely dominant in volume at the dawn of the twentieth century. A second group consists of the other social sciences—political science, economics and sociology—each with about 4600 to over 5000 students enrolled per year. A third group with smaller numbers than the other two groups consists of literature, anthropology and philosophy which, in Sweden, would all be considered more humanities disciplines than social sciences (Fig. 7.9).

Student-to-Teacher Ratios

Increases and decreases in student numbers only highlight changing balances between disciplines and do not say anything about the demographic crises that occur within the disciplines at times of over- and under-population. These are almost inevitable since there is always a time lag between a wave of expansion and the hiring of teachers to handle it. At these times, the student to teacher ratio goes up. Since these expansion waves are partly due to population demographics in a narrower sense, e.g. baby boom generations, one might experience the opposite problem with student to teacher ratio going down.

[6]The history of business studies in Sweden has been well documented by Engwall (2009).

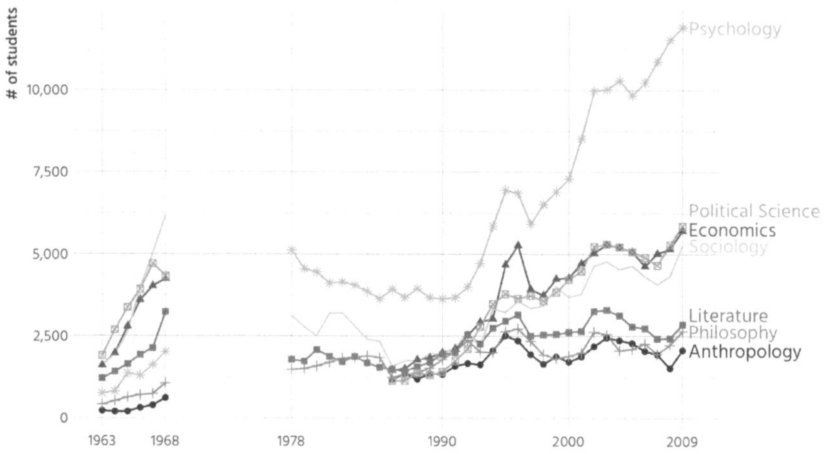

Fig. 7.9 Number of students 1963–2009 (*Sources* Register data from SCB (1978–2014); Statistiska meddelanden Serie U (1963–1968))

The rapid expansion of the 1960s led to overpopulation—in the sense above—particularly in political science and sociology. For a couple of years, there were between 140 and 164 students per teacher in these two disciplines. A sociologist teaching at this time described it as a "shocking experience," having "to teach large groups of students in restaurants or movie theatres" (Björnberg 2014: 199). Economics seem to have been under less strain, with around 77 students per teacher. Psychology, literature and philosophy managed to keep this ratio at 30 or less students per teacher, while anthropology experienced a crisis of the opposite kind with about 9 students per teacher.

To be an undergraduate student, establishing some kind of relation with the professor of your discipline might never have been a simple task. In political science and sociology of the 1960s, it must have been next to impossible as the ratio between students and professors were 841 and 655 to 1 in each of these disciplines respectively.

During the stagnation phase of the 1980s (see section "Organization and Expansion of Higher Education and the Social Sciences and Humanities" above), the previous order was reversed, albeit at a lower magnitude. Economics, political science and sociology now all had more manageable student to teacher ratios at the same level as literature, philosophy and psychology experienced in the 1960s. The latter three disciplines had all increased the number of students per teacher; philosophy had even doubled this ratio. Yet, while the literature student of 1985 had more classmates than their 1965 counterpart, there was still a far greater chance of meeting a professor in 1985 as the student to professor—or professor to teacher—ratio had decreased.

After the third wave of expansion in the 1990s (see section "Organization and Expansion of Higher Education and the Social Sciences and Humanities" above), the ratio between students and teachers ranged between 16 and 30 in all disciplines, i.e. the size range of a normal school class. Sociology only had 16 students per teacher, suggesting a coming demographic crisis. Anthropology, economics, philosophy and political science all had roughly the same ratio of students per teachers in 2005, but the teacher corps were not constituted with the same relative amount of different teacher categories. There were more professors in economics and philosophy than in anthropology and political science, suggesting a different weight of teaching versus research activities in these pairs of disciplines (Table 7.1).

Table 7.1 Ratio between students and professors

Numbers	All teachers (1)			Professors (2)			Students (3)		
	1965	1985	2005	1965	1985	2005	1965	1985	2005
Anthropology	21	18	107	4	9	20	198	1167	2285
Economics	36	62	225	13	22	81	2784	1460	5072
Literature	57	34	–	7	11	35	1643	1483	2784
Philosophy	21	20	96	8	9	27	627	1242	2102
Political science	24	41	244	4	13	44	3362	1130	5071
Psychology	50	83	329	5	16	58	1364	3917	9839
Sociology	16	44	284	4	11	65	2619	1579	4646
Ratios	Professors/all teachers (2/1)			Students/all teachers (3/1)			Students/professors (3/2)		
Anthropology	0.2	0.5	0.2	9	65	21	50	130	114
Economics	0.4	0.4	0.4	77	24	23	214	66	63
Literature	0.1	0.3	–	29	44	–	235	135	80
Philosophy	0.4	0.5	0.3	30	62	22	78	138	78
Political science	0.2	0.3	0.2	140	28	21	841	87	115
Psychology	0.1	0.2	0.2	27	47	30	273	245	170
Sociology	0.3	0.3	0.2	164	36	16	655	144	71

Conclusions

The unprecedented expansion in student enrolments in the social sciences and humanities was partly determined by the absence of *numerus clausus* in those parts of the higher education system. The number of Ph.D. candidates did not rise at the same pace. Such mechanisms created significant differences in the relation between research and education. Medicine and sciences became more research oriented than the humanities and social sciences, as is illustrated by the ratio of undergraduates to Ph.D.s. When comparing different disciplines, using the number of students per professor as an indicator, research carries more weight in economics and philosophy than in political science or anthropology, for example.

At the beginning of the new millennium, the social sciences have unquestionably gained a dominating position *vis-à-vis* the humanities in terms of enrolled students, employed staff and research resources. Thus, the order that lasted until the mid-twentieth century has been reversed. Even with almost every discipline expanding since then, most of the humanities failed to keep pace with the social sciences, as could already be seen to some degree in the great expansion wave in the 1960s, only to be cemented by the next wave in the 1990s.

This rise of the social sciences took off with the institutionalization soon after World War II, when professorships, departments, dedicated Swedish journals, research councils and so on were in place. The main institutionalizing principle was one chair, one discipline, one department; an arrangement that allowed each discipline to function as an autonomous field. The close links between most of the emerging social sciences and the contemporaneous welfare state project did not necessarily imply weak autonomy in relation to the political fields and state administration since. In many cases, social scientists were significant agents in the societal transformation.

In contrast, the more recent transformation of Swedish social sciences, and also humanities, from the late twentieth century onwards has entailed deinstitutionalization and loss of autonomy. Departments ceased to be centered around one discipline with one or, later on, perhaps a few chairs. The social sciences (to a lesser extent the humanities) lost their Swedishness, a development that is mirrored in the fate of the journals. Those publishing contributions in the Swedish language—such journals had been decisive vehicles in the early formation of Swedish disciplines—were "outflanked" by Swedish or Nordic journals in which the English language was used. Those were, in turn, relegated to second-rate positions when so-called international journals gained supremacy (see Table 7.2).

Today Swedish economists, philosophers, political scientists, psychologists or sociologists tend, if they are in the position to choose, to favor publishing in journals regarded as international. Among anthropologists, and some other disciplines, publishing in international fora was always the case. However, scholars in literature or history still publish in Swedish and their national journals have not been anglicized.

As already mentioned, a particularity of Swedish (and, to a somewhat lesser extent, other Nordic) social sciences during the founding period immediately after World War II was the tight links to the welfare state project. To write in Swedish signified to write for politicians, state officials and concerned citizens. The turn to foreign journals—most often edited in English speaking countries and behind paywalls—implies the decoupling of social sciences from the welfare state project. This development—together with the general erosion of the Swedish welfare project and the related weakening of specialties such as the previously most dominant "welfare sociology"—might, on the one hand, in some respects present new and autonomous opportunities to the social sciences—unless the state defined research problems are replaced by objectives originating from other external powers such as the economic field or the media and entertainment industry. On the other hand, the weakened links to stakeholders within the political and administrative fields will no doubt contribute to the ongoing deinstitutionalization of Swedish social sciences.

Table 7.2 Institutional patterns 1945–2015

Discipline	Chair	Degree	Journals	Organisations	Sequence
Anthropology	Allmän och jämförande etnografi Gothenburg (1924–1932), Stockholm/Etnogr. mus (1928–1969), Uppsala 1962	1948	*Ymer* (1881–) *Ethnos* (1936–)	SSAG (1878–)	Org-Journal-Chair-Journal-Degree
Economics	Lund and Uppsala (pre-1900) Nationalekonomi och sociologi Gothenburg (1903) Nationalekonomi och socialpolitik Stockholm (1904)	1909 (faculty of philosophy)	*Ekonomisk tidskrift* (1899–), changed name to *The Swedish Journal of Economics* (1965) and *The Scandinavian Journal of Economics* (1976)	Nationalekonomiska föreningen (1877–)	Chair-Org-Journal-Degree
Literature	Stockholm (1899), Gothenburg (1901), Lund (1920), Uppsala (1922)	1908	*Samlaren* (1880–) *Tidskrift för litteraturvetenskap* (1971–)	Svenska litteratursällskapet (1880–)	Org/Journal-Chair-Degree
Philosophy	Lund and Uppsala (pre-1900) Gothenburg (1893) Stockholm (1937)	pre-1876	*Theoria* (1935–) *Häften för Kritiska Studier* (1969–) *Filosofisk tidskrift* (1980–)	Föreningen för filosofi och specialvetenskap (1935–)	Chair-Degree-Journal

(continued)

Table 7.2 (continued)

Discipline	Chair	Degree	Journals	Organisations	Sequence
Political science	Uppsala (pre-1900) Lund (1909), Gothenburg (1901), Stockholm (1935)	1870 independent degree 1935	*Statsvetenskaplig tidskrift* (1897–) *Scandinavian Political Studies* (1966–)	Statsvetenskapliga föreningen i Uppsala (1919–) Swepsa (1970)	Degree-Chair-Journal-Org
Psychology	Uppsala (1948), Stockholm (1953). Lund (1955), Gothenburg (1956)	1948	*Psyke: tidskrift för psykologisk forskning* (1906–1920) *Arkiv för psykologi och pedagogik* (1920–1929) *Studia psychologica et paedagogica* (1947–) *Nordisk psykologi* (1949–) *Scandinavian Journal of Psychology* (1960–)	Sveriges psykologförbund (union, 1955–) Specialised, e.g. Swecog (2012)	Journal-Degree-Chair-(Org)
Sociology	Uppsala (1947), Stockholm (1954), Lund (1956), Gothenburg (1959)	1948	*Acta sociologica* (1955–) *Sociologisk forskning* (1964–)	Sociologiska föreningen i Lund (1936–1940) Sveriges sociologförbund (1962–)	Chair/Degree-(Journal)-Org-Journal

References

Unpublished Sources

Riksarkivet (RA).
Statens råd för samhällsforskning [Social Science Research Council].
Rådets verksamhetsberättelser [Annual reports] (B III).
Vol. 1 (1947–1967).
Vol. 2 (1967–1977).
Statens humanistiska forskningsråd [Humanities Research Council].
Anslagsframställningar, verksamhetsberättelser [Annual reports] (B III).
Vol. 2 (1972–1977).
Riksantikvarieämbetet/Enheten för arkiv och bild (RAÄ).
Humanistiska fondens arkiv [Humanities Research Council].
Verksamhetsberättelser [Annual reports] (B 4).
Vol. 1 (1929–1958).
Vol. 2 (1948–1959).

Printed Sources

Prop. 1992/93:1 *Regeringens proposition 1992/93:1 om universitet och högskolor: frihet för kvalitet*. 1992. Stockholm: Riksdagen.

Prop. 2004/05:162 *Regeringens proposition 2004/05:162: ny värld - ny högskola*. 2005. Stockholm: Riksdagen.

SOU 1946:74 Socialvetenskapliga forskningskommittén. 1946. *Betänkande angående socialvetenskapernas ställning vid universitet och högskolor m.m.*. Stockholm: Nordiska bokh. i distr.

Statistiska meddelanden. U [=Statistical reports]. 1963–1968. Stockholm: Statistiska centralbyrån.

Statistisk årsbok för Sverige = [Statistical yearbook of Sweden]. 1914–2014. Stockholm: SCB.

Sveriges statskalender. 1945–2005. Stockholm: Fritzes offentliga publikationer.

Universitet & högskolor: Högskoleverkets årsrapport. 2015. Stockholm: Högskoleverket.

Secondary Literature

Agevall, O., and G. Olofsson. 2013. The emergence of the professional field of higher education in Sweden. *Professions & Professionalism* 3 (2): 1–22.

Allardt, Erik, Aage Bøttger Sørensen, and Sverre Lysgaard. 1988. *Sociologin i Sverige: vetenskap, miljö och organisation: en utvärdering av svensk sociologi utförd på uppdrag av Humanistisk-samhällsvetenskapliga forskningsrådet och Universitets- och högskoleämbetet.* Stockholm: HSFR.

Askling, Berit. 2012. *Expansion, självständighet, konkurrens: vart är den högre utbildningen på väg?* Göteborg: Göteborgs universitet.

Bauer, Marianne, Berit Askling, Susan Marton, and Ference Marton. 1999. *Transforming universities: Changing patterns of governance, structure and learning in Swedish higher education.* London: Jessica Kingsley.

Björnberg, Ulla. 2014. Lokalsamhällen och familjeliv i omvandling. In *Det personliga är sociologiskt. 14 professorer om svensk sociologi*, ed. Gunnar Andersson, Thomas Brante, Christofer Edling, and Sverre Wide. Stockholm: Liber.

Blomqvist, Göran. 1993. Akademiska visioner under 1800-talets tre sista decennier. *Scandia* 59 (2): 205–256.

Börjesson, Mikael. 2011. Studentexplosionen under 1960-talet. Numerär utveckling och orsaker. *Praktiske Grunde. Nordisk tidsskrift for kultur- og samfundsvidenskab* 4: 11–27.

Börjesson, Mikael, and Donald Broady. 2016. Elite strategies in a unified system of higher education: The case of Sweden. *L'Année Sociologique* 66 (1): 115–146.

Börjesson, Mikael, Emil Bertilsson, and Tobias Dalberg. 2014. Sweden. In *Enrolment Patterns in Nordic Higher Education, ca 1945 to 2010*, ed. Mikael Börjesson, Sakari Ahola, Håvard Helland, and Jens Peter Thomsen. Oslo: Nordisk institutt for studier av innovasjon, forskning og utdanning.

Bourdieu, Pierre. 1984. *Homo academicus.* Paris: Minuit.

Bourdieu, Pierre. 1989. *La noblesse d'État.* Paris: Minuit.

Engwall, Lars. 2009 [1992]. *Mercury meets Minerva: Business studies and higher education: The Swedish case*, 2 [extended] ed. Stockholm: Economic Research Institute, Stockholm School of Economics (EFI).

Heilbron, Johan. 2004. A regime of disciplines: Toward a historical sociology of disciplinary knowledge. In *The dialogical turn: New roles for sociology in the postdisciplinary age*, ed. Charles Camic and Hans Joas. Lanham, MD: Rowman & Littlefield.

HSV. 2006. *Högre utbildning och forskning 1945–2005: en översikt*. Stockholm: Högskoleverket.

Jordansson, Birgitta. 1999. *Jämställdhetspolitikens villkor: politiska intentioners möten med den akademiska världen: exemplet "Thamprofessurerna"*. Göteborg: Nationella sekretariatet för genusforskning.

Kock, Karin (red.). 1941. *Våra yrken: en handbok i yrkesvägledningens tjänst. D. 1*. Stockholm: Kooperativa förb.

Lebaron, Frédéric. 1997. La dénégation du pouvoir. Le champ des économistes français au milieu des années 1990. *Actes de la recherche en sciences sociales* 119 (1): 3–26.

Lundgren, Anders. 1993. Discipliner och institutionalisering inom 1900-talets svenska vetenskaper. In *Från hermetism till rationell distribution*, redaktör Bosse Sundin, 99–110. Umeå: Umeå universitet.

Lundin, Per, and Niklas Stenlås. 2015. The reform technocrats: Strategists of the Swedish welfare state, 1930–60. In *Scientists' expertise as performance: Between state and society, 1860–1960*, ed. Joris Vandendriessche, Evert Peeters, and Kaat Wils, 135–146. London: Pickering & Chatto.

Markusson Winkvist, Hanna. 2003. Som isolerade öar: de lagerkransade kvinnorna och akademin under 1900-talets första hälft, dissertation, Umeå University, Umeå.

Mauss, Marcel, and Paul Fauconnet. 1901. Sociologie. In *Grande Encyclopedie*, vol. 30. Paris: Société anonyme de la Grande Encyclopedie.

Odén, Birgitta. 1991. *Forskarutbildningens förändringar 1890–1975: historia, statskunskap, kulturgeografi, ekonomisk historia* = [Changes in postgraduate studies, 1890–1975]. Lund: Lund University Press.

Rigné, Eva-Marie. 2002. Profession, science and state: Psychology in Sweden 1968–1990, dissertation, University of Gothenburg, Göteborg.

Slaughter, Sheila, and Leslie, Larry L. 1997. *Academic capitalism: Politics, policies and the entrepreneurial university*. Baltimore: Johns Hopkins University Press.

Strang, Johan. 2010. Theoria and logical empiricism: On the tensions between the national and the international in philosophy. *The Vienna Circle in the Nordic Countries*, 69-89. Dordrecht: Springer.

Trow, Martin. 1974. Problems in the transition from elite to mass higher education. In *General Report on the Conference on Future Structures of Post-Secondary Education*, 55–101. Paris: OECD.

Verger, Jacques, and Christophe Charle. 2012. *Histoire des universités. XIIIe-XXIe siècle*. Paris: Presses Universitaires de France.

Wisselgren, Per. 2015. *The social scientific gaze: The social question and the rise of academic social science in Sweden*. Farnham, Surrey: Ashgate.

Electronic Sources

Johansson, Leif. 2009. Förteckning över svenska doktorsavhandlingar i statsvetenskap 1890–2009. http://swepsa.org/doktorander/avhandlingar.html. Accessed Jan 2017.

Libris webbsök [Elektronisk resurs]. 1997–. Stockholm: Kungliga biblioteket. http://libris.kb.se/. Accessed Jan 2017.

Vetenskapsrådets projektdatabas [Swedish Research Council project database]. 2001–. http://vrproj.vr.se. Accessed Mar 2013.

8

Institutionalization and Professionalization of the Social Sciences in Hungary Since 1945

Victor Karády and Peter Tibor Nagy

Introduction

For a meaningful discussion of our topic, one cannot dispense with a short reminder of the sociopolitical and cultural settings. This part of the story is heavily marked by at least three paradoxes of the post-feudal modernization of Hungarian society since the late nineteenth century.

From the early sixteenth century, the Hungarian Kingdom was part of the Habsburg Empire, but governed by the legislation of its own, regularly voted on by the ruling nobility. The society was heavily fragmented by its denominational division as well as by ethnic multiplicity, to the extent that it was the only would-be nation state in Europe without ethnic or religious majorities. In the lower gentry—one of the largest in scale in Europe (some 4–5% of the population), besides those

V. Karády (✉)
Central European University, Budapest, Hungary

P. T. Nagy
John Wesley Theological College, Budapest, Hungary

C. Fleck et al. (eds.), *Shaping Human Science Disciplines*, Socio-Historical Studies of the Social and Human Sciences, https://doi.org/10.1007/978-3-319-92780-0_8

in Poland and Croatia—the ethnic balance showed 85% of Magyars against only 34% in the titled aristocracy (Janos 1982: 19). In the rank and file population, Magyar speakers represented less than 40% in the early nineteenth century (Katus 2012: 397).

The second paradox concerns the educational provision and its quite remarkable achievements in this otherwise underdeveloped country. After the 1867 Compromise with Austria, when the Hungarian state gained full autonomy for its internal affairs within the so-called Dual Monarchy, a vast program of development of the educational system was undertaken. This resulted in the growth of the number of primary schools by some 21% (from 13,643 to 16,510)—which lagged behind the increase of the population (cc. 35%), while the stock of secondary schools leading to elite training more than doubled, increasing by 120% (from 103 to 227). In 1872, a new university in Kolozsvár/Cluj was added to the previously unique university of Budapest (founded as a provincial Jesuit institution in the seventeenth and nationalized in the eighteenth centuries). Two others were added in 1913 (in Pozsony/Bratislava and Debrecen) to the academic network, which also comprized a Polytechnic University in Budapest and a large number of vocational academies and colleges; altogether 64 secular institutions of higher education by World War I. Hungarians could and often did take higher education degrees abroad, mostly in nearby institutions of advanced learning of the Dual Monarchy, Germany or Switzerland. As a consequence, the production of men endowed with higher educational credentials—with some 1.4–1.5% of the young age groups concerned—reached by the early twentieth century Western European standards, comparable to that of France or Germany.

The third paradox is concerned with the decisively Western orientation of the emerging creative intelligentsia in the country. In the classical gymnasiums, German was the third or fourth most important subject (following Latin, Hungarian and Math), while in *Realschulen* (representing the 'modern' track) German and French shared similar positions around 1900 (Mészáros 1988: 103). Those of Jewish and German background often made up a qualified majority of Hungarian students both abroad and inside the country, strengthening the

West-oriented intellectual dispositions of the new (non-noble) sectors of the upcoming educated elites.[1]

These preliminary observations refer to basic characteristics of the social circles that gave rise to the first institutional initiatives for the professionalization of the social sciences. By the late nineteenth century, the sizable intelligentsia in the country followed intellectual developments in the rest of Europe closely having adopted—thanks to language skills and studies abroad, international congresses, and transnational academic contacts—the intellectual standards of their Western counterparts. Their social fragmentation in terms of ethnic origin, religious background and educational mobility was strongly related to scholarly dispositions to engage in new disciplinary tracks and develop critical stances to classical types of study and inherited cognitive assets.

The Emergence of Social Sciences

The social sciences and modern humanities (henceforth SSH) started to be organized in the country under three types of institutional umbrellas.

Following earlier initiatives going back to the *Vormärz* (the 'Reform era' of Hungarian nation-building, before 1848), a small number of national scholarly institutions were founded, notably for the production of knowledge about Hungarian society, civilization and language in the past and the present, besides other, more classical intellectual pursuits. The Hungarian Academy of Science (henceforth HAS) was brought to life in 1825, thanks to the efforts of enlightened aristocrats (preceding its equivalent in Vienna by some 20 years). The Academy's six initial classes were reduced to three in 1869, the earlier 'historical' class being henceforth dedicated to the study of Hungarian 'society and

[1]Though a global assessment of the ethnic composition of the student body is difficult due to their dispersion in some 60 odd institutions of higher learning concerned (outside theologies), for the main training centers data are available with a good approximation as follows. Around 1900 among students born or residents in Hungary those of Jewish and German background represented 50% in the University of Budapest, 62% in the Budapest Polytechnic, 65% in institutions of higher education in Vienna and as much as 67% in those of Germany. See Karády (2012).

history,' while the 'legal' class included scholars of the constitutional system and contemporary 'state sciences' as taught in Law faculties and colleges. The social statistician Elek Fényes was elected member of the HAS as early as 1837, after having published the first volume of his comprehensive report on the multiethnic population. Fényes became the founding director of the first national statistical service set up by the revolutionary government in 1848. This embryonic institution served as the prototype of the two major sources of the production of high quality demographic, economic, educational and social-statistical data after 1867; the Central and the Budapest statistical bureaux. Later, in the interwar years, they were instrumental in the institutional professionalization of statistics with ramifications in demography, educational science, economics, human geography, political science or sociology.

The HAS was later also eager to admit statisticians of different sorts as well as protagonists of the classical social sciences endowed with university chairs by the outgoing nineteenth century, especially when their development contributed to the nation-building process. Cartography, geography, social and political history or ethnology were cases in point, just like 'public economics.' They could be considered 'national' disciplines on several counts. Increasingly they produced knowledge in Hungarian (thus ending the hitherto dominant position of Latin and German) and thereby created the relevant national vocabularies of erudition in Hungarian. Generally, they formulated their findings in harmony with the political interests of the Magyar ruling elite. The prestige of the Academy and university positions helped mobilize new clienteles for these disciplines mostly, if not exclusively, from educated sectors of the nationalized (or 'assimilated') German, Slav and, above all, Jewish middle class.

Thus, in the same year as the conclusion of the political Compromise with Vienna (1867), a *Hungarian Historical Association* was founded, not long before the *Hungarian Geographic Society* (1872). These were followed by a set of similarly discipline-based learned gatherings for philosophy (1882), ethnography (1889), pedagogy (1891), economics (1894), experimental psychology (1889–1902), literary studies (1911), and psychoanalysis (1913). In the interwar years, these served as a model for a Statistical Society (1923), and a Psychological Association (1927). Most were equipped with specialized journals as well as book series. Some of them earned significant subsidies from the HAS or the government proper. Their activities comprised publications, public debates, lectures, conferences, mostly representative of contemporary mainstream scholarship, but in pursuit of a nationalist agenda in their respective fields. Their leaders were often integrated in the staff of universities and other 'national' institutions (museums, libraries, archives, etc.), and/or associated with the Academy in Budapest. Rank and file members of the other associations came from circles of educated amateurs, such as secondary school teachers, civil servants, free professionals (medical doctors, lawyers) or the clergy.

In the early twentieth century, however, new types of learned societies emerged with ideological or professional agendas of their own, adopting a critical stance *vis-à-vis* contemporary social conditions and guided by different projects of social, political and cultural therapy and/or reform. One could cite various intellectual initiatives, some of them rather conservative—like St Stephen's Academy, a gathering of Catholic academics (1915)—but it was not connected to any 'properly' professional project. Two other scholarly associations, however, have long-standing intellectual heritage that is still effective.

The main initiative of this sort was the first 'workshop of Hungarian sociology.' It derived from intellectuals of the liberal camp strongly connected to Freemasons' lodges. They brought about a *Society for Social Science* in 1901 and later the *Galileo Circle* (1908) as its radical avant-garde, following the secession of the more moderate and nationalist wing (1906) from the initial group (Litván and Szűcs 1973). The membership of the remaining *Society for Social Science* consisted essentially of Magyarized Jewish students and professionals (up to 81%) with a relative majority (40%) active in medicine and in law[2] (27%), while the moderate nationalist sector's membership originated almost exclusively from Christian milieus and various strata of employed

[2]Estimation based on a prosopographical survey of members of the radical wing, the Galileo Circle, for the years 1912–1915.

intellectuals, mostly jurists in the provinces.[3] Though no one among the members could boast of a specialized training in sociology, social history, demography, etc., which was not available at that time in the Hungarian academe. Both associations developed scholarly agency in a vast array of studies, books and other publications in their journals, especially in the vastly circulated *Huszadik század* (Twentieth Century) (1900) and belonged to the radical wing, touching upon the main social issues of the time. These included, among other things, rural misery on feudal properties, urban pauperism, industrial relations under early capitalism, maintenance of gentry prevalence in the administration, Magyar hegemony in the multiethnic state, crass political inequalities due to suffrage restrictions. In 1918, a special issue was dedicated to 'the Jewish Question' with several dozen contributions (Hanák 1984).

The second important initiative was much more professionally organized. It was the outcome of the important local psychoanalytical movement, rooted early on in the urban middle classes of Jewish background, like elsewhere in Central Europe (Erős 2016: 51–52). The *Hungarian Psychoanalytical Society* (1913) was headed by Sándor Ferenczi, a close associate of Freud. A privately organized and selective agency, its set-up was, from the outset, grounded on the association of specialists only. This did not prevent it exerting influence in all kinds of new artistic and scholarly circles of Budapest and bigger cities. Its impact is well documented in the central organ of modern Hungarian art and literature *Nyugat* (West—1908–1941).

Though differing in their professional targets, the two associations had a lot in common. Both remained extramural in academia, unlike other learned societies. They represented new-born scholarly ventures at odds with established ways of thinking and investigation. They were united in their dual opposition to the Catholic Church (hostile to secularization and liberalism) and the dominant pre-1918 conservative-liberal nationalism represented in government circles (allied with the Jewish high bourgeoisie). All this prepared many of their

[3]With 66% jurists and only 3% of Jews, according to estimations. Vasvári (2007): 101–102.

activists for active participation in the two revolutionary regimes after the loss of the war: the liberal experience, marking the end of the Dual Monarchy and the historic kingdom (October 1918–March 21, 1919) and the Bolshevik one (March 21–August 1, 1919). In early 1919, due to a successful petition filed by students of the Faculty of Arts, Ferenczi was appointed, temporarily, to a chair of psychoanalysis at the University of Budapest (Erős 2016: 61). With the ensuing counter-revolutionary backlash of the extreme right, starting with the murderously anti-Semitic White Terror in 1919–1920, many outstanding *Galileans* engaged in social studies – e.g. Arnold Hauser (1892–1978), Oszkár Jászi (1875–1957), Georg Lukács (1885–1971), Karl Mannheim (1893–1947), and Karl Polányi (1886–1964) – were forced to emigrate. Bódog Somló (1873–1920), the first appointed university professor to lecture on contemporary sociology (in Kolozsvár/Cluj), committed suicide (Johnston 1983: 178). Apart from protagonists of the political Left, during the White Terror and the ensuing anti-Semitic regime many Jewish intellectuals fled the country under duress. Out of 18 members of the *Psychoanalytical Society*, only 6 completed their career in Hungary (Erős 2016: 62). With the numerous émigrés, the newly emerging critical social and human sciences lost their best minds (Frank 2009).

The change of regime in 1919 started with the national disaster of the dismantlement of the Dual Monarchy due to the Trianon Peace Treaty (1920). This left Hungary with some two-fifths of its former territory and population, surrounded by hostile new states of the 'Petite Entente' and governed by an openly anti-Semitic and antiliberal new elite. It was supported by the Christian Churches and the propertied classes including (initially) the Jewish high bourgeoisie (with many apostates). Universities and public institutions were purged of liberals and Jews, all 'revolutionary' appointments were abolished, the Freemasonry and the *Society of Social Science* outlawed and the remaining social scientific organizations were watched over closely. After the introduction of the infamous *numerus clausus* law in universities (1920—the first anti-Jewish legal act in modern Europe—Nagy and Karády 2012) more balanced policies were implemented under the 'consolidation government' of count Bethlen (1921–1931).

The self-proclaimed 'Christian regime' embarked upon a program of cautious modernization to strengthen its social bases and compensate with a program of 'cultural superiority' for the dismemberment of the erstwhile 'great Hungary.' This included an embryonic land reform (partly at the expense of Jewish owners), a *modus vivendi* with the social democratic party (whose activities were limited to big cities), implementation of universal male suffrage, but without secrecy outside the capital (securing the stable rule of government majorities), promotion of literacy thanks to primary school constructions, support for elite training inside the country (with four universities and a set of new secondary schools) and abroad (creation of research centers in Berlin, Rome and Vienna, filled with holders of state scholarships), as well as the extension of publicly sponsored medical insurance programs.

All this was accompanied by several parallel movements in the field of the social sciences.

The first event that must be mentioned is the further professionalization of economics, particularly due to the foundation of an Economic Faculty in Budapest (1920). Thus, for the first (and only) time, a social discipline achieved fully fledged institutional independence and canonization, securing its educational reproduction via special university curricula and degrees. In communist times the faculty would be developed (after 1948) into a full-scale university. As a follow-up, connected to publicly employed or freelance statisticians a *Hungarian Institute for Economic Research* was founded privately (1928–1949) publishing quarterly reports on the state of the country (Ungváry 1913: 291–296).

The reorganization of the statisticians' profession (including demographers) should also be noted. In 1922, the *Hungarian Statistical Society* was born, beginning the publication of its specialized journal both in Hungarian and French. This led to a remarkable production of new social data on educational inequalities, estimations of current national income and fortune, tax returns by various criteria, social stratification, unemployment, women active in economic markets, trends of depopulation, among others, as well as the first ever countrywide survey of professional mobility in the 1930 census.

If other social disciplines could not boast comparable institutional success because of the 1919 halt and rupture, some further achievements on the road to professionalization can be registered. Through most of the period, several scholarly journals offered their pages to studies from various social sciences—even empirical ones—though few were specialized. The disciplinary journals and associations, founded in the late nineteenth and the early twentieth centuries, also continued to operate (except those of politically liberal or socialist orientation).

More importantly, some upcoming core social sciences obtained new appointments in the Arts Faculties. Experimental psychology was taught continuously from 1918 in the Budapest University, where Leopold Szondi (1893–1986), the initiator of 'fate analysis,' was offered a teaching and research position (1927–1941), until he was dismissed as a Jew. Another chair in psychology was founded at the University of Szeged (1929), where a first chair in 'Hungarian folklore' was also put up in the same year, preceding another in Budapest (1934). For sociology, a lectureship in Pécs was created at the end of the period (1941) for István Weisz (1889–1973), a professional civil servant and sharp minded observer of contemporary realities. At the Arts Faculty of Budapest, the first chair in sociology (1942) was also founded for a conservative social philosopher István Dékány (1886–1965). Last, but not least, Béla Kovrig (1900–1962) a Catholic sociologist and prominent promoter of the Hungarian social security system, was appointed professor—later dean—at the Faculty of Law of the transitorily re-Magyarized University of Kolozsvár/Cluj (1942–1944).

The most significant institutional development at the end of the period remains the *Teleki Institute* (1941), named after the known geographer and rightist-conservative politician Paul Teleki. It was formed by the government out of three formerly established state agencies, the *Institute of State Science* (1926), the *Research Institute of the Landscape and the People* (1938) and the *Transylvanian Scientific Institute*, set up after the reannexation of Northern Transylvania by Hungarian troops thanks to the infamous 'Vienna Dictate' passed by Hitler and Mussolini (August 30, 1940). Visibly, the last governments of the neoconservative and authoritarian old regime undertook on a restricted scale the project of developing social scientific studies in their own key.

But the most decisive steps in the field of actual social research were made outside academia, thanks to both politically left and right activists of the so-called 'populist' movement; both were equally critical of the ruling political establishment. A number of largely circulated, and much discussed, 'sociographic' studies appeared without any institutional support, focusing on rural pauperism and malnutrition, high infant mortality, ethnic inequalities and discrimination, conflicts and violence in village life, feudal type relations of power and authority in latifundia, the spread of 'nativist' sects at the expense of historic Churches in the rural proletariat. Some of these studies, among them several masterly works of participant anthropological observation, also contributed to enhance and exacerbate the division in the intellectual opposition of the regime between 'populists' (often of provincial and Christian lower-class background) and 'urbanists' (with mostly Jewish middle-class ties from Budapest and bigger cities). A few of the 'populists' joined the mounting tide of anti-Semitic campaigns of the 1930s and endorsed the process of nazification that ensued. The imprint of this division can still be traced today in the fragmentation of various sectors of the creative intelligentsia in the country.

Indeed after 1938—the date of the first anti-Jewish law (after the 1920 *numerus clausus*)—the regime entered into a decisive phase of radicalization in alliance with the *Third Reich*, accompanying it in the disastrous (and, for Hungary, totally pointless) adventure of the war against Soviet Russia. This led to enormous losses in human life (close to 10% of the national population) and infrastructure, not to speak of the historically unprecedented moral decay and havoc generated by the transitory Nazi rule in 1944–1945. The *Shoah* in Hungary and the heavy destruction of intellectual capital inflicted considerable losses to the burgeoning extramural social sciences too.

The Regime Change and the Stalinist Disruption of Classical Social Studies (1945–1960)

The catastrophic collapse of the old regime as the last Nazi satellite brought about a transitory multiparty democracy under the close surveillance of the Red Army, allowing a rather quick economic recovery.

In spite of, or in part thanks to, purges of the politically compromised academic staff, a number of new academic positions, journals and research centers were quickly put up or developed. Among these was the renewed Teleki Institute (until 1948), a new chair of sociology—conferred to Sándor Szalai (1912–1983), a prominent leftist social democrat—, the refurbishment of the Economic Faculty, an *Institute of Popular Culture* (1946–1948), an Institute for public opinion studies attached to the services of the state radio (1945–1949), a *National Institute of Educational Science* (1948–1950). All of the latter were in Budapest, to which a few others may be added (such as a lectureship in the history of statistics at the University of Szeged). Most of these had a short period of prosperity, which provided the opportunity to initiate the elimination of the heavy ideological heritage of the old regime: chauvinist historical memory (directed against neighbor states); anti-Semitism; the so-called 'urbanist' versus 'populist' controversy; the nationalist narcissism of the ('neo-baroque') 'Christian regime' (Szekfű 1934/1989); and the unrealistic policy of territorial 'revisionism' (the 'Trianon syndrome'). Some prominent authors—such as the historian Gyula Szekfű (1883–1955) on pitfalls of expansionist nationalism or the political scientist István Bíbó (1910–1978) on collective responsibility for anti-Jewish persecutions (Bíbó 1994)—must be credited with essential contributions to this 'intellectual revolution' of sorts, a short-lived harbinger of Western type development of the social sciences in the country.

The academic emancipation of women must be counted along the same lines as the major novelty of the democratic transition. After primary education, secondary schooling started to be coeducational. Half a century after the admission of women to universities (1895) all academic degrees were made accessible to women (even the earlier prohibited legal and technical studies) together with appointments in academia (from which they had been almost completely excluded in the old regime).

To this must be added a systematic policy of positive discrimination to promote qualified offspring of the lower classes. This was exemplified first in the publicly sponsored movement of 'popular colleges' hosting students of peasant and proletarian extraction. Since this was a grassroots organization with close to 10,000 students in its dorms at its apogee, after a while the communist authorities suppressed it formally due to its political independence (1949), while formalizing the program of preferential enrolment in higher education of lower-class secondary school graduates at the expense of middle-class youngsters in the shape of formally fixed admission quotas 'by social background.'

The related resurgence of the social sciences was short lived. In 1948, the 'Year of the Turn' quickly canonized in Communist historical memory, marking the seizure of exclusive power of the highly personalized communist leadership (the infamous 'Troika'), proved to be fatal for most social disciplines and their institutional infrastructure. Starting in 1948, the triumph of Stalinism equaled an all but complete State control and forced ideological '*Gleichschaltung*' of all publicly supported intellectual activities together with the prohibition of pursuing private ones. It also involved the suppression or curtailment of publications in all forms of classical social science. The publishing market, the scholarly press and the whole institutional structure followed suit under strict 'Party censorship,' forcing some top-level professionals to emigrate, silencing other, even depriving many of their livelihood (e.g. by withdrawing pensions) and—in the worst but not infrequent cases—sending them to prison (including the sociologists István Weisz or Sándor Szalai). The whole spectrum of the educational ladder was invaded by the regimes' propaganda brochures, the Soviet system serving as a mandatory model in scholarly communications. The academic staff appointed before 1945 were severely purged (Péteri 1998: 94).

Following the prohibition of privately organized cultural and social movements, agencies and publications—except those adopted and strictly controlled by communist authorities—most learned societies in social or para-social studies, regarded as 'ideologically suspicious,' were outlawed by 1950, together with their journals. For a varying number of years, the ban concerned the *Society for literary studies* (1948–1963), the *Philosophical society* (1949–1956), the *Psychological society* (1948–1961), the French edition of *Statistical review* as well as the *Statistical society* itself (both eliminated in 1949), the *Pedagogical society* (1950–1967), the *Geographical society* (1949–1959), the *Ethnographic society* (1948–1958)—with the exception of its journal *Ethnographia*—and even the *Economic society* (1950–1959), promoting a discipline carrying official favor with the communist regime. The long-established journal of the *Institute for Economic Research* was abolished to leave space for a *Hungarian-Soviet Economic Journal* (1948–1954). After 1948, for almost a decade, only skeletal statistical yearbooks in duly numbered copies, and with restricted circulation, were published (Péteri 1998: 135). The population data of the 1949 census remained largely unedited for several decades. Moreover, many of the hitherto commonly applied and most efficacious variables in the study of the population (religion, mother tongue, residence, professional stratification, etc.) were dropped or (as with denominational or ethnic diversity) tabooed by socialist statistical services.

The ban frequently concerned the disciplines themselves, stigmatized as 'alien to socialism.' This coincided with drastic political purges of university staffs as well as the HAS. The purges responded to the logic of forced elite change from above. It followed a complex scheme of constantly shifting power relations in the period of 1945–1949 at the expense of the transitory coalition parties and to the benefit of the communist leadership, which finally assumed absolute preponderance in 1948. A good part of the emerging scholarly elite (up to around 68–94% of doctoral students in the social and historical sciences in 1951–1955) hailed from communist Party staff (Péteri 1998, 179–180). From this point forward, party membership represented an important asset for joining the new academic establishment which, by 1950, comprised not less than 80 recently elected members of the Academy and another hundred university chairholders. Law schools and faculties were the big losers in the forceful recomposition of educated elites (Biró and Székelyi 1996). New promotions were made in Medicine, the Technical and the Science Faculties to the detriment of legal training, social studies and Western type humanities.

Still, each social science discipline had its own particular conditions of negotiating its fate under duress in the Stalinist regime. In principle, all social sciences were re-qualified as 'bourgeois sciences'—like sociology, political science, demography, psychoanalysis—and forbidden. Most of the time their messages could be referred to critically only in lectures on 'scientific socialism' or official 'Marxism-Leninism' (also called

'Dialectical materialism'), which became a mandatory academic subject in all tracks of higher education. This was part of the major Stalinist project of ideological reeducation in order to shape the 'new socialist man.' No branch of scholarship escaped the obligatory ritual to begin specialized publications with a homage paid to the classics of official Marxism, including Stalin, the quasi-sanctified 'Father of progressive humanity' (Vorisek 2011, 38–39). However, apart from 'political economy' of Stalinist style, some applied disciplines or those having close equivalents in Soviet Russia—pedagogy, folklore and psychology—could survive the political exile of the others in conformity to preset Stalinist patterns, especially if they could count on prestigious specialists with sufficient 'political capital' to protect them. Thus, folklore studies were maintained due to efforts of well established political 'fellow travelers' like Ferenc Erdei (1910–1970) and Gyula Ortutay (1910–1978). In 1950, even the publication of a multilingual journal—*Acta ethnographica*—was authorized for the discipline (Kuti 2011). The famous composer Zoltán Kodály (1882–1967), one of the cultural heroes expropriated by the regime, could organize a Folk Music Research Group (1949), which soon passed under the auspices of the HAS (1953).

Still, repressive as it was, Stalinism was instrumental in four types of innovations that ultimately served the rebirth and later development of several social sciences.

As low as the scholarly credibility of dogmatic 'Marxism' proved to be, it put in the focus of public opinion some major themes of modern social disciplines: social class systems; property and power relations; problems of economic development and modernization; historical continuities and discontinuities in the latter; the impact of mass movements and revolutions; interrelations between social and intellectual structures (culture as 'superstructure' grounded upon socioeconomic foundations), etc.

The new regime consented to considerable state investment in higher education such as the abolition of tuition fees, numerous student grants, upgrading of many training tracks, construction of new academic premises. This included the foundation (as in Miskolc in 1949) of new university campuses, the upgrading or the expansion or available institutions of higher learning together with the establishment of a number of new study tracks, mostly in applied sciences. All this produced an historically unprecedentedly large and ever-growing number of graduates, both men and women, and this student body contained within it would-be clients for training in the social sciences (when this became accessible two decades later). In 1937/38, 11.747 students were enrolled in higher education. By 1953/1954 student numbers had reached 53,330; a multiple of close to five.[4] This expansion was taking place not without significant intellectual counterselection. This was due to the drastically applied 'social discrimination' (both positive for lower-class offspring and negative for alleged 'class enemies') in the admission process, accompanied by politically biased training conditions under heavy censorship. Still, it brought about a large cluster of young scholars liable to be converted later into social scientists. Philosophers and specialists of '*tudszoc*' (sci-soc = as 'scientific socialism' was nicknamed) were notable cases in point.

Thirdly, the Stalinist era accomplished a fundamental reform of the market for creative scholarship by the parallel development of universities and academic research facilities, with new functions conferred to the Academy of Sciences—according to the Soviet model. After 1950, the HAS became the main authority that awarded scholarly degrees based on original research achievements, due to a new hierarchy of degrees and positions of its own: 'aspirants' under training; academic 'candidates'; 'academic (or 'big') doctors' and corresponding and full members of the Academy. To stimulate research activities, academic degree-holders received regular monthly payments. Political bias in the election of 'aspirants' and the choice of research subjects, as well as the complex relationship between university and 'academic' careers, was occasionally detrimental to scholarly progress. Nevertheless, even with ups and downs, this parallel structure proved to be a feasible one. It has maintained itself, with minor modifications (notably the progressive replacement of 'candidature' by PhDs since 1993), to the present.

Finally, the HAS was not only endowed with new entitlements as a degree-granting institution. It was also enriched progressively with

[4] *Magyar statisztikai évkönyv* [Hungarian statistical yearbook], 1949–1955, 332.

new endowments and competences, exemplified in a growing set of its research centers offering life-long careers. Such disciplinary study centers were founded for philosophy, history, linguistics, geography and law as early as 1951. Parallel to this, public administrations also started to organize research institutions of their own to procure expert knowledge in their fields of decision making. An *Institute for the study of the workers' movement* had already been established in 1948, to later become an *Institute for Party history* (1957). Similar institutions were created for research on problems of education, economics, cooperative agriculture and mass culture. All this generated a quasi-new category of intellectuals; researchers with full-time employment in specialized research institutions. By 1953, one counts 81 of the latter, with a scholarly staff of 2416. By 1959, this number grew to 6669 out of which 425 were even classified as social scientists, including 108 women. The HAS alone had 33 such institutions by that time.[5]

From Deinstitutionalization and Repression to Reinstitutionalization and State Controlled Development (cc. 1960–1989)

The 'thaw period' after Stalin's death (1953) and the 1956 national uprising and anti-communist revolution gave rise to 'revisionist' state policies, often qualified as 'soft dictatorship.' Parallel to the ghastly repression of proven involvements in 'the October events,' there was an immediate alleviation of former Stalinist restrictions. International relationships were eased, legal emigration (at least to Israel) allowed (in spite of the outflow of over 150,000 refugees to the West in the very months after the uprising), victims of trumped-up political trials were rehabilitated, police practices started to follow legally prescribed procedures. The lifting of the ban on Western type social sciences (though only outside universities up to the late 1960s) was part of this process of relative

[5] *Magyar statisztikai évkönyv* [Hungarian statistical yearbook], 1959, 329.

liberalization of the otherwise reaffirmed one-party rule. The transition was facilitated by both the emigration of many would-be scholars and the political demise of the Stalinist ruling staff. Many of the latter were in search of career changes in the direction of intellectual disciplines endowed with scholarly legitimacy and liable to free them from their compromising ideological loyalties.

Hungarian communism, in spite of its formal alignment with Soviet political positions, continued to maintain a cautiously moderate course in matters ideological, educational, cultural and academic, but all of this came with several reservations. On the one hand, more than lip service was paid to ideological conformism—censorship was not fully lifted, self-declared political opponents were deprived of public employment and the right to publish, but only exceptionally condemned by tribunals. On the other, the post-1956 official course did its best to stress its positive distinction—indeed radical demarcation—from Stalinist policies. The admission of modern social science initiatives was part of the new trend. A price had to be paid; a level of ideological conformism. This forced several first-class scholars to temporary emigration in the late 1970s, like the philosopher Ágnes Heller (1929–), the sociologists Iván Szelényi (1938–) or István Kemény (1925–2008). Thus the situation of the social sciences, most particularly those liable to challenge the taboos, fictions and lies of doctrinaire Marxism–Leninism—sociology, political science, philosophy, social and political history—did not cease to be ambiguous till 1989 (Vorisek 2011, 49). Though sponsored by the regime, most staff members of the reemerging critical social sciences remained (or became) opposed—with a large variety of arguments—to 'real socialism.' The novelty of the situation after 1956 was that this was integrated in the very dynamics of the readmission of modern social sciences with a measure of effective tolerance.

A paradigmatic case in point can be identified in the career of the dismissed, and politically disappointed, Stalinist prime minister András Hegedüs (1922–1999), who had been entrusted with the organization of a sociological study group (1963) of the HAS, which would later become a research institute proper. The political credentials of Hegedüs served here as a guarantee for the expected level of 'ideological conformism,' at least for a time. He was dismissed in 1968, due to his disagreement with the crushing of the 'Prague Spring' by the armies of the Warsaw Pact (Huszár 2015: 284). Among similar new foundations, special research groups were set up in demography (1963), pedagogy (1967) and folklore (1967) after the post-1956 relaxation of political pressures marked by the amnesty of political prisoners in 1963 and 1964. Some prominent victims of the repression were set free at this time, such as the famous political scientist István Bibó, former minister of the revolutionary coalition government in October 1956. The ensuing three or so socialist decades constitute a period of 'refoundation' for the SSH, when the main disciplines involved could achieve a measure of professionalization and (re)institutionalization under state sponsorship. In 1959, there were only 425 social science fellows officially employed in research centers (a mere 9.4% of the total in research). Their numbers grew to 1352 by 1970 (18.6%) and as many as 5990 by 1990 (20% of the total number of researchers).[6]

Psychoanalysis is a rather singular, almost borderline, case in this respect. Its professional practice, pursued though in a quasi-clandestine manner in the 1950s, never actually ceased, nor did the training of practitioners. The stages of its reinstitutionalization started in the 1970s when the international psychoanalytical association admitted several Hungarian members on an individual basis. In 1980, a *Hungarian Psychiatric Society* was officially recognized with a therapeutic section made up by psychoanalysts. In this change, the action of prestigious specialists—among them Ferenc Mérei (1909–1986), another former political prisoner after 1956—proved to be decisive. But a new *Psychoanalytical Association* was not set up in Hungary before the end of the socialist era in 1989.

An interesting transformation occurred in the vast network of 'philosophical' chairs and institutes teaching official Marxism by a staff often trained in Soviet Russia. In the years when 'Marxism' remained mandatory in the curriculum of higher educational institutions, some of them started to look for career alternatives in modern social science research. The *Institute of the History of the Party* (1948), founded as a

[6]Data from the *Magyar statisztikai évkönyvek* [Hungarian statistical yearbooks] of relevant years.

classic instrument of communist propaganda, was renamed *Institute of Political History* and moved toward the support of realistic historical, social and even political investigations (while maintaining basic taboos imposed on public reflection: relationship with the Soviet Union; 1956; Jews and anti-Semitism, etc.). Following earlier initiatives (1963) of the state radio and television agency, a *Research Institute of Mass Communication* (1969) was founded to promote public opinion polls. Another contemporary foundation of the Political Bureau of the Communist Party was the *Institute of Social Science* (1966), which started to commission significant research projects on important public issues. After 1989, it would be taken over by the HAS as its modern *Political Science Research Institute.* Training in the same spirit was the charge of the *Political Academy* (1968), another Party foundation, which acted always as a reform institution.

More or less moderate versions of 'political revisionism' were implicitly built into the working of these new institutions from the outset. Studies emanating from them regularly confronted increasingly objectively official Marxist statements with Western achievements and experiences in the social sciences. Indeed, the translation of major Western publications of the sort—including political science—were regularly commissioned by the Party hierarchy and distributed confidentially among members of the communist *nomenklatura* (and their relations), without being officially published. Thus, many would-be social scientists and interested intellectuals could get first-hand information about mainstream currents of their disciplines in the West. Paradoxically, even in the last phase of the one-party regime, this was conducive to the foundation of a grassroots organization that promoted studies in modern political science: the *Hungarian Political Science Association* (1982). Introductory training in this 'politically sensitive' field was soon taken up by a proto-department, of sorts, at the Budapest University (1984) (Szabó 2002). Courses were organized, but there was not a full study cycle culminating in a degree. It is understandable that publicly canonized Marxism-Leninism could not permit such an intellectual competition. However, this initiative served as the matrix for the future academic reception of the discipline following the fall of the regime. Modern political science in Hungary has its own journal only since 1992.

In the years of the post-1956 political thaw, new publicly supported publications also started to report on recent developments in Western social sciences. The 'generalist' monthly *Valóság* ('Reality,' 1958) was the first journal—widely circulated among the educated public—that regularly published research reports on initiatives in the social sciences as well as reviews of important Western innovations in these fields. *Világosság* ('Light,' 1960)—a monthly journal for matters religious, philosophical and ideological—followed suit and was equally important. Much later *Medvetánc* ('Bear dance', 1981–1989), which was regarded as a tolerated, semi-dissident journal, contributed to breaking official taboos in all matters, challenging consensual dogmas and going as far as to give voice to emigrant scholars in the West whom it was otherwise forbidden to publish in the country. In the 1980s, political dissidents also produced several *samizdat* journals—among them the *Beszélő* ('Visiting hours in prison,' 1981)—bringing out original or translated studies on political and social science issues that were excluded from the official press. All of these organs played a role in the normalization of the position of Western-oriented social sciences in a country where, officially, Soviet-type Marxism remained dominant. When publications were censored by intellectual watchdogs of the regime, Western publishers in Hungarian (such as the *Cahiers hongrois* in Paris) helped to bring out—and smuggle back to the country—the works of victims of political ostracism. Such was the case for many prominent 'dissident' social scientists, who produced an influential discourse of applied political science on the true nature of 'real socialism' in privately organized 'flying university' lectures in Budapest.

In one way or another, most social sciences thus acceded to a level of autonomy conducive to their reinstitutionalization and professionalization in the second, post-1956 phase of communism, while their scholarly agency remained under official ideological censorship.

1968 was a year of a significant economic reform in Hungary, the preparation of which goes back to 1965 in the wake of political changes of the post-Khrushchev era in the Soviet Union. The proposition was to combine a measure of continued State governed economy with elements of a market system, including petty private entrepreneurship. Although the reform itself was later officially canceled under post-Stalinist counterattacks, in Hungary it brought about a new political course that led, ultimately, to the liberalization of the regime in the 1980s and its negotiated fall in 1989 (Tőkés 1996). Even before this, the political thaw combined with reform efforts of the economic planning system enhanced the importance of training economists and other social scientists. An official *Institute of Economic Research* (1966–1992), designed to secure scholarly support for economic planning, was one of the early outcomes of this movement. Internationally recognized critics of the socialist economic regime—such as János Kornai (1928–) or Tibor Liska (1925–1994)—were hosted either in this Institute for part of their career, or in the Budapest Economic University. In late socialist times, Budapest Economic University became a prestigious central institution for all kinds of social research, with similar training schemes begun later in provincial universities, such as Miskolc whose scheme dates to 1987.

The fate of sociology deserves particular attention in this respect (Némedi and Róbert 2002). The creation of a special training course in 1969, in form of evening classes in sociology in the Arts Faculty of the University of Budapest by Tibor Huszár (1930–), an earlier member of the communist establishment, must be regarded as a breakthrough in this sense. Interestingly, one can find a trace of the change dating back to 1967 in the official *Statistical yearbook* of that year which, for the first time, names 'sociology and statistics' in its thematic categories that describe book production of the year, indicating the appearance of two such books.[7] Two years later the same happened in the roster of categories presenting periodicals—signaling five journals with an average circulation of 15,600.[8] Between 1970 and 1980, both the number of journals in the category 'sociology, statistics' doubled, journals from five to ten and copies printed from 92,000 to 161,000. The growth was even more spectacular for copies printed in the category 'philosophy, psychology,' going from 16,000 to 110,000.[9] All this might be attributed to a special Party decision aimed to secure international

[7] *Magyar statisztikai évkönyv* [Hungarian statistical yearbook], 1967, 379.
[8] *Magyar statisztikai évkönyv* [Hungarian statistical yearbook], 1969, 464.
[9] *Magyar statisztikai évkönyv* [Hungarian statistical yearbook], 1980, 458.

recognition of a 'cultural modernity' of sorts for the post-Stalinist regime, though not independently from the fact that sociology, along with other social disciplines, were readmitted in academe in the Soviet Union at the same time. The course was later transformed into a regular professional training program, to which others were added in the 1980s. In 1984, Rudolf Andorka (1931–1997)—a forced labor serviceman in the 1950s and victim of the post-1956 repression—was appointed chair of the Sociology Department of the Economic University. At the end of the period, all full universities were endowed with sociological programs and the first specialized journal—*Szociológiai szemle* (Sociological observer)—was published in 1985 by the Sociology department of the University of Budapest. All HAS research groups—in sociology, demography, ethnology, etc.—became fairly well-funded research institutes, added to others attached to various ministries.

Despite the various ups and downs over time, state sponsoring allowed the funding of several highly original collective scholarly ventures in charge of the new research centers, sometimes even in contradiction to doctrines of official 'Marxism.' Such Marxism would not allow poverty be seen to exist in state socialism, but around 1970 István Kemény (another political prisoner after 1956) was commissioned to carry out large-scale surveys on 'economically handicapped groups,' especially Gypsies. Socialist society was supposed to be egalitarian, but several projects were funded for the empirical study of inequalities of education—by Zsuzsa Ferge—socio-professional stratification and mobility—Ferge, Andorka—industrial leaders—Kemény—housing—Szelényi—and birth control and family life—László Cseh-Szombathy—among others. Andorka published a countrywide study of occupational mobility in the pre-socialist regime based on the data of the 1930 census (a unique document of its kind in Europe). Problems of stratification and mobility received large-scale scholarly attention in population censuses, with ensuing statistical reports, which secured a prominent place for Hungarian sociologists on the world map of this research orientation.

From the 1970s onwards, a new scheme of funding via centrally fixed 'main scholarly directions' offered research institutes the opportunity to develop new research programs. Political-ideological control of such projects appeared to be increasingly relaxed over time. In the outgoing period of the socialist era, even the formerly enforced exclusiveness of state institutions to carry out research in matters economic and social was tacitly abandoned, which brought forward the first privately funded and market-oriented research agencies. This was the case for *TÁRKI* (1985), which became a major institutional actor of the study of contemporary Hungarian society. Another state-run institution, the *Economic Research Institute* started a new market-oriented career, only to then convert itself into a private enterprise after the regime change (1992). The same happened to another notorious research center launched by the government under the 1968 movement for economic reform, the *Institute for Financial Studies* (1968). It was privatized as early as 1987, only to become as a *Financial Research Centre Ltd.* a major nonacademic scholarly entrepreneur in social studies (Csaba 2002).

The Postcommunist Expansion and Professional 'Normalization'

The smooth regime change in 1989 eliminated all previous obstacles to professionalization for the already active, but often under-institutionalized, social sciences. But the consequences for their further development were mixed.

Before 1989 most, although certainly not all, actors within the renovated social disciplines could be counted among active or passive members or sympathizers of opposition circles. After, however, many of them were drawn into positions of political or administrative responsibility, far away from their scholarly commitments. At the same time, intellectual debates about the social changes that the emerging generation experienced made social studies more popular than ever.

The response to this growing interest was the perfection of the hitherto incomplete professional self-organization of the disciplines concerned, the extension of academic teaching and research facilities, the partial reshuffling of the training system and the reorganization of research funding.

Thanks to the disappearance of prefixed ideologically motivated constraints and restrictions, the 1990s were a refoundation period for independent professional organizations in most branches of the social sciences, including the liberation from political influences and bureaucratic hindrances and the passage to free self-management and development. Democracy and free enterprise brought about the blossoming of institutional innovations, initiatives and creativity in many areas, including scholarly markets. New journals came to life, competing with each other in every discipline. The boom of new publications had already started before the regime change. *Replika*, a new interdisciplinary social science journal appeared in 1990, followed by a number of disciplinary publications of various intellectual gatherings and schools. In 1980, there were 264 periodical titles. By 1990, 361 titles were published in the core social disciplines (as cited in official statistics).[10] Summer schools, topical conferences, study groups and project-based networks were multiplied, as well as subsections of professional associations. International relations in the form of collaborations, mutual invitations, study trips, temporary or definitive expatriation as staff members in foreign universities went from being extraordinary opportunities to ordinary in career building (Karády and Nagy 2018). The expansion of study options, and the growing demand for research, apparently favored the social sciences more than other disciplines. From 1990 to 2010 the staff in the statistical category 'research and development' increased from 30,256 to 35,700, within which the share of social science scholars swelled from 20% to 28%.[11] This is a clear indication of the substantial enhancement of the weight of the social disciplines within research markets in the country.

The teaching of various new currents of the social sciences was now openly promoted in Hungarian universities. Moreover, the liberation from communism triggered the general expansion of both demand and supply of advanced studies, the social sciences included. There was a statistically demonstrable 'boom' of the student population in economics

[10] *Magyar statisztikai évkönyv* [Hungarian statistical yearbook], 1990, 286.

[11] Ibid., 1990, 276 and ibid., 2010, 222.

and law, but also—though more modest—in most other social science disciplines (Csaba 2002, 86). After Miskolc, all provincial universities opened economics faculties or departments, while the number of students in law faculties—limited under communist times—experienced steady growth. New types of study units—notably private ones—were added to the existing universities and vocational colleges. Several business schools—both publicly and privately funded—started to provide executive training so that the post-socialist economy could be open to private entrepreneurship.

A special case must be made here of the *Central European University* (CEU) in Budapest, founded by the philanthropist George Soros in 1991; a postgraduate campus with accreditation in the USA and entirely dedicated to studies in the SSH. It is the only fully international university in the country by the recruitment of both its staff and students. It was integrated into the local higher educational network by a law, voted on in the Hungarian parliament, in 2006. Over time, the CEU appears to have become a major research university, by far the most successful academic unit to attract European research funds compared to any other institution of higher education in any formerly Sovietized society.

The regime of higher education underwent significant changes, the biggest of which was the adoption of the Bologna system of the organization of bachelor's, master's and PhD degrees (1993). Though Hungarian universities continued to grant until 1949/50, and reestablished in 1959 the classical 'small' doctorates inherited from the old regime, this degree was not prepared via specialized training. Hence it was not much in demand, being regarded as a purely decorative distinction without promotional value in scholarly professions, as opposed to 'academic degrees' (like 'candidacy').[12] The new Western type PhD at first tended to compete with the 'academic candidacy,' replacing it altogether by 2003. This entailed the setting up of a profusion of doctoral schools with their own staff, conducive to the birth of numerous

[12]This is why several highly qualified 'old regime doctorates' of established academics were officially converted into 'academic candidate' degrees after 1989 without any further procedure.

new specializations (Karády 2016). The number of specialties (chairs or departments) in the faculties of Arts and Law in Hungary grew from a mere 65 in 1950 to 143 in 1990 and as much as 215 in 2000, and has continued to increase since then.[13] Table 8.1 shows an overview of the distribution of doctoral schools' staff, together with other scholarly clusters, by discipline as of 2010.

Institutionalized funding also continued to be developed in various ways, following Western models. The regime of applications for targeted or proposed funding offers became the rule. Hungarian candidates could compete, or enter into collaboration, with their European counterparts for research support. The HAS, local, national and international administrations launched increasingly generous and diversified funding schemes, especially after 2004 when the country joined the European Union. The role of the 'OTKA' (*Országos Tudományos Kutatási Alapprogramok* (basic national programs for scientific research), was established in 1986 to fund research projects as a main state program. It can be more closely analyzed using the list of its beneficiaries together with the topics of their various projects. This was the first state-run funding scheme of scholarly research open to individual or institutional applications in Hungary. OTKA offered a complement to the Soros Foundation, which had been operating since 1984, as a result of a deal between the HAS and the American-Hungarian philanthropist George Soros. The Soros funds served as an abundant source of scholarships for studies in the West for individuals until the 2000s, through which even well-known political dissidents—otherwise deprived even of passports—benefited from extended study trips in America or Western Europe in the late 1980s.

The simple list of projects financed by OTKA allows for important observations as to the institutional, residential and disciplinary allocation of funded research potential in the Hungarian social sciences in the postcommunist transition period (1994–2015),[14] especially if

[13]Data drawn from the prosopographical lists of scholars active in the fields of Arts and Social studies gathered in the framework of the Interco-SSH project.

[14]http://www.otka.hu/en.

Table 8.1 *OTKA* projects and qualified scholars by disciplines in the SSH after 1990

	1	2	3	4	5	6
	All OTKA projects 1994–2015	% of OTKA projects in Budapest 1994–2015	Distribution of OTKA projects 1994–2015	Distribution of staff of doctoral schools 2010	Distribution of academic candidates 2003	Distribution of academic doctors 2003
Philosophy	48	89.6	4.8	4.8 (153)	7.3 (266)	5.5 (40)
Literature	105	78.1	10.6	12.0 (382)	10.0 (365)	13.9 (101)
Ethnology, folklore	60	88.3	6.0	1.4 (45)	1.6 (58)	2.2 (16)
Psychology	39	100.0	3.9	4.7 (150)	3.7 (135)	3.0 (22)
Political science	39	84.6	3.9	1.4 (45)	2.4 (88)	1.2 (9)
Sociology	120	100.0	12.1	4.7 (150)	5.3 (193)	5.1 (37)
Demography	10	100.0	1.0	?	?	?
Economics	55	100.0	5.0	17.4 (556)	25.2 (920)	16.5 (120)
History	229	84.7	23.0	13.7 (438)	15.2 (555)	19.2 (140)
Law	13	100.0	1.2	9.0 (288)	6.7 (244)	9.6 (70)
Pedagogy, education	33	75.8	3.3	5.4 (173)	6.6 (241)	3.2 (23)
Other SSH disciplines	244	95.5	24.5	25.5 (814)	16.0 (584)	20.7 (151)
Altogether SSH disciplines		90.1	100.0	100.0	100.0	100.0
N	995		995	3194	3649	729
Unknown	107	86.9	107			1

comparedvto other social science practitioners. The databank exploited here comprises only those institutions (together with the projects funded) which were awarded at least five grants. Among 1778 such projects having received support, some 1102 could be classified as emanating from the SSH, though the discipline of one out of ten projects could not be unambiguously defined.

This snapshot of the disciplinary composition of the active staff of the social sciences in broadly the same period refers to the importance of project-based research and the prevalence of the capital city in obtaining research funding as well as the weight of doctoral staff and 'academic candidates' in the various disciplines under scrutiny.

The sheer number of OTKA funded projects already provides a summary indication of the rank order of disciplines following their liability to be funded. A full explanation of these figures cannot be attempted here. Such an endeavor should take into account the 'lobbying power' of different scholarly schools, the internal fragmentation of disciplines as to the share of scholars and other (managerial, therapeutic, etc.) practitioners, as well as the technical need of the former to be financed. Obviously, the funding requirements of classical legal studies are not comparable with the needs of large-scale surveys in political science or sociology. Research focused on societal issues in non-applied studies appears to have been funded more often than in other disciplines. History, sociology, political science and ethnology absorbed almost half (46%) of OTKA budgets, while constituting only 21% of the staff concerned. If economics benefited relatively rarely from OTKA, it could count on several other public and, more recently, private sources of funding.

OTKA funding benefited the institutional network located in the capital city preferentially. In several disciplines, only Budapest institutions obtained funding. If one observes, the list of the 91 institutions have received at least five grants, the first 32 are all in Budapest. Only 19 provincial institutions can be detected at the bottom of the list. Among the 20 institutions funded most often (with 16 projects or more for each), 14 were HAS research centers, 5 belonged to the University of Budapest and 1 was a scholarly publisher. The academic preponderance of the capital city can be globally characterized by two observations.

In 2010, Budapest hosted both the absolute majority of doctoral school staff and those obtaining OTKA funding. Despite this, the share of Budapest among established scholars followed the rise in the academic hierarchy. Members of the HAS and celebrities cited in encyclopedias appeared to be systematically concentrated in the capital.

Comparing columns 3, 4, 5 and 6 of Table 8.1, one can gain insight into the structural weight of different SSH disciplines in the three scholarly clusters analyzed for the last period under scrutiny.

The first observation concerns the relative intensity of funding in various disciplines. Manifestly some 'research intensive disciplines' stand out, such as ethnology, sociology, political science and, to some extent, history and demography, which all received much larger shares of OTKA grants than their proportion of all academic staff. In the opposite cluster we find law, educational science (pedagogy) and economics. The presence of economics in this group appears paradoxical, given the well-founded reputation of the economic discipline for requiring large-scale databanks, with the relevant investigations, for its progress. The explanation of this apparent anomaly lies in the fact that economic research could claim, and has always profited from, funding from other institutional sources (like national statistical services) or from private means (mobilized by economic actors and firms), all outside the scope of OTKA. Moreover, many 'academic candidates' or even members of doctoral schools in economics are engaged in practical management and advisory activities rather than in research proper.

Columns 4 and 5 of Table 8.1 help evaluate the present day 'power relations' among disciplines in terms of the quantified size of two relevant clusters, the staff of doctoral schools and 'academic candidates.' The two sets of figures are strongly correlated and even overlap to a large extent, although some differences can also be detected.

On top of this ladder in terms of size, economics prevails by far, followed by history and literary studies. The first is the big modern research discipline endowed with a special university, several faculties of its own, together with a number of recent training centers for management studies. The other two have belonged to the classical study tracks of the Arts Faculties since the eighteenth century, if not earlier. The difference in their share of doctoral school staff and 'candidates' in economics can be interpreted by the fact that many such 'candidates' are engaged in day-to-day economic functions and not in academia. Among other important disciplines, philosophy and educational science are also 'old' disciplines, as is law, which has always represented a professional study track with a faculty of its own since the dawn of the university system. Among the relatively new SSH, sociology appears to be the strongest, followed by psychology, political science and ethnology. But, on the whole, the weight of these 'new' tracks still remains modest at a mere 12–14% of available positions, much less than economics alone. As will be illustrated below, there was a breakthrough of sorts of the 'modern' SSH disciplines both in and following the last phase of the socialist regime, but they did not fundamentally disrupt the established hierarchy of disciplines.

Differences between the share of disciplines among academic 'candidates,' doctoral staff and 'academic doctors' could be interpreted in terms of the standing of the discipline in academe. When the proportion of 'doctors' is significantly higher than that of 'candidates' and doctoral staff—as is the case for history, literary studies and ethnology—one might suppose that the discipline has a strong position in universities, if only because appointments of professors were preferentially (if not mandatorily) made for laureates of the 'academic doctorate.' In the opposite case—for economics, pedagogy, political science or psychology—one might presume that a larger part of its personnel was active outside academe, where an advanced academic title was not necessary for professional self-assertion and promotion.

The Dynamics of the Institutionalization Process Over Time. Inequalities of Growth and Feminization of SSH Disciplines (1945–2015)

In conclusion, three fundamental features will be emphasized as marking the development of the SSH in Hungary.

First, aspects of the growth—perhaps best objectified by data on what could be qualified as a Hungarian 'state nobility' of research

(Bourdieu 1987)—are exemplified by the changing numbers and proportions by disciplines of laureates of the HAS (Table 8.2).

The absolute number of 'candidates' and 'doctors' (the penultimate lines of Tables 8.2 and 8.3) in 1962 can be estimated from 'researchers' in the official statistics (554 in 1959, as cited above). Thus, after modest beginnings in the 1950s, growth subsequently accelerated for both clusters. Following other estimations, one out of three or four 'candidates' achieved the prestigious degree of 'academic doctorate,' a stepping stone for nomination in university chairs or for further careers as members of the HAS. These relative proportions do not appear to have changed much over time. The growth of the two academic degrees is similar (Table 8.3).

The position of disciplines changed significantly over time, in spite of some continuities. Economics and history maintained almost permanent two digit figures in the share, making up around one-third of academic degree-holders during the decades under scrutiny. In a different key, these were the two master disciplines of the socialist age

Table 8.2 'Academic candidates' at various dates by social science disciplines (1962–2003)

	1962	1970	1980	1990	2003
Philosophy	4.3	6.9	7.9	8.0	7.3
Literary studies	13.9	11.4	10.4	9.2	10.0
Ethnology, folklore	1.4	1.5	1.8	1.8	1.6
Psychology	1.7	3.0	3.8	4.0	3.7
Political science	–	–	–	1.9	2.4
Sociology	1.2	1.7	2.8	4.5	5.3
Economics, incl. management	18.9	20.1	22.8	24.3	25.2
Law and state sciences	8.9	10.3	10.0	8.5	6.7
Geography, earth sc.	6.7	5.6	4.5	4.7	4.7
Art history	6.0	2.9	2.8	2.3	2.7
Educational science	2.5	3.7	6.2	6.3	6.6
Linguistics	13.9	9.5	7.3	7.8	7.8
History	18.7	19.5	18.7	15.7	15.2
Religious studies	1.2	0.2	0.1	–	–
Musical studies	0.7	1.2	1.0	0.9	0.8
All in the SSH	*100.0*	*100.0*	*100.0*	*100.0*	*100.0*
Numbers	418	861	1869	2730	3649
Relative growth	*100*	*206*	*447*	*653*	*873*

Table 8.3 Date of nomination of 'academic doctors' in social science disciplines (1951–2003)

	Before 1971	1971–1980	1981–1990	1991–2000	2001–2003
Philosophy	3.2	5.7	5.8	5.4	5.0
Literary studies	6.4	11.3	11.9	17.7	15.4
Ethnology, folklore	1.3	0.9	2.7	3.1	2.5
Psychology	3.8	3.0	1.5	3.1	3.5
Political science	–	–	0.3	2.7	1.5
Sociology	1.9	1.7	4.6	4.2	7.5
Economics, incl. management	18.6	19.1	18.6	13.1	9.2
Law and state sciences	10.9	15.2	14.0	5.0	3.5
Geography, earth sc.	5.7	3.4	4.9	8.1	11.4
Art history	7.1	5.2	1.2	2.7	4.5
Educational science	3.2	2.2	3.0	2.3	3.5
Linguistics	16.7	9.6	8.5	11.2	9.5
History	17.9	21.7	20.7	20.0	20.9
Religious studies,	1.3	–	–	–	–
Musical studies	1.3	0.9	2.1	1.2	2.0
Other	0.7	–	–	0.4	–
All in the SSH	*100.0*	*100.0*	*100.0*	*100.0*	*100.0*
Numbers	156	230	328	260	201
Relative average yearly growth	*100*	*287*	*410*	*325*	*885*

and—for different reasons—also the post-socialist period. Among classical branches of study Law, and especially religious studies, endured a visible decline, while the importance of the no less classical disciplines of the Arts Faculties—art history, literary studies, linguistics and philology, geography—held or consolidated their positions on various levels. The figures of both tables on 'official academics' show that the 'new' core social sciences—sociology, political science—broke out from earlier ghettoized positions, but on the whole remained marginal compared to the classical disciplines. The same fate befell those SSH—ethnology, educational science—that were preserved, at least institutionally, from the Stalinist disruption.

The second major process implicit in the general growth of staff, publications and institutional diversity concerns the feminization of actors across the field. Though data are available in our prosopographical surveys, only the final results are worth displaying, given both the small number and dispersion of cases among disciplines during the first decades immediately following 1950 when the new 'academic degrees' were created.

The scale of feminization varies considerably over time. It is clear from the penultimate line of Table 8.4 that contemporary doctoral schools present the highest proportion of women, followed by 'academic candidates' since 1950 and—at some distance—by 'academic doctors.' The 'glass ceiling' of female career chances is well demonstrated in these figures, but also important advances over time in most disciplines (last two columns of Table 8.4). Increase in the share of women is particularly manifest in marginal classical disciplines (with few practitioners) connected to earlier feminized artistic occupations like art history or musical studies, but also in some 'new' core SSH such as ethnology, psychology, sociology or political science, though this is less prominent. In some other, rather classical branches of scholarship—educational science, literary studies, linguistics and philology—the contrast is strongly marked between the relatively high degree of feminization at the lower echelon of academic hierarchy and the weak presence of women in the higher ranks ('academic doctors' and members of the HAS). Among the classical disciplines, history and philosophy remain in the middle of the scale. Law, economics, religious studies and geography seem to have resisted pressures for feminization. There again, for the interpretation of these large-scale disparities, discipline-specific intellectual conservatism, the structure of the occupational market connected to professional research (private practice, public employment besides academic positions), moreover the historically transmitted and inherited representations of gender roles and attributions—despite the impact of Western academic patterns—must also be taken into account.

On the whole, the paradoxical conclusion of all these findings might be stated as follows. Obvious efforts at a forcefully imposed Soviet-Russian intellectual colonization under socialism had only limited effects and was ultimately doomed to failure. From the 1960s onwards,

Table 8.4 Share of women in Academe in the social sciences and the humanities

	Among all 'academic candidates' 1950–2003 (%)	Among all 'doctors' of the HAS 1950–2003 (%)	Among 'doctors' of the HAS 1951–1990 (%)	Among 'doctors' of the HAS 1991–2003 (%)	In the staff of doctoral schools (2010) (%)
Philosophy	20.1	14.8	14	17	18.5
Literary studies	28.4	12.5	5	21	36.5
Ethnology, folklore	36.9	19.2	8	31	43.5
Psychology	39.2	21.2	6	40	50.7
Political science	20.2	18.2	–	20	13.6
Sociology	30.6	20.8	13	27	36.4
Economics	21.6	12.8	13	13	23.5
Law,	11.2	7.6	5	–	19.2
Geography	17.3	5.1	–	9	19.6
Art history	36.4	32.6	33	31	24.4
Educational science	34.8	6.1	5	8	50.3
Linguistics, philology	30.2	13.4	14	13	41.6
History	20.5	12.9	12	15	20.9
Religious studies	16.7	–	–	–	9.8
Musical studies	20.9	16.7	–	43	22.9
All	*24.2*	*13.3*	*12.8*	*16.9*	29.2
Total	*4729*	*1175*			*3183*

it did not exclude the growing impact of Western SSH in the reconfiguration of scholarly markets in Hungary, in spite of the brutally imperial nature of early communist rule, as recalled by collective memory (Karády and Nagy 2018). The smooth political transition in 1989 appears to have involved more continuities than ruptures in the established SSH, but also gave them a new lease of life. This can be observed in terms of better funding, access to European resources (especially after 2004, when the country joined the European Union), facilities with transnational exchanges and inter-state cooperation, multiplication of institutional or informal scholarly chapels with publications and a decisive reorganization of training for research via PhDs.

A new story appears to have started after 2010 when the self-proclaimed conservative and archaically nationalist ruling elite embarked upon policies detrimental to academic freedom. This included the recent attack on the *Central European University* (2017), the discriminatory ('clientelist') distribution of research funds (a 'déjá vu' from communist times), political intervention in academic elections and appointments, legal onslaught on humanitarian and scholarly foundations subsidized from abroad, the insidious demonization of George Soros (a uniquely generous promoter of the regime change as well as sponsor of the SSH in the country), and authoritative governmental statements on moot questions of historical memory. Further analysis of these recent developments must be reserved for a later study.

Bibliography

Bíbó, István. 1994. *Misère des petits états d'Europe de l'Est*. Paris: Albin Michel.

Biró, Judit, and Mária Székelyi. 1996. A tudomány újjáépítése, 1945–1950 [The reconstruction of science, 1945–1950]. *Szociológiai Szemle* 3–4: 81–95.

Bourdieu, Pierre. 1987. *La noblesse d'État: grandes écoles et esprit de corps*. Paris: Les Éditions de Minuit.

Brunnbauer, Ulf, Claudia Kraft, and Martin Schulze Wessel. 1911. *Sociology and ethnography in east central and south-east Europe: Scientific self-description in state socialist countries*. München: Oldenburg.

Csaba, László. 2002. Economics—Hungary. In *Three social science disciplines in Central and Eastern Europe, handbook on economics, political science and sociology*, ed. Max Kaase and Vera Sparschuh, 83–101. Bonn and Budapest: Social Science Information Centre and Collegium Budapest.

Csunderlik, Péter. 2017. *Radikálisok, szabadgondolkodók, ateisták. A Galilei-kör története* [Radicals, freethinkers, atheists. A history of the Galilei-Circle]. Budapest: Napvilág.

Erős, Ferenc. 2016. *Psyche és hatalom* [Psyche and power]. Budapest: Kalligram.

Frank, Tibor. 2009. *Double exile: Migration of Jewish-Hungarian professionals through Germany to the United States, 1919–1945*. Oxford: Peter Lang.

Hanák, Péter. 1984. *Zsidókérdés, asszimiláció, antiszemitizmus. Tanulmányok a zsidókérdésről a huszadik századi Magyarországon* [Jewish question, assimilation, anti-Semitism. Studies on the Jewish question in twentieth century Hungary]. Budapest: Gondolat.

Huszár, Tibor. 2015. *A magyar szociológia története* [History of Hungarian sociology]. Budapest: Osiris.

Janos, Andrew. 1982. *The politics of backwardness in Hungary, 1825–1945*. Princeton: Princeton University Press.

Johnston, William M. 1983. *The Austrian mind: An intellectual and social history, 1848–1938*. Berkeley: University of California Press.

Kaase, Max, and Vera Sparschuh (eds.). 2002. *Three social science disciplines in Central and Eastern Europe, handbook on economics, political science and sociology*. Bonn and Budapest: Social Science Information Centre and Collegium Budapest.

Karády, Victor. 2012. Les Allemands dans l'intelligentsia moderne émergeante en Hongrie à l'époque de la Double Monarchie. *Austriaca* 1: 193–221.

Karády, Victor. 2016. The academic profile of doctoral school staffs in Hungarian universities in the social sciences and humanities. A comparative study of disciplines with special reference to educational science (2000–2010). *CIAN-Journal of History of the University* (Madrid) 19: 71–89.

Karády, Victor, and Peter Tibor Nagy. 2018. A case of state controlled Westernization. Foreign impacts in the Hungarian social sciences (1945–2015). In *The social and human sciences in global power relations*, ed. Johan Heilbron, Gustavo Sora, and Thibaud Boncourt, 297–332. Cham, Switzerland: Palgrave Macmillan.

Katus, László. 2012. *A modern Magyarország születése, 1711–1914* [Birth of modern Hungary. The history of Hungary, 1711–1914]. Pécs: Kronosz kiadó.

Kuti, Klára. 2011. Hungarian ethnography. Science of the nation and/or socialist society. In *Sociology and ethnography in East Central and*

South-East Europe: Scientific self-description in state socialist countries, ed. Ulf Brunnbauer, Claudia Kraft, and Martin Schulze Wessel, 197–211. München: Oldenburg.

Litván, György, and László Szűcs. 1973. *A szociológia első magyar műhelye* [The first workshop of sociology in Hungary]. Budapest: Gondolat.

Mészáros, István. 1988. *Középfokú iskoláink kronológiája és topográfiája, 996–1948* [Chronology and topography of our secondary schools, 996–1948]. Budapest: Akadémiai kiadó.

Nagy, Peter Tibor, and Viktor Karády (eds.). 2012. *The numerus clausus in Hungary. Studies on the first anti-Jewish law and academic anti-Semitism in modern central Europe*. Budapest: Pasts Inc., Centre for Historical Research, History Department of the Central European University.

Némedi, Dénes, and Péter Róbert. 2002. Sociology—Hungary. In *Three social science disciplines in Central and Eastern Europe, handbook on economics, political science and sociology*, ed. Max Kaase and Vera Sparschuh, 437–451. Bonn and Budapest: Social Science Information Centre and Collegium Budapest.

Péteri, György. 1998. *Academia and state socialism: Essays on the political history of academic life in post-1945 Hungary and Eastern Europe*. Boulder, CO: Social Sciences Monographs. Distributed by Columbia University Press.

Szabó, Máté. 2002. Political science—Hungary. *Three social science disciplines in central and eastern Europe, handbook on economics, political science and sociology*, ed. Max Kaase and Vera Sparschuh, 258–274. Bonn and Budapest: Social Science Information Centre and Collegium Budapest.

Szekfű, Gyula. 1934/1989. *Három nemzedék és ami utána következik* [Three generations and what follows afterwards]. Budapest: AKV-Maecenas Reprint.

Tőkés, Rudolf. 1996. *Hungary's negotiated revolution: Economic reform, social change and political succession*. Cambridge: Cambridge University Press.

Ungváry, Krisztián. 1913. *A Horthy-rendszer mérlege. Diszkrimináció, szociálpolitika és antiszemitizmus Magyarországon* [Balance sheet of the Horthy regime. Discrimination, social policy and anti-Semitism in Hungary]. Pécs and Budapest: Jelenkor kiadó.

Vasvári, Ferenc. 2007. A Magyar Társadalomtudományi Egyesület megalakulása [The birth of the Hungarian Association of social science]. *Acta Sociologica* 2 (1): 90–105.

Vorisek, Michael. 2011. Sociology in Soviet Europe. In *Sociology and ethnography in East central and South East Europe: Scientific self-description in state socialist countries*, ed. Ulf Brunnbauer, Claudia Kraft, and Martin Schulze Wessel, 29–58. München: Oldenburg.

9

Arduous Institutionalization in Argentina's SSH: Expansion, Asymmetries and Segmented Circuits of Recognition

Fernanda Beigel and Gustavo Sorá

Argentina is an interesting case with which to explore the institutionalization and professionalization of the social sciences and humanities outside the traditional "centers of excellence." It is a field that, in the last ten years, has experienced an impressive expansion of its research capacities, leading to rapid scientific development and increasing internationalization. As in other historical periods (1958–1966, for example), recently

F. Beigel (✉)
Instituto de Cienicas Humanas y Ambientales (INCIHUSA), Consejo Nacional de Investigaciones Científicas y Técnicas (CONICET), Mendoza, Argentina
e-mail: mfbeigel@mendoza-conicet.gob.ar

F. Beigel
National University of Cuyo, Mendoza, Argentina

G. Sorá
Instituto de Antropología (IDACOR), Consejo Nacional de Investigaciones Científicas y Técnicas (CONICET), Córdoba, Argentina

G. Sorá
National University of Córdoba, Córdoba, Argentina

C. Fleck et al. (eds.), *Shaping Human Science Disciplines*, Socio-Historical Studies of the Social and Human Sciences, https://doi.org/10.1007/978-3-319-92780-0_9

Argentina's academic institutions grew to a size that led it to become a significant international academic node. Nevertheless, as has happened in the past, expansion phases are interrupted by the change of political and economic conditions for scientific activity. The new government, a right-wing coalition, has questioned the expansion of full-time researchers at the main public agency, the National Council for Scientific and Technical Research (CONICET, its Spanish acronym) and has made a major cut in the budget for science and technology. In order to understand the singularity of this type of "peripheral-center", it is pertinent to describe the *cycles* of expansion and contraction, institutionalization and de-institutionalization. It is also crucial to observe the impact of these cycles on the development of circuits of production and circulation of SSH knowledge, some clearly internationalized and others rather nationally-oriented.

The main task of this chapter is to describe the current state of research capacity in the SSH, resulting from the recent expansion, in order to analyze the coexisting and diverse principles of legitimation. The participation of the Argentinian scholars in increasingly segmented publishing circuits is defined by different rewards in terms of recognition: (a) international (mainstream) journals, indexed in WoS (now Clarivate) or Scopus, are highly valued for tenure and promotion at CONICET, especially in the exact and natural sciences; (b) Transnational journals indexed in directories such as DOAJ and Latin American journals, indexed in SCIELO,[1] REDALYC,[2] LATINDEX,[3] are mostly rewarded in the SSH for tenure and promotion and CONICET, while; and (c) local, non-indexed journals, are mainly used for teaching career-building at national universities. These different *circuits of recognition* that exist in the country can be understood by observing institutionalization along with internationalization, combining field analysis with a circulation approach, and paying special attention to interactions and the publishing landscape (Beigel 2014b).

[1]Created in Brazil in 1998, SCielo nowadays includes journal collections from 14 Iberoamerican countries and an indexing service. Recently it was acquired by WoS-Clarivate.

[2]Created in Mexico in 2003, it had a collection of 932 journals by December 2017.

[3]Created in Mexico in 1995, it is the biggest catalogue of Latin American journals evaluated based on 25 criteria, but only part of these journals are available on line. Latindex does not offer an indexing system. Latindex-C is the most inclusive and representative repository of evaluated journals, while SciELO and RedALyC have more restrictive range collections, but they are available in full text and offer citation services.

Two major trends involve circulation in Argentina's SSH, and both are discussed in this chapter. First, the polarization between the local and international orientations in production and circulation of knowledge in the country. While natural and exact sciences participate dominantly in mainstream publishing circuits, the SSH tend to feed the national journals and the Latin American circuit, while it is only an internationalized elite that gets involved in the mainstream circuit. Most of the journals belonging to the SSH are still edited in paper and, accordingly, distributed in reduced circles; almost half are not indexed in any international database nor in the national index CAICYT (Beigel and Salatino 2015). Second, the paradox of the "nationalist" orientation of the public policies for Ph.D. formation and the requirement of international publications for tenure as a full-time researcher at CONICET. The majority of the new researchers at CONICET have obtained their doctoral degree in Argentina, but the agency inclines researchers to publish in foreign journals. A key question will be raised: If the Ph.D. formation is no longer the privileged path for internationalization, what other factors explain the circulation of Argentinian CONICET researchers?

In the first section of this chapter, we review the historical process of institutionalization of the SSH in Argentina in the tense atmosphere between CONICET and the national universities. We focus on these two institutions because they are the main pillars of Argentinian higher education and research. As will be shown, together with the expansion, the concentration of resources at metropolitan University of Buenos Aires (UBA) has reinforced geographical and institutional asymmetries, deepening what Beigel (2014b) describes as the structural heterogeneity of the field. Geographic inequalities are shown by highlighting the morphology of eight academic regions. We outline the main features of pre-graduate and graduate studies in SSH and describe the current state of SSH research capacities in terms of personnel and institutes, focusing on institutional inequalities and differences between disciplines. Finally, we describe the publishing circuits and the particular role of the SSH in the universe of Argentinean scientific journals. Our aim is to examine the recent expansion and consider its impact in historical structures along with genetic processes in order to describe the state of the field in the new context.

Institutionalization and Professionalization of SSH in Argentina: Politicization, Discontinuities and Specific Forms of Prestige

The disciplines assembled here under the umbrella SSH emerged early in Argentina. The first chair of Sociology was created in 1898, Psychology followed in with its first chair in 1902 and Anthropology in 1906. The country has a long history of distinguished scientists—including five Nobel Prize winners—who play an active role in academic diplomacy and transnational networks. During the second half of the twentieth century, the circulation of prestigious scholars was assisted by political exile, reinforcing an internationalization based more on individual trajectories than on stable institutional policies (Beigel 2010). Argentina's higher education system has been mostly public, and therefore largely dependent on public expenditure, as well as political and institutional stability. These last two factors have been threatened by military interventions that have repeatedly affected academic freedom, reduced higher education budgets and have therefore had a strong impact on the structure of the academic field. Public education is part of the country's social self-image and the "national" [i.e. public] universities are sustained by a strong tradition of university autonomy. Their institutional autonomy is granted by the Constitution and is defined in terms of preventing State interference in universities' "internal affairs", offering those inside the universities self-government, autarchy and "teaching freedom" [*libertad de cátedra*]. Therefore, the relations of each university with the State are tense and public policies have to deal with a fragmented higher education system containing 50 powerful and demanding institutions. They are self-governed by professors, alumni and students. They are gratuitous (no fees are paid by the students), and most of the undergraduate programs have unrestricted entry. Student movement is highly politicized and rectors (university presidents) play an active role in defining university policies.

The institutionalization of SSH was directly associated with the modernization, expansion of matriculation and feminization of the higher education system, a phenomenon that occurred in all major

Latin American countries during the 1950s and 1960s, changing the social composition of student body. As in many other countries, careers with their roots in SSH—such as Education, Social Work, Psychology, Sociology, Anthropology, Economics—were expanded during this cycle of growth, innovation and democratization. Vast external resources were geared toward "university development." The elite of Argentina's first cohort of professionally educated SSH scholars participated actively in international networks and created institutes and professional associations capable of capturing foreign aid for research and teaching (Blanco 2010; Sorá and Blanco 2017).

The creation of several inter-governmental agencies for research and teaching—such as the Latin American Economic Commission (ECLA) and the Latin American Faculty of Social Sciences (FLACSO), mostly based in Santiago de Chile—promoted a *regional academic circuit* that stimulated the emergence of a region-wide form of academic prestige (Beigel 2010). In acquiring institutional positions in the new undergraduate programs, institutes and regional centers those who wished to play a role in the definition of the new disciplines' profiles developed institutional power, the most significant capital at stake. Academic disputes were as important as the struggle to dominate institutions, therefore university politics shaped Latin America's professionalization process. Most of Argentina's renowned figures have been simultaneous founders of the first research institutes and of undergraduate schools. Even during periods of political radicalization of SSH scholars, seizing "intramural" institutional power was at the core of their concerns (Beigel 2013).

During the 1950s, the impetus for the modernization of the higher education system was the creation of new universities organized in the form of Departments, in a similar way to the organization of US Universities of the time. It was in this context that CONICET was created, in 1958, along with the full-time research career, marking a milestone in Argentina's academic professionalization. This public organization was the result of an intense debate on the institutional form of the activities of science and technology that took place from the beginning of the 1950s, when scientific research was concentrated within national universities and private associations (Feld 2009).

The SSH had only a marginal space at CONICET during this early period with its development occurring more within the national universities and a few private institutes, until the 1980s.

Attempts to extend this "academic modernization" to the old-established universities were rejected by the student movement and became increasingly politicized, advancing the message that Departmentalization would never fit within Argentina's traditional universities. The old system of chairs still prevails, even today. Influenced by the Cuban Revolution (1959), and in the context of the cultural cold war, an *anti-academic illusio* became dominant in the SSH when militant practices prevailed within professors and students. Political engagement triumphed and the social essay was the preferred mode to interpret this society in revolt.

The dictatorship of 1976–1983 had a violent impact on public universities, particularly in the SSH undergraduate schools and research institutes. After the elimination of democratic rule, hundreds of professors and students were killed or imprisoned, and programs of Sociology, Anthropology, Psychology, Social Work and Journalism were closed down over the whole country. This de-institutionalization had strong effects on these disciplines; some are still visible today. As has been demonstrated by Bekerman (2013), during these dark years, public expenditure on research increased, surprisingly, as a result of an external loan given by the Inter-American Development Bank. CONICET received a huge amount of money for expansion, while at the same time, the research budget for the national universities was reduced to a minimum. Academic life at the universities was severely diminished and research was concentrated at CONICET, where more than 100 new research institutes were created. The SSH received substantial funding for research institutes detached from the former politicized agenda. This policy also benefitted some individuals, directors of 15 institutes who had no scientific prestige but were loyal to the military regime. They were later prosecuted for fraud. As a result, a deep rift between research and teaching was created that largely transcended the end of the dictatorship, becoming a structural feature of the scientific-academic system. In this way, the share of SSH researchers in CONICET increased during the dictatorship.

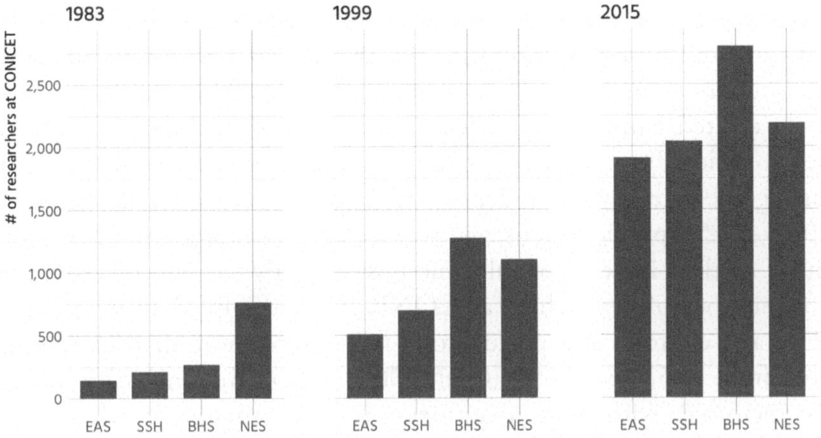

Fig. 9.1 Number of researchers at CONICET by scientific area, 1983–1999–2015. *Note* EAS: Engineering and Agricultural Sciences; SSH: Social Sciences and Humanities; BHS: Biological and Health Sciences; NES: Natural and Exact Sciences (*Source* Beigel et al. (2018))

After the end of the military rule (1983), SSH departments were re-established and new research institutes were created, primarily at the public universities. The return to democracy brought with it an attempt to reverse the rift between CONICET and the national universities; the latter got some incentive to regain their role in the research tasks. By that time, less than 15% of university teachers participated in research activities, and only a tiny portion of professors held doctoral degrees and listed scholarly publications in their CVs. Teaching was the dominant activity at the universities and research was almost exclusively undertaken by CONICET researchers. As can be seen in Fig. 9.1, the share of full-time SSH researchers was 15%, while Natural and Exact Sciences prevailed. During the 1990s, a period of economic crises, the budget for universities was cut and the competition for tenured positions at CONICET reduced to minimum.

After the last, and most brutal, social and financial crisis that affected the country (2001), an expansive cycle started with the Presidency of Néstor Kirchner (2003–2007), as part of the leftist turn and the economic recovery that took place in Latin America at this time.

This was a strong period for the academic field, based on the increase and diversification of the budget, the creation of new public universities, and the promotion of scientific and technological agencies. The newly created Ministry for Science and Technology (2007) promoted the professionalization of research and emphasized internationalization. During the next two Presidencies of Cristina Fernández de Kirchner (2007–2015), public expenses in science and technology were enlarged with a marked increase of full-time positions. By the end of 2015, the total number of researchers at CONICET had tripled from 3579 to 9236, and doctoral and postdoctoral fellowships more than quadruplicated from 2351 to 10,092 (CONICET 2016). The university budget also increased, with huge investments in infrastructure, newly created research centers in different parts of the country, and several new Ph.D. programs, especially in the SSH. In terms of internationally comparable data, a rate of 2.64 researchers per thousand labor force was reached by 2014 (UNESCO 2016) and the public plan established by the Ministry was to increase it to 5 per 1000 by 2020.

Figure 9.1 shows the evolution of researchers at CONICET from 1983 to 2015 and the increasing number of full-time positions for SSH, reaching 22.1% of the total. The growth of the personnel was not yet met by an equal expansion of financial resources for research projects. The National Agency for Promotion of Science and Technology (ANPCYT) concentrated its funds for scientific projects mainly in areas such as technology, health, and exact and natural sciences. The share of funding for SSH projects was somewhere in the range of 10–12% and these disciplines were not considered priority areas (Rovelli 2015).

The Evolution of Teaching in the SSH

Argentina is a traditionally public higher education system although currently composed of equal quantity of private and public universities. The state finances undergraduate programs at public universities and enrollment is much higher at public institutions: by 2014, 78.5% of undergraduate enrollment and 75.5% of graduate enrollment were in public institutions. Ph.D. holders increased in all scientific areas and a particular move toward professionalization of research was observed in the national universities. In previous studies (Beigel 2014b; Beigel et al. 2018), we have shown that this professionalization during the recent expansion reinforced the existence of competing cultures of evaluation and diverse principles of legitimation. Table 9.1 shows the synthesis of the structural features of the field with specific observations for the SSH.

The SSH in Argentina have a consolidated tradition of undergraduate studies that are developed in most parts of the country, with separate faculties or departments for the humanities and the social sciences across all national universities. The evolution of the SSH faculty in the public universities shows periods of stagnation and recent growth. In 1993, the Incentive Program for Teachers-Researchers (PIDI, its Spanish acronym) was created. This had an important resonance and impacted the professionalization of research in the SSH.

Pre-graduate student enrollment in the social sciences matches the rise in the humanities. But concerning certification, the former grew stronger than the latter. Figure 9.2 shows that alumni in the social sciences were doubled in absolute terms from 2001 until 2014, while the humanities were almost stable. However, the distance between enrollment and certification is higher in the social sciences, due to higher dropout rate.[4] During the last decade, the National Council of Deans of SSH Faculties was a very active agent in building public policies for the SSH and obtaining funds for increasing certification rates, intra-national mobility, full-time teaching posts, fellowships for Latin American students and other programs to reinforce professionalization.

Concerning graduate programs, until 2004, the SSH in Argentina had a stronger emphasis on Master's programs and the "academic Ph.D." was restricted to an individual's thesis. During the last 15 years, Ph.D. schools for Social Sciences, Sociology, Political Sciences and Anthropology were created and accredited. They expanded due to the

[4]The distance between enrollment and certification in SS and H may be explained by considering differences in social origin and positions of the agents, a crucial issue that has not been studied yet.

Table 9.1 The SSH in Argentina's academic field

Institutional setting	• Dominantly public funded • Major public specialized agencies for scientific research such as CONICET • 52 autonomous national universities with diverse regulations concerning requisites for professor tenure, university government, student credits and curricula • 50 private universities, with scarce research activities • 79% of higher education enrollment belongs to public universities
Public policies and institutionalization of research	• Equitable growth of full-time researchers at CONICET among scientific areas. SSH is currently represented at 22.1% of the total personnel • There is a national (public) agency for funding scientific research (Agencia Nacional de Promoción Científica y Tecnológica, ANPCYT). Research grants are not distributed equally among areas, SSH less favored • Disconnection between public policies implemented by the Ministry of Education (addressed to national universities) and scientific policies led by the Ministry of Science and Technology (addressed to CONICET and specialized research organizations)
Teaching	• 3/4 of the total higher education enrollment in public universities belong to the SSH • There are 69 SSH Faculties within 47 national universities. Mostly separated between the label of **social sciences** (political sciences, sociology, psychology, social work, anthropology) and **humanities** (history, philosophy, literature and education) • Teaching at universities is not mandatory for CONICET researchers, although 78% of the total SSH researchers are also professors and the majority of the SSH institutes are located at the national universities
Distribution of research capacities	• High concentration of CONICET researchers at 3 metropolitan universities (UBA, UNLP, UNC) 29% of total SSH researchers at CONICET work at UBA • The capital city and metropolitan area of Buenos Aires concentrates 60% of total researchers at CONICET
Evaluation cultures	• A unique public National Agency for Evaluation and Accreditation of universities (CONEAU) • Diverse prestige-building paths between international and nationally oriented SSH scholars at national universities and CONICET • Diverse evaluation cultures and regulations at each national university, mainly centered in teaching background and endogenous recruitment

(continued)

Table 9.1 (continued)

Circulation of scientific output	• Segmented circuits of publication, separated by language, institution and discipline • Strong local journal landscape fueled by non-indexed university journals, mostly from SSH, many still published only in paper

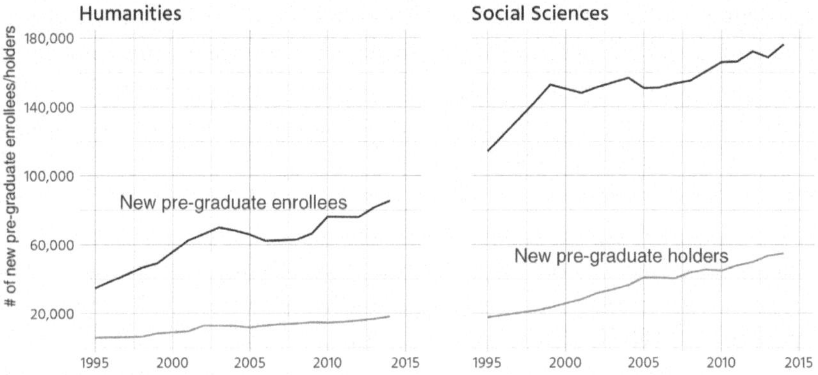

Fig. 9.2 New pre-graduate enrollees and new pre-graduate holders, Social sciences/humanities, per year (1995–2014) (*Source* Sistema de Consulta de Estadísticas Universitarias http://estadisticasuniversitarias.me.gov.ar/#/home/1)

stimulus of CONICET doctoral fellowships, and other public policies such as the "DoctorAR" Program, which contributed to the institutionalization of Doctoral programs. Between 2006 and 2012, the number of doctoral students in these disciplines nearly exploded. While the natural and exact sciences experienced an increase of 27% in the doctoral enrollment, the SSH doctoral students grew by 217%. In 2012, the SSH reached a 35% share of the total doctoral students in the country. From 2001 to 2015, new Ph.D.s grew from 37 to 637 in the social sciences, and from 74 to 447 in the Humanities; 56% of the new Ph.D. holders were women and 75% of the total degrees were given by public universities (SCEU 2016).

There are two different national careers for researchers, one regulated by the universities with five categories (I–V), and the other at CONICET with five stages, from Assistant to Superior. The majority of the increasing number of new Ph.D. holders in SSH are employed by CONICET, despite the competition for tenure in this organization having always been harsh. The number of full-time teaching positions has also increased at public universities during this period, but each university has its own regulations for tenure. In general, in the universities, a doctoral degree is not a prerequisite for a teaching position. According to the last data available, only 9.9% of all university staff have a doctoral degree (SPU 2013). At many provincial universities, international publications still are not considered important nor are a determinant factor in the periodical evaluations at the universities by CONEAU. They do not have an impact on government funding either. The traditional chair system prevails in the traditional universities and the "temporal" power of head professors and deans has great influence in the selection process. As a consequence, young doctors aspire to become CONICET researchers and feel discouraged by the impossibility of gaining a teaching post at universities. Accordingly, an elite group of SSH researchers shaped by CONICET's evaluative culture has evolved based on papers published in international indexed journals. These people participate in academic mobility, attend congresses and collaborate internationally, but more frequently within the Latin American region (Beigel 2017).

An Increasingly Nationalized Graduate Formation for an Internationalized Academic Elite: The Social Science Researchers at CONICET

Of the total number of researchers at CONICET by 2014 (7905 for all disciplines), we collected data for the formation of 7343 individuals, the rest representing mostly senior researchers of a generation in which many of them did not earn a Ph.D. degree. Of those with a doctoral degree, 6470 obtained it in Argentinian institutions (6293 at public universities) and only 873 abroad. For the natural and exact sciences, national Ph.D. degree represents 92% of the total, and in Biology and health 95%. The SSH disciplines have a larger share of Ph.D. degrees earned outside the country: 21.6% of all its researchers.

This can be attributed to the historical de-institutionalization, exile and the late-coming of SSH doctoral schools in the country. Gallardo (2015) has analyzed the evolution of the relation between the national and foreign doctoral degrees in the SSH researchers at CONICET between 1989 and 2013, showing that the tendency has been a decrease in the Ph.D. abroad and a downward slope of researchers without doctoral degree.

The morphological distribution of the doctoral degrees of SSH researchers at CONICET demonstrates a generational lag marked by the times in which undergraduate schools were closed (during the dictatorship), leaving a void space between those currently in their 70s and older and those in their 40s and 50s. Its composition shows the existence of a big, young faculty with doctoral degrees gained in Argentina, a small intermediate generation partly with doctoral degree gained abroad, and a core of "big names" coming from the pre-dictatorial generation, a great many of whom obtained their doctorates abroad, though relevant parts are without a doctoral degree. This composition is typical in sociology and anthropology, but is different in disciplines not subject to de-institutionalization, particularly economics and political science. These disciplines also have more doctoral degrees gained abroad than other disciplines. For the seven INTERCO-SSH disciplines, the proportion of degrees gained in Argentina is lower than the universe of researchers at CONICET (66.9%) and a degree gained abroad is higher, reaching 29%. This number changes greatly if analyzed by gender: 22.6% of women have a doctoral degree abroad, compared to 36.6% of male researchers for the seven disciplines. The principal destinations, in order of importance, are Spain, the United States, France and Brazil.

Figure 9.3 shows the particular composition of the place of doctoral degree in the seven INTERCO-SSH disciplines. Psychology has the highest number of researchers who obtained a doctorate in Argentina (85%), followed by anthropology (78.1%) and literature (77.7%). Interestingly, these three disciplines are also those with the highest degree of feminization. For psychology, this is (86.3%), anthropology (77.5%) and literature (71%), and these figures are mostly composed of people under 50 years old. In Political Science and Economics, less than half of its researchers finished their doctoral degree in the country (47.2

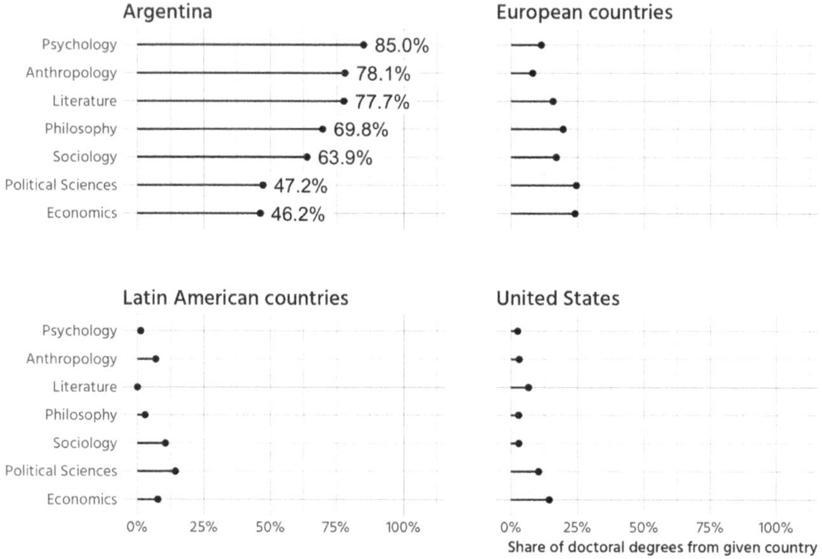

Fig. 9.3 SSH researchers at CONICET by discipline and country of doctoral degree, 2014

and 46.2%, respectively). Spain is the main destination for Psychology and Philosophy, though German universities are also a major attraction for the latter. The United States prevails in Political Science, Economics and Literature, followed by UK. In the case of Sociology, the main destinations are France and Spain. Only in the case of Anthropology does a Latin American country, Brazil, appear as the main place to gain a doctorate abroad (Gallardo 2015).

Of the researchers with a national Ph.D. degree (66.9%), more than half of them were obtained at the UBA. In previous studies, Beigel (2017) has shown that the UBA played a crucial role in the formation of the internationalized academic elites at CONICET. This dominance is not capitalized in the prestigious title, but more likely the possession of certain institutional knowledge (*savoir-faire*, *savoir-dire* and *savoir-vivre*) given by specific skills raised in long existent research teams and the participation in international networks. The ability to write a research project or an application for a fellowship is revealed as a sort of social capital forged at prestigious institutes that have long existed at UBA. On the other hand, researchers formed at UBA concentrate institutional power in CONICET because they have a dominant presence in the evaluation committees where the criteria for entrance and promotion are determined (Beigel 2017: 840–843).

Morphology and Geography of the SSH Researchers

In order to account for the overall SSH population of researchers in the country, we must compare the CONICET researchers with the non-CONICET scholars accredited by the national classification made by the Incentive Program for Teachers-Researchers (PROINCE, its Spanish acronym). The total SSH researchers at CONICET for 2014 was 1711, while total professors categorized by PROINCE rose to 4559. Some of these are researchers at CONICET, but the available statistics do not account for this subpopulation with two different posts. Our own survey indicates that 1327 out of 1711 researchers at CONICET hold a university post in addition to their researcher post. Only 384 are researchers exclusively at CONICET and should be added to the total professors categorized at the universities. Accordingly, the number of SSH researchers in the country is 4943. An additional group of young SSH professors could soon be added because the census of accredited "professors-researchers" is now old (2009) and many have recently asked for categorization in 2015. The results are not yet available.

Research positions at CONICET are completely independent from competitions for teaching positions in Argentina's higher education system. To apply for tenure at CONICET, it is necessary to present an application describing one's research background and a scientific project. There are currently five positions a researcher can hold at CONICET and these are, in order of rank: Assistant, Adjunct, Independent, Principal and Superior. The lowest position (assistant) is reserved for young researchers under 35; adjunct, ages 36–40;

independent, ages 41–45; and principal, ages 46–50. The highest position, superior researcher, is assigned through a special selection process. The institution accepts applications for promotions once a year; application is voluntary and is evaluated by special committees that consider production and scientific contributions. The full-time positions are tenured for life, as long as the researcher approves his/her annual or bi-annual research report. There is a moderate salary difference among these categories.

The five categories of professors-researchers at the national universities differentiate only symbolically because there are virtually no differences in terms of salary. Whereas CONICET researchers are accustomed to being examined by national committees with increasingly internationalized evaluation criteria, many of the university professors, working at autonomous institutions, resisted the implementation of external evaluations in the 1990s. This favored a populist discourse where "the people's" professors were inclined against CONICET researchers who were viewed as living in an ivory tower. This was impulsed by the fact that each university has its own regulations for tenure and the selection committees are composed according to internal power relations.

The differences observed between the two career-paths—CONICET and national universities—are substantial. A great many non-CONICET researchers at the universities are nationally oriented—inclined toward the production of social essays and more engaged in university politics—and a large share of them do not hold a Ph.D. As for SSH researchers at CONICET, doctoral degrees are required for tenure and the evaluation culture is strongly oriented toward an internationalized profile. But the fact that most of the CONICET researchers have their workplace at national universities, and that both styles of career-building live together in that locus, makes these two principles of legitimation an arena of permanent disputes. The most conspicuous representatives of both styles co-habit at the national UBA.

The morphology of this population evidences important differences comparing SSH CONICET researchers and professors-researchers. Table 9.2 describes SSH scholars in relation to demographic and academic indicators: CONICET researchers are highly concentrated in the metropolitan area of Buenos Aires (see column 5), while the SSH professors-researchers are better distributed throughout the country (see column 3). In previous works (Beigel et al. 2018), we classified 8 academic regions on the basis of the higher education regions defined by the national statistics.[5] It is critical to differentiate areas inside this huge Argentinean province that includes the capital of the country. Actually, 3 different regions can be distinguished within "Buenos Aires": (a) the Autonomous City of Buenos Aires ("CABA", for its Spanish acronym), capital of country; (b) the metropolitan area of Buenos Aires ("Gran Buenos Aires"); and (c) other zones of the province of Buenos Aires (the "Bonaerense" region).

The concentration of CONICET researchers in the CABA node is particularly visible, and higher in the SSH than in other scientific areas, in spite of the efforts made in this organization to prioritize other regions. The prominence of the UBA as a workplace for a significant proportion of researchers from CONICET is noteworthy. But this is intensified in the SSH: CABA accumulates 31.5% of the total researchers at CONICET, but virtually half of the total SSH researchers. In contrast, the distribution of the professors-researchers shows concentration first in the center-west region, with 26.4%, followed by Great Buenos Aires (18.4%) and CABA is only in the third place with 14.2%. The rest of the regions virtually double their percentage in PROINCE compared to the distribution of CONICET SSH researchers. Accordingly, seen from a spatial perspective, the SSH at CONICET are highly concentrated in the capital city and the professors-researchers are more fairly distributed in the national universities across the country.

Regarding the geographical distribution of the SSH by level of education, in Table 9.2, we can see that pre-graduates are significantly concentrated in CABA (32.7%). The available data do not allow a difference to be drawn between Great Buenos Aires and the Bonaerense

[5]These regions are 7 and they are called Councils of Regional Planification for Higher Education (CPRES for its Spanish acronym). Buenos Aires is divided in two CPRES: "bonaerense" and metropolitan. But in order to compare regional inequalities without forcing concentration data, it is necessary to observe the capital city as a separated region from the area that includes the big and prestigious Universidad Nacional de La Plata (here called Gran Buenos Aires).

Table 9.2 Demographic indicators, graduate degrees, researchers, professors and fellows, by academic region

Region	1	2	3	4	5	6	7	8	9
	Population	PROINCE Total Categorized professors-researchers	SSH Categorized professors-researchers	CONICET researchers all scientific areas	CONICET researchers SSH	New SSH total pre-graduate holders, 2014	New Ph.D. holders, all scientific areas 2014	New Ph.D. holders, SSH 2014	SSH Doctoral fellows at CONICET, 2015
CABA	2,890,151 (7.2%)	3405 (15.1%)	1127 (14.2%)	2489 (31.5%)	849 (49.6%)	23,880 (32.7%)	3790 (41.6%)	429 (46.5%)	933 (35.6%)
Great Buenos Aires	9,916,715 (24.7%)	3351 (14.8%)	1460 (18.4%)	1353 (17.1%)	231 (13.5%)	15,081 (20.6%)	1243 (13.6%)	122 (13.2%)	617 (23.5%)
Bonaerense	5,708,369 (14.2%)	2603 (11.5%)	842 (8.7%)	887 (11.2%)	135 (7.8%)		552 (6.1%)		
Center- West	6,161,170 (15.4%)	5564 (24.6%)	2099 (26.4%)	1321 (16.7%)	226 (13.2%)	13,747 (18.8%)	1849 (20.3%)	161 (17.4%)	476 (18.2%)
Center-East	4,430,531 (11%)	2471 (10.9%)	572 (7.1%)	729 (9.2%)	78 (4%)	10,112 (13.8%)	851 (9.3%)	185 (20.0%)	199 (7.6%)
North-West	4,911,412 (12.2%)	2765 (12.2%)	939 (11.8%)	413 (5.2%)	95 (5.5%)	5266 (7.2%)	469 (5.1%)	8 (0.9%)	206 (7.9%)
North-East	3,679,609 (9.2%)	987 (4.4%)	364 (4.6%)	143 (1.8%)	35 (2%)	3687 (5%)	171 (1.8%)	17 (1.8%)	83 (3.2%)
South	2,419,139 (6%)	1429 (6.3%)	574 (7.2%)	570 (7.2%)	62 (3.6%)	1179 (1.6%)	188 (2.1%)	1 (0.1%)	102 (3.9%)
Total	**40,117,096 (100%)**	**22,575 (100%)**	**7977 (100%)**	**7905 (100%)**	**1711 (100%)**	**72,952 (100%)**	**9113 (100%)**	**923 (100%)**	**2616 (100%)**

Elaborated by the authors on the basis of the following sources: For population: National Census 2010 (National Institute for Statistics and Census, INDEC). PIDI Professors all categories I-V (SPU 2013). CONICET Researchers: PIDAAL database, Dec. 2014. New SSH pre-graduates (SCEU 2016, http://estadisticasuniversitarias.me.gov.ar/#/home/1); New Ph.D. holders Anuario SPU 2012; New Doctoral fellows at CONICET, 2015 http://www.conicet.gov.ar/becarios/

region (both concentrating 20.6%) but, along with Center-West, these four regions accumulate an impressive 72.1% of the total pre-graduates in SSH, while the South region's share is a paltry 1.6%. The distribution of Ph.D. holders in all scientific areas is concentrated in CABA (41.6%) because of the dominant role played by UBA, as we have already mentioned. The distribution of SSH doctoral fellows granted by CONICET shows a structure similar to the distribution of pre-graduates; high metropolitan concentration (35.6% of the doctoral fellows belong to institutions placed in CABA) and similar shares for the Buenos Aires area and Center-West region. But it is within Ph.D. holders in SSH where the concentration at CABA is the highest of all variables: 46.5% obtained their title in the capital city of the country, almost all coming from UBA. Buenos Aires and Center-West diminish their share when pre-graduates and doctoral fellows are compared, highlighting in this variable the Center-East region with a share of 20% of the new doctoral holders. This can be attributed to the dynamical National University of Rosario with several consolidated doctoral schools in Political Science, International Relations, Social Work and Social Communication.

Let us now delve into the CONICET organization in order to analyze disciplinary differences, gender, age, hierarchy and geographical asymmetries. Compared to other scientific areas, the global SSH's share of full-time researchers is pretty equal to the other 3 scientific areas, as we have seen before (22%). Looking at the status hierarchy, the situation for SSH is less promising. As can be seen in Fig. 9.4, the "hard" sciences occupy higher positions, whereas the SSH show the inverse picture.

The SSH CONICET researchers tend to be young (average age 46) and a high proportion are in early career positions (Assistant and Adjunct); both features are the result of the expansion during the last decade. Of the 1072 researchers from the seven INTERCO-SSH disciplines, 272 entered CONICET before 2003, while 800 did so from 2004 onward. Researchers aged 30–49 years represent more than 60% of most disciplines. Psychology is the youngest discipline, with 86.3% of members within that age range. Table 9.3 shows the distribution of researchers among disciplines, and hierarchy. Philosophy and History exhibit a similar distribution, mainly due to the fact that these existed

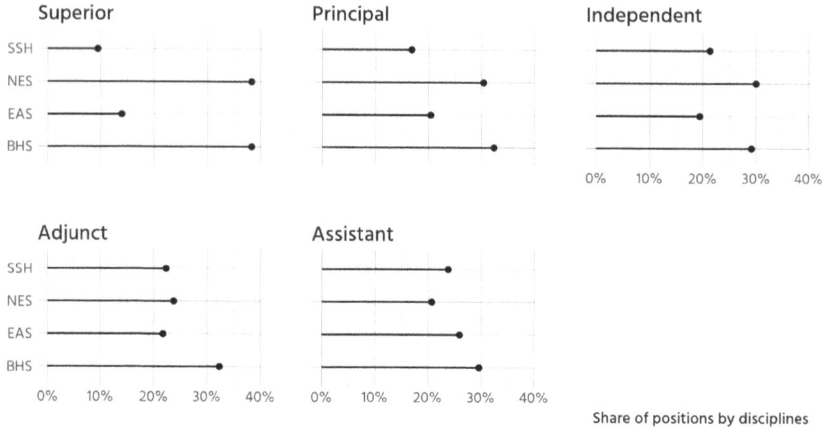

Fig. 9.4 CONICET researchers, by category and scientific area (n = 7905), 2014. *Note* EAS: Engineering and Agricultural Sciences; SSH: Social Sciences and Humanities; BHS: Biological and Health Sciences; NES: Natural and Exact Sciences (*Source* Beigel et al. (2018))

longer in the country and were not de-institutionalized during the dictatorship. Economics follows a similar pattern of hierarchy but with more presence in the adjunct position. Sociology and Psychology have nearly half of their total agents within the lower post (assistant). Social Anthropology, Political Science and Literature have most of their effectives in between the assistant and independent categories.

The relation between men and women has remained relatively constant for the total CONICET researchers. Women occupied 54.3% of the positions in 2003, and 55.1% in 2014. The most balanced fields are philosophy and sociology. Male researchers account for more than 65% of researchers in economics, and 60% in political science, while the strongest feminized disciplines are psychology, literature and anthropology.

Inter-institutional asymmetries are visible when observing the distribution of the workplaces for CONICET researchers. While the exact and natural sciences have higher percentages of researchers at institutes with exclusive administrative dependency of CONICET, the dominant workplace of SSH researchers is in national universities (59.5% of the total), as can be seen in Fig. 9.5. The exception is Economics,

Table 9.3 CONICET SSH researchers by discipline and hierarchy n = 1711 (2014)

Discipline	%	*n*	Superior (%)	Principal (%)	Independent (%)	Adjunct (%)	Assistant (%)	No data (%)	Total (%)
Social Anthropology	9.4	160	0.0	6.3	20.0	35.0	38.8	0.0	**100**
Political Science	6.2	106	0.9	10.4	16.0	33.0	37.7	1.9	**100**
Economics	6.1	104	1.9	5.8	18.3	45.2	26.9	1.9	**100**
Philosophy	11.9	203	2.5	8.9	26.1	34.0	26.1	2.5	**100**
Literature	7.1	121	0.0	5.8	28.1	31.4	33.9	0.8	**100**
Psychology	4.7	80	0.0	6.3	15.0	32.5	46.3	0.0	**100**
Sociology	17.3	296	0.7	5.7	18.6	26.4	45.9	2.7	**100**
History	13.5	232	1.3	6.9	27.6	31.9	30.2	2.2	**100**
Others	23.9	409							
Total	**100**	**1711**							

Source Beigel and Gallardo (2014)

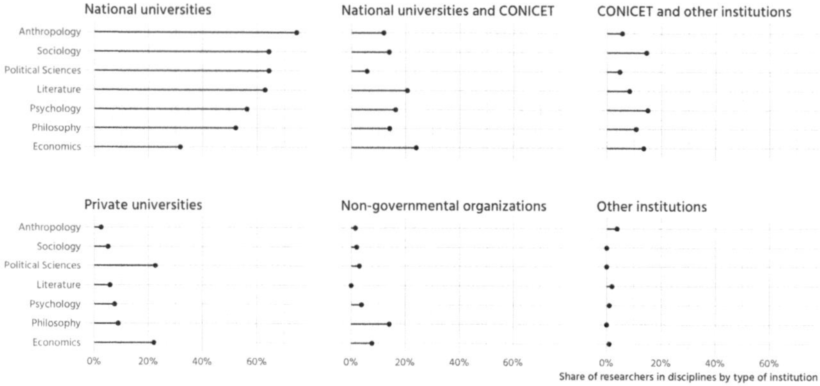

Fig. 9.5 CONICET researchers 7 SSH disciplines per workplace, (n = 1072)

where the workplace in national universities is slightly over 30%. SSH researchers at exclusively dependent CONICET institutes represent 10.7% of workplaces. As for double dependency (CONICET-university institutes), higher values are given in Literature and Economics, with over 20% each. Compared to other scientific areas, in the SSH private universities reach a relatively high number of workplaces (9.1%). Low values are seen in most disciplines—around 5%—except in Economics and Political Science, where over 20% of workplaces are located in private universities. Finally, the two cases in which workplaces of exclusive dependency from CONICET are higher are Psychology (15%) and Sociology (14.5%).

Regarding the preponderant role of the national universities as workplace for the SSH researchers at CONICET, two main asymmetries must be highlighted. First, the institutional concentration and the overwhelming weight of the UBA, which brings together 45.9% (426/928) SSH researchers based in universities. Far behind come the other three universities located in the Greater Buenos Aires area (San Martin, Quilmes, La Plata), with the rest dispersed around the country (in 29 different places, together comprising 37% of the total researchers). For the major and more prestigious universities, Gallardo (2015) has observed a strong association between the pre-graduate and doctoral degree institution and the same university as a workplace at

CONICET; 8 of 10 researchers with workplace at UBA or UNLP did their undergraduate degree at the same institution, and 65% of them also obtained their Ph.D. in the same university (2015: 142–143). Second, we can observe the geographical concentration of research capacities in the metropolitan area, which is a historical feature of the structure of Argentina's scientific field. The concentration is even higher because many private institutes and other types of institutions are also located in the capital city.

The Research Institutes

Focusing now on the institutional setting, the most extended scientific loci are the institutes, whose main activity is research. Some of these are integrated within teaching departments, while others develop within 14 big regional centers created by CONICET exclusively for research. At national universities, there are institutes of all sizes, with diverse degrees of consolidation; research is also done in traditional chairs, frequently led by isolated individuals. Institutes belonging to CONICET have standardized features. Many efforts have been made in the last decade to bridge the old split between CONICET and the national universities, although disconnection still prevails in many provinces. By 1999, 15% of total researchers had their workplace in an institute depending exclusively on CONICET; 22% at institutes of double dependency (National University + CONICET) and 44% at institutes belonging exclusively to national universities. In 2014, almost half of the total researchers worked at double dependency institutes (47.26%), only 6.59% at institutes depending exclusively on CONICET, and 29.5% at institutes belonging exclusively to national universities. Most of these latter belong to UBA and UN Córdoba.

Bekerman (2017) has collected primary data on the universe of the total research institutes in Argentina for all scientific areas: 813 institutes belong exclusively of the national universities and 233 institutes are organized under CONICET (or double dependency CONICET + national universities). The exact and natural sciences are mostly based in the CONICET orbit: 54% of these institutes and

76.1% of the researchers. Biology and health are more equilibrated in terms of institutes: 53% belong to the national universities while 47% belong to CONICET. However, 75.8% of the total researchers from this area work in a CONICET institute. Engineering and Agricultural Sciences have more institutes in the orbit of the national universities (87%), but the bigger ones belong to CONICET and 65% of the total researchers work at these institutes. The social sciences and humanities are in an opposite situation, with a marked trend toward the universities: 91% of the institutes depend exclusively on a national university and 65% of the researchers at CONICET work at an institute dependent upon a national university. As we have seen, most of these researchers work at one specific university: UBA.

Observed by discipline, the distribution of institutes and researchers is different at CONICET and national universities; 35% of CONICET institutes belong to the Natural and Exact Sciences and 31% to Biological and Health Sciences. Concerning the university institutes, 51% belong to the SSH and 30% to Engineering and Agricultural Sciences. If we now isolate the SSH institutes, we see that an astonishing majority (91%) depend exclusively on a national university while 9% are part of the CONICET system. This particular nature of the SSH institutes can be understood by returning to the historical features of its institutionalization in Argentina. During the last dictatorship (1976–1983), virtually no new SSH institutes were created at the universities, while there was a moderate growth at CONICET, an agency that experienced expansion during those years, as we have shown. SSH institutes at the universities started growing in 1984 while, at CONICET, this was a period of stagnation. In 2003, both spaces show an increase, although this is much higher at the universities. Between 2008 and 2014, 87 institutes were created at the national universities against only 13 at CONICET (Beigel et al. 2018).

If we look at the distribution of the SSH institutes in universities, a relatively fair geographic distribution arises. Seen through the quantity of researchers per institute, however, a concentrated structure emerges in the old and prestigious universities. In absolute numbers, 16% of the institutes belong to the National University of Rosario, followed by Cuyo and Tucumán (11%) and San Martin. But the share of the UBA (9%) does not reflect the real distribution of research capacities. UBA's institutes are the largest in terms of quantity of full-time researchers and resources. Only the Instituto Gino Germani had 203 researchers, 142 research assistants and 248 fellows by 2015 (Memoria IIGG, 2014–2015). Of the total 457 SSH institutes, 27% are multidisciplinary while the strongest disciplinary institutes are in Economics (11%) Literature (10%), Education and Arts (9%). At CONICET, the interdisciplinary institutes are dominant and traditional discipline institutes are in anthropology, economics, philosophy, history, geography and literature. At the national universities, there is more diversity, with the presence of arts, communications, psychology and public administration (Bekerman 2017).

The Output and the Publishing Circuits

The structural asymmetries of the field and the principles of legitimation in conflict observed between the evaluative culture at CONICET and the national universities have direct consequences for the circulation of knowledge produced locally. In other works, Beigel (2014b, 2015, 2017) has observed that publishing circuits have a determinant role in academic career-building and that these circuits are segmented according to 3 factors: language of writing (English being hypercentral), discipline and the institution of affiliation. In the case of Argentina, the universe of researchers whose publications we are observing gain different types of local academic rewards according to the evaluative cultures dominant in their institutions. Table 9.4 illustrates the distribution of these circuits of recognition, how they are related to disciplines and language, although institution is the determinant factor. Starting a career as a researcher at CONICET depends mainly on publishing articles in indexed journals and researchers highly value international recognition. This dynamic is prevalent in all scientific areas; however, SSH are oriented to the regional circuit while the exact and natural sciences to the mainstream.[6]

[6]For a thorough discussion on the integrity of the field and the institutional stake, see Beigel et al. (2018).

Table 9.4 Segmented circuits of academic recognition in Argentina

Circuit	Institution	Prevalent disciplines	Language
Local	Non-metropolitan National Universities	SSH	Spanish
Regional–Transnational	Metropolitan National Universities and CONICET	SSH and agrarian sciences	Spanish Portuguese
Mainstream	CONICET and internationalized institutes at UBA-UNLP-UNC	Exact and natural sciences, Medicine and Biology, Archaeology, Engineering, technologies	English

Two circuits are strongly segmented and representing the extremes of opposite evaluative cultures (local/mainstream).

There are few studies on the styles of publication and the impact of these circuits for the SSH career-building in Argentina. Gantman's (2011) study on the CVs of Argentinian researchers of political science, sociology, anthropology and psychology argues that the articles published in Spanish and local journals are prevalent, while very few publish in English and/or SSCI indexed journals. A recent study by Baranger (forthcoming) based on the CVs of the universe of the SSH researchers at CONICET shows a different picture. Around 23% of the SSH CONICET researchers follow the standards of the natural and exact sciences, 43% respond to a group featured by lower rates of productivity and a national orientation, publishing only in Spanish. Another 34% is represented by a cosmopolitan group with a high rate of productivity featured by publishing in languages other than English. The distribution of these styles of production within disciplines is revealing. Sociology and anthropology appear to be less dependent on WoS standards and journal rankings, while resorting to cosmopolitan and local orientations. Economics and psychology have a majority of researchers with a publishing style following the standards of the natural and exact sciences. Regarding languages, Baranger points out that "Anthropology, History and Sociology have the lower share of researchers that have published in English at least once, at the same time with relatively high percentage of publishing in French".

The survey on the evaluation committees for tenure (Beigel 2014a, b) and the study on the *career-best publications*[7] selected by the CONICET researchers to apply for promotion (Beigel 2017) show that an internationalized evaluative culture is prevalent among CONICET researchers, along with a strong tendency to publish papers in English and in mainstream journals. The SSH researchers are part of a general shift showing a preference for non-Argentinian journals and replacing the evaluation of the originality of published papers in favor of the acceptance of judgment on the journal made by international indexes. However, one difference is remarkable: in the SSH, Latin American indexing services are valued, while in the exact and natural sciences, the SCI (WoS-Clarivate) is the unique gold standard. Similar to other parts of the globe, the use of SCI or Scopus promotes the evaluation of individual authors via the impact factor of the journal in which they publish and assign factions in cases of joint authorship; it goes without saying that this routine drives off the real assessment of the quality of individual contributions. The general tendency in the younger generations of SSH researchers at CONICET is to select publications in the format of papers instead of books, and to point to publications in English.

Let's now delve into the relation between these conflicting, evaluative cultures among CONICET and the national universities and the landscape of Argentinian scientific journals. Beigel and Salatino (2015) built a database covering all active Argentinian SSH journals. The most challenging task was to identify non-indexed journals edited at faculties or departments across the country. In a survey performed in every national university addressed to elaborating a policy to digitalize journals that were edited only in paper, two numbers give a rough picture: 349 out of 716 journals are not indexed and 345 out of 716 appear in paper only. The majority of the indexed journals are covered by regional

[7]When applying for promotion, a CONICET researcher must select his 5 career-best publications and submit them into his online presentation. We built a data base with 23,852 career-best publications in order to assess tendencies regarding styles of production (books, articles and presentations to congresses) and prevalent publishing circuits. It is relevant to note that in the comparison with the complete CV, the trajectories are less attached to the patterns observed in the 5 career-best publications (see Beigel 2017).

systems, such as LATINDEX, REDALYC and SCIELO. As argued by Beigel (2010) a regional academic circuit exists in the SSH since the 1960s and was particularly dynamic up to the time when the military dictatorship in Chile in 1973 disarmed its base in Santiago de Chile. This regional orientation was manifest in the name of relevant "Latin American" journals, such as the Revista Latinoamericana de Sociología, edited by the Torcuato Di Tella Institute and directed by Juan Marsal (Sorá and Blanco 2017).

Most of the Argentinian journals that are indexed are published in the metropolitan area of Buenos Aires, emphasizing the centralist tendency already observed for researchers and institutional capacities. The journals that are indexed in transnational open access indexing services—(DOAJ) or repositories such as DIALNET—are published in digital format and stand out as thematic or interdisciplinary. In comparison with the journals indexed in the regional circuit, these journals are edited in different provinces in Argentina with at least 10 journals per region, a large number in Cuyo, Northeast and Northwest regions.

Only 25 journals (5% of all SSH journals currently in circulation) are indexed in WoS-Clarivate or secondarily in SCOPUS. In general, these journals are old-established, 40% of them were founded between 1943 and 1980 and they belong to political science, economics and psychology (for example, the *Journal of Applied Economics*, *Intersecciones en Antropología*, *Estudios migratorios latinoamericanos* and *Salud Pública*). The inclination for open access of scientific communication in the social and human sciences in Argentina is clearly seen in the fact that even the journals indexed in restricted access systems (WoS, Scopus or Ulrich) are also indexed in other systems belonging to the regional circuit; journals are opting for the democratizing momentum of the Argentinian scholarly publication market. Among these are publications such as *Intersections in Anthropology*, *Mundo Agrario*, *Runa*, *Synthesis*, *Andes*, *Cuadernos del CILHA*, *La Trama de la Comunicación*. These journals belong to literature, sociology, philosophy, history and psychology; they are published in the most prestigious National Universities and were created in the 1990s.

In contrast to these journals, publications restricted to local circulation are mostly edited in national universities from the Northwest and Northeast regions. Of the 222 non-indexed journals, 68% are published only on paper, which greatly reduces their circulation and effective access outside the geographical limits of the publishing institution. In this circuit, we considered the journals that, despite being indexed in Latindex, are not available in full text, limiting their access and circulation (Beigel and Salatino 2015).

The disciplinal morphology of the Argentinian journals is evidence of what Heilbron and Gingras (2015) call the "resilience of disciplines." A significant group of the SSH journals can be classified as interdisciplinary and make up 34% of the total. From the 1990s onward, journals addressing specific issues based on interdisciplinary approaches flourished. However, as can be seen in Fig. 9.6, numerous new disciplinary journals have been created since 2001 and these are strongly driven to imitate international styles of publication and circulation. This was particularly noticeable in history, literature, psychology and political science. Illustrating the historical development of SSH journals, we divided all the active journals from eight INTERCO-SSH disciplines plus interdisciplinary journals into the time periods of their founding.

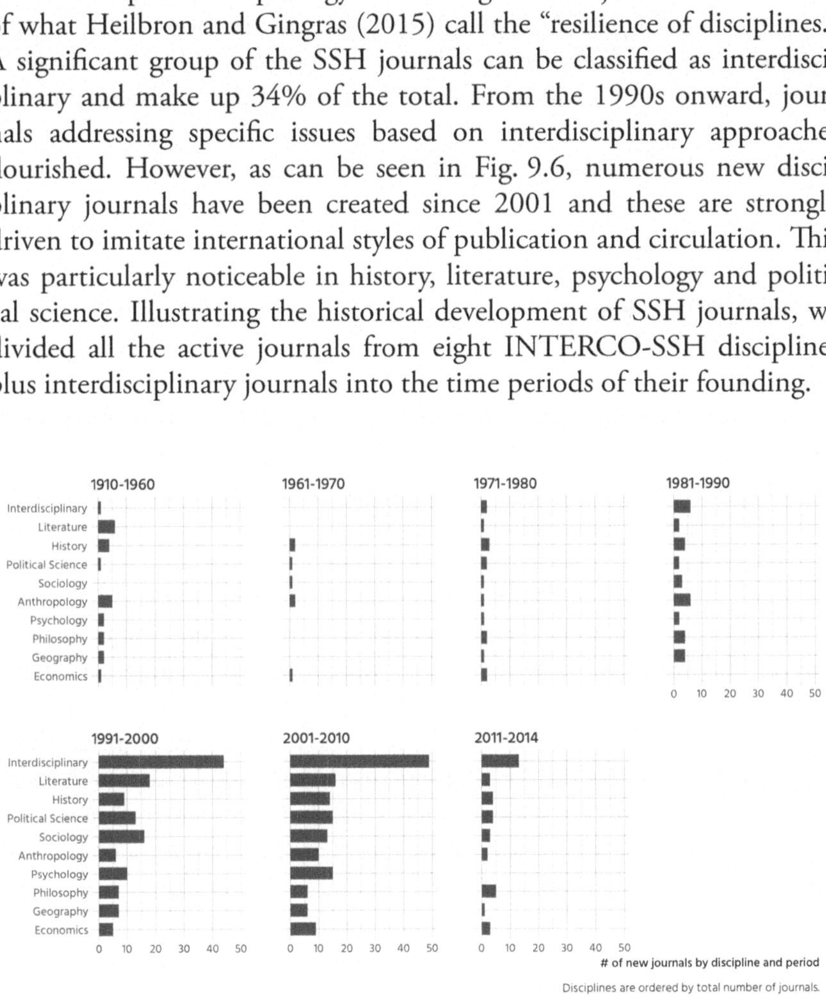

Fig. 9.6 Founding period of SSH journals 1917–2015, for selected disciplines (n = 559) (*Source* Beigel and Salatino (2015))

Anthropology, economics and literature are represented in the earlier decades—up to 1970—with half a dozen journals. History is somewhat different with a few more journals enduring to the present. Such is the case of the *American History Bulletin Dr. Ravignani* and the *Journal of Argentine and American History*. Economics is particularly strong, with a precursor journal like *Economic Development*, founded in 1958, which remains a leading publication. Also worth mentioning is the unique *Journal of Applied Economics*, published in English and indexed in WoS. Between 1960 and 1980, many social science and humanities journals were created, but few survived. A noteworthy feature of the last two decades is the increasing emergence of interdisciplinary or thematic journals.

Final Remarks

This chapter sought to objectify the historical setting of the academic field in order to characterize the structure and dynamics of the SSH in contemporary Argentina. We analyzed the recent institutional expansion and growth in terms of research capacities in 2004–2015, along with the structural and emergent geographical and institutional asymmetries. The analysis was condensed in the last cycle of expansion and we have attempted to understand the differences between national universities and CONICET largely because the tensions and potentiality of this scientific field lie in the "hinge" between these two major institutions.

Focusing on the geography of the SSH population of researchers, we show that the CONICET researchers are concentrated in a single university UBA while the professors-researchers are, for the most part, better distributed. This is due to the fact that the autonomy of the universities prevents the imposition of international standards for tenure. Regarding styles of production and circulation, CONICET researchers tend to produce increasingly more papers than books. Publishing in Argentina is not generally rewarded in career terms and these researchers have made an important effort to increase publishing in English. The dominant circulation for SSH CONICET researchers is within Latin

American journals indexed in regional systems while the social scientists with teaching positions (not CONICET researchers) tend to publish in Argentinian journals, frequently edited in their own institutions.

Accordingly, along with the expansion, the gap between Argentine scientists versed in the dominant production styles of the world academic system and those with a more endogenous agenda has widened. The dynamic of these styles of production is connected to the symbolic capital at stake in the field and, in particular, the role played by publishing in a field with diverse principles of legitimation in tension. These principles can be observed in the existence of different *circuits of recognition*, all locally valuable, but with different rewards according to the institution. We have argued that these circuits are fueled by struggling evaluative cultures and the particular features of the publishing landscape in the country. To publish in international journals is highly valued for tenure at the national research agency, the CONICET. To publish in national or local, non-indexed journals, is mainly rewarding for teaching career-building at non-metropolitan national universities.

In a higher education system such as Argentina's, with a strong tradition of university autonomy and politicization, the local circuit of recognition is a very dynamical space with hundreds of journals edited by university departments, distributed mainly in paper, where local scholars publish their works, far from the standards of international publishing. Given that these large and extensive local circuits have not yet been studied, we cannot presume its scientific worth, but we can indeed affirm that this "national/istic" orientation still prevails in many institutions, and especially in the SSH. In contrast, we have observed the evaluative culture at CONICET, where there is a replacement of the assessment of quality and originality in favor of indexation, impact factor or h-index—bibliometric data whose link to defining quality is currently under scrutiny. In the midst of such diverse (even opposite) evaluative cultures, the Argentinian social scientists with a local orientation and those with an international agenda live together in conflict, given the existence of two national research careers (one at CONICET, another at the national universities), along with autonomous regulations for tenure at each university.

Currently, this recently expanded field shows an uncertain future and a contractive cycle is envisioned under the neoliberal orientation headed by the right-wing new President Mauricio Macri. The growth of the scientific national system was suddenly slowed down. CONICET experienced a budget cut-off that affected the full-time positions annually offered, diminishing in 50% by 2016. The orientation of previous scientific policies, concerned by equitable growth of all disciplines is today in dispute. Basic research is relegated while applied research is awarded with half of the available positions. This affects directly the social sciences and humanities which had experienced an impressive recovery that seemed to finally make up for de-institutionalization performed by the last dictatorship.

References

Baranger, D., and F. Niño. Forthcoming. *El espacio de las disciplinas sociales en el CONICET*. PISAC.

Beigel, Fernanda. 2010. Social sciences in Chile (1957–1973). A laboratory for an autonomous process of academia-building. In *Academic dependency in the social sciences: Structural reality and intellectual challenges*, ed. Syed Farid Alatas and Kathinka Sinha-Kerkhoff, 183–212. New Delhi: Manohar.

Beigel, Fernanda. 2013. The politics of academic autonomy in Latin America. In *The politics of academic autonomy in Latin America*, ed. Fernanda Beigel, 1–27. London: Ashgate.

Beigel, Fernanda. 2014a. Current tensions and trends in the World Scientific System: Alternative circuits and new forms of peripheriality. *Current Sociology* 62 (5): 743–765.

Beigel, Fernanda. 2014b. Publishing from the periphery: Structural heterogeneity and segmented circuits: The evaluation of scientific publications for tenure in Argentina's CONICET. *Current Sociology* 62 (5): 617–625.

Beigel, Fernanda. 2015. CULTURAS [evaluativas] ALTERADAS. *Política Universitaria*, N°2, IEC-CONADU: 11–21.

Beigel, Fernanda. 2017. Peripheral scientists, between Ariel and Caliban. Institutional know-how and circuits of recognition in Argentina. The "career-best publications" of the researchers at CONICET. *Dados–Revista de Ciências Sociais, Rio de Janeiro* 60 (3): 825–865.

Beigel, Fernanda, and Maximiliano Salatino. 2015. Circuitos segmentados de consagración académica: Las revistas de ciencias sociales y humanas en Argentina. *Información, Cultura y Sociedad* 32: 7–32.

Beigel, F., and O. Gallardo. 2014. *The SSH researchers in Argentina*. Interco-SSH Report N°2. Mendoza: PIDAAL.

Beigel, F., O. Gallardo, and F. Bekerman. 2018. Institutional expansion and scientific development in the periphery: The structural heterogeneity of Argentina's academic field. *Minerva: A Review of Science, Learning and Policy*. https://doi.org/10.1007/s11024-017-9340-2.

Bekerman, Fabiana. 2013. The scientific field during Argentina's latest military dictatorship (1976–1983): Contraction of public universities and expansion of the National Council for Scientific and Technological Research (CONICET). *Minerva: A Review of Science, Learning and Policy* 51 (4): 253–269.

Bekerman, Fabiana. 2017. Distribución desigual de las capacidades de investigación en las ciencias sociales argentinas: una mirad relacional [Unequal distribution of research capacities in the Argentine social sciences: A relational perspective]. *Revista Iberoamericana de Ciencia, Tecnología y Sociedad* 13 (37): 257–288.

Blanco, Alejandro. 2010. Ciencias Sociales en el Cono Sur y la génesis de una nueva elite intelectual (1940–1965). In *Historia de los intelectuales en América Latina*, ed. Carlos En Altamirano, vol. II, 606–629. Buenos Aires: Katz.

CONICET. 2016. http://www.conicet.gov.ar/acerca-de-conicet-en-cifras/.

Feld, A. 2009. El Consejo Nacional de Investigaciones: estado y comunidad científica en la institucionalización de la política de ciencia y tecnología argentina (1943–1966). In *Conocer para transformar. Producción y reflexión sobre Ciencia, Tecnología e Innovación en Iberoamérica*, ed. H. Vessuri, et al., 133–152.

Gallardo, O. 2015. Trayectorias de formación de investigadores del CONICET. *Sociedad* 34: 129–147.

Gantman, E. 2011. La productividad científica argentina en Ciencias Sociales: Economía, Psicología, Sociología y Ciencia Política en el CONICET (2004–2008). *Revista Española de Documentación Científica* 34 (3): 408–425.

Heilbron, J., and Y. Gingras. 2015. La résilience des disciplines. *Actes de la recherches en sciences sociales* 210: 5–10.

IIGG. 2015. Instituto Gino Germani, Universidad de Buenos Aires. Memoria institucional.

Rovelli. 2015. Un modelo para armar: áreas prioritarias e investigación en Universidades Nacionales. *Revista Ciencia, Docencia y Tecnología* 26 (51): 26–53.

SCEU. 2016. *Sistema de Consulta de Estadísticas Universitarias.* Secretaría de Políticas Universitarias. http://estadisticasuniversitarias.me.gov.ar/#/seccion/2.

Sorá, Gustavo, and Alejandro Blanco. 2017. Unity and fragmentation of social sciences in Latin America. In *The social and human sciences in a global perspective*, ed. Johan Heilbron, Gustavo Sorá, and Thibaud Boncourt. Basingstoke: Palgrave Macmillan.

SPU. 2013. *Anuario de Estadísticas Universitarias* [University statistics yearbook]. Buenos Aires: Secretaría de Políticas Universitarias, Ministerio de Educación de la Nación. http://portales.educacion.gov.ar/spu/investigacion-y-estadisticas/anuarios/.

UNESCO. 2016. International Institute for Statistics: Indicators for science and development. http://uis.unesco.org/indicator/sti-rd-hr-res.

10

Concluding Remarks

Christian Fleck, Matthias Duller and Victor Karády

These concluding remarks are not executive summaries of the eight chapters but do five things. First, we discuss why the nation state molds any generalizable knowledge claims before looking into conditions supporting international collaboration in different scholarly disciplines. We will then try to make sense of the variety of nation states in Europe with regard to the systems of higher education and research before analyzing how disciplines developed differently during the last seven decades in European countries. Finally, we sketch an explanatory model for the forces behind the rise and decline of scholarly fields.

C. Fleck (✉) · M. Duller
University of Graz, Graz, Austria
e-mail: christian.fleck@uni-graz.at

M. Duller
e-mail: matthias.duller@uni-graz.at

V. Karády
Central European University, Budapest, Hungary

C. Fleck et al. (eds.), *Shaping Human Science Disciplines*, Socio-Historical Studies of the Social and Human Sciences, https://doi.org/10.1007/978-3-319-92780-0_10

The Unsurmountable Methodological Nationalism

Since Martins (1974) and Smith (1983) introduced the label "methodological nationalism" for the fact that social scientists and scholars from the humanities often select the nation state as their "natural" object due to the availability of data by state agencies, a kind of consensus has been reached about such a strategy in deficit (cf. Wimmer and Schiller 2002). Without elaborating on the foundations on which this practice settles, or the factors that support it, we want to underscore one aspect favoring nation state-based analyses in the study of the SSH. Whereas most natural (and formal) scientists can claim universal applicability of their findings, social scientists and other students of what have been sometimes disparagingly called the "inexact sciences" (Helmer and Rescher 1959; Hausman 1992; Boumans 2014) create bounded knowledge. Ethnomethodologists label this feature "indexicality", which means that any statement reveals its meaning only if there is a local referent assigned. Physical laws, biological mechanisms, and certainly mathematical formulae are not affected by whether they are applied by someone in Oxbridge or in Transnistria. Similar universality of knowledge is rare in the fields of SSH; there, the more universal a proposition, the less informative it often is. This is, of course, very old news. The inventor of sociology, Auguste Comte, made the same point in 1839 and it has concerned methodologists and philosophers of the SSH ever since. For the present purpose, it is sufficient to remind the reader that the overwhelming portion of SSH knowledge is bound to conditions that are not the same across space and time. The universal applicability even of such well-established SSH claims as the Philipps Curve in economics, or the demographic transition, does not stand up to close scrutiny. Neither does increasing wealth affect fertility unidimensionally, nor does the inverse relationship between the rate of unemployment and the level of inflation work everywhere and at anytime. This insight holds true on levels far less aggregated than nation states. For example, findings about the level of satisfaction with gratification systems in industry are not easily transferable to other businesses, not to mention exporting them abroad.

This is not to insinuate that *all* SSH knowledge is true only locally. However, one should bear in mind that there is a reasonable part of knowledge in the SSH that is, in one way or another, anchored in the societal and historical territory to which it applies and where it holds particular meaning; unchained knowledge loses its heart.

Neither is this a claim that knowledge produced in SSH is ontologically different from knowledge in the natural or "hard" sciences. John Searle emphasized the difference between a dollar and gravity. A dollar works only because actors agree on its function and meaning. Gravity, however, doesn't give a damn about the opinions of those living under its effect (Searle 2005). Nevertheless, even if all SSH knowledge is based on convention, some elements are applicable worldwide while others lack this universality. Money, marriage, and chess exist only because of our mutual agreement about what they mean and how they should be handled. These objects are culturally bounded, but their global usability is not affected by the fact that they are based on convention. If people accept Bitcoin as money, it works as long as people stick to their persuasion. If someone fails to acknowledge that marriage is primarily a juridical regulation but holds it as a realization of religious obligations, the same person could not accept the extension of the right of marriage for partners of the same sex. Nevertheless, SSH scholars can compare different modes of marriage transgressing cultural and conceptual borders. Systems of rules and instructions created and kept alive by humans exist as long as some humans stick to them. But people in one corner can follow different conventions to those populating nearby territory. The different systems of business accounting represent different systems of knowledge that are "true" only if one accepts the frame. The United States Generally Accepted Accounting Principles (US-GAAP) and the International Financial Reporting Standards (IFRS), for example, could be used in parallel, but they work only if the frame is considered. How rich a firm is depends on what system of accounting it uses; remarkably, in the business world one can play different versions of chess. From a sociological perspective, the simultaneous existence of two mutually exclusive systems of doing—in this case accounting—is a much more interesting topic than the widely overused constructivist claims.

Our discussion of types of knowledge produced by SSH does not follow the popular distinction between fundamental vs. applied disciplines, or its offspring Mode I and Mode II sciences (Gibbons 1994; Nowotny et al. 2001). Only very local knowledge, produced at the level of factories or communities, for example, resembles applied knowledge in this sense. We should keep in mind, however, that sometimes scholars from SSH emerge from their small fields of specialization and report something of worth outside of their usual place. This could even happen for micro-sociological insights, by definition the most local type of knowledge. The rest of the knowledge also produced within SSH echoes the particularities and conditions of their socio-cultural environment which is, due to convention, simultaneously a territory defined by the institutions of the given nation state. In some cases this societal embeddedness might appear to have negligible impact on the knowledge produced in other cases, but the nexus is undeniable close. A proposal that practically all knowledge produced in the SSH is indexical might not be an over exaggerated.

It would be misleading to expect that the SSH will ever be in the position to compete with the knowledge producing branches of scholarship that do not show comparable bounded types of knowledge claims. The nature of the SSH disciplines is their embeddedness in cultural and social environments, which molds their results in a lasting way. While we hold that locality and embeddedness is thus inherent to practically all SSH knowledge, how that locality is defined depends very much on the questions one is interested in studying. Nation states are only one of a number of possibilities with which to draw the epistemological boundaries within which a phenomenon receives significance.

Cooperating Internationally

One could look also at the same debate from the angle of cooperation within and beyond the borders of nation states. Although recognition for academic work at its highest level is dominated by international practices (prizes, medals, membership in transnationally elected positions of excellence, invitations to dominant places of scholarship), most researchers need to establish their reputation on several levels simultaneously.

If science and scholarship are, besides other characteristics, a "recognition game" then it is a bundle of interlocked games. For research induced activities these are teaching, academic administration and what is now called the organization of social impact. Each of them asks for different types of effort and no scholar is encouraged to participate in all of them at the same level of intensity. In addition to this horizontal differentiation, we have to take into account a vertical differentiation too. Depending on the size and build-up of the particular nation state to which scholars belong, they strive for recognition not only on the national level and regional forums (e.g. in federalist or big states) but also in local academic micro-environments. Attempts to secure the continuous opportunity to work and contribute to their field of choice could happen on one or more of these vertically integrated arenas.

There are no strong hints that the international or European level is the dominant one. Nevertheless, diverse readings of internationalization are continuously becoming part of the game that need to be followed by new entrants to the field if they strive for success in disciplines beyond their national contexts, be it the publication in a lingua franca in contrast to the national language, attendance at international conferences, training or teaching jobs abroad or cooperation with disciplinary peers on an international level.

Given the particularities of national traditions and the functioning of academic fields, international cooperation—in the form of joint research and publication endeavors—comes at a cost and thus needs justification. One source for justification is the need for the SSH to study increasingly transnational political processes, such as European integration, because these provide a common object of interest for scholars from different countries. This is an area that the European framework programs have supported in the SSH since at least 1994 (Schögler and König 2017). Another justification might be awareness of problems that exist in different countries in the same or similar kind, which makes cooperation fruitful for mutual exchange of knowledge and ideas. One of the most promising projects in the EU schemes is the search for "best practices," where international comparison is mandatory. Finally, cooperation can be desirable if researchers at a particular place possess knowledge, information, or skills that potential partners abroad want to acquire. Cooperation of this kind is more likely to emerge where national disciplinary fields are fragmented and thus small communities exist with little opportunity to maintain sufficiently dynamic exchange practices within their countries (Fleck and Hönig 2014).

All three conditions have become more salient throughout the past 50 years, although to different degrees in the respective disciplines. Linguistic subtleties, explicit or implicit references to cultural particularities and shared historical understandings are among the factors that make literature much harder to transfer into other national contexts. On the other end of the cognitive spectrum there are universal analytical forms heavily reliant on mathematics. Econometrics' algorithms, for example, travel smoothly through cultural and linguistic spaces and need virtually no translation of differences in meaning. Only slightly more challenging are knowledge forms that rely on well-established abstract concepts as in, for example, parts of analytic philosophy, not to mention the quintessential universalistic logic. Though dependent on verbal language, such meanings are largely decoupled from local contexts and variations and thus travel easily. Natural science knowledge principally consists of these knowledge forms. The differences in the relative importance of particular and universal knowledge forms explain why the humanities have been much less shaped by internationalization than the social sciences. But even in the humanities, we can identify heavily internationalized research branches, archaeology, for example, with its move towards a scientific type of digging as in the case of Ground Penetrating Radar, or Classical Studies with their astonishing cosmopolitan research agenda.

Among visible signs of these differences is the unequal inclination to publish primarily in English, as is the case in economics or psychology. Accordingly, publication markets and mechanisms of scientific recognition work differently for these disciplines (Heilbron et al. 2017). What is more, many SSH disciplines are internally differentiated in a similar way. Sociology, to take the most obvious example, hosts communities that produce highly contextualized literature close to what has been characterized as knowledge typical for humanities, for which there is little or no incentive to publish in foreign languages since not much international interest or cooperation can be expected. On the other hand, particular parts of quantitative sociology show very similar publication practices to economists and psychologists, because there is a broad international consensus on the methodological standards and tools they use. Their final results need little contextual clarification because they are liable to reach trans-lingual similarity (e.g. Eurobarometer, European Social Survey, International Social Survey Programme, Kropp 2017). Methods and topical foci are thus crucial elements in differentiating scientific fields that are likely to engage in international cooperation, favoring quantitative over qualitative approaches. However, other factors also need to be taken into account.

In the next section we raise the question of why states, especially nation states, are the most important frame of reference for our analyses of the SSH in postwar Europe.

How Different Are the Nation States in Europe?

Europe has about three dozen states, mostly nation states proper (with Switzerland as the only significant federal exception), depending where one draws the outer line of the continent and whether mini-states such as Andorra and Monaco and state-like entities as The Holy Sea are counted or not. Following Norman Davies, one could draw several lines across the map of Europe to indicate splits into subunits which were in force for some time: Catholic (Latin) vs Orthodox (Greek) Christianity; the Roman Limes and the Iron Curtain; industrialized Western vs peasant Eastern Europe etc. (Davies 1996). More recently, social scientists have developed a number of typologies for comparative analysis. Welfare state systems, varieties of capitalism and media systems are just three, substituting the Cold War bipolarity with a more differentiated classification of macro units (cf. Esping-Andersen 1990; Hall and Soskice 2001; Hallin and Mancini 2004). A tripartite division of systems of higher education prevailed in the nineteenth century, governed by the Prussian, Napoleonic French and British models. During the Cold War, the opposition between Western and real socialist Eastern

Europe affected higher education too in important respects. Beyond these outdated or rough differentiations, however, there are no widely used typologies for the system of higher education, research and innovation in contemporary Europe. Suggested comparative analyses tend to focus primarily on teaching methods and the range of disciplines but less often on governance and institutional regulations (Ziegele 2013; van Vught et al. 2010; Bartelse and van Vught 2007). OECD reports, as well as university rankings, focus either on the level of the basic institutional units (e.g. university) or the nation state. Groupings of states into larger types has occured more frequently since the fall of Communism, without consciously constructed typologies but rather arbitrarily, in the style of 'winners vs. losers'.

Categories and distinctions established in international comparative research so far do not make sense when studying the trajectories of a handful of SSH disciplines. Let us first consider differences at the level of universities. As regards the funding and governing of universities, the opposition between private and public institutions, which is of such crucial importance in the United States, does not play a comparable role in Europe.[1] Putting the United Kingdom aside because of the formal survival of its historic private universities (which are now financed to a large degree by public money such that their qualification as "private" is a misnomer of sorts), there are only few privately managed European universities whose fame transgress the borders of the country in which where they are located, like, for example, the Central European University in Budapest or Bocconi University in Milan. Most private universities do not play a big role in the competition between universities or the nation states, but we can identify differences in the funding regimes.

The heavily debated "economization" of universities, a.k.a. "neoliberal revolution," is of minor relevance when drawing differences between countries.

[1]According to data from UNESCO the share of students enrolled in private institutions of Higher Education increased in the first decade of the 21st c: In France from 15 to 17%, in Hungary from 13 to 15, in Italy from 6 to 8 and in Argentina in 2007 to 25%. The other countries from our sample did not report any numbers of students enrolled in private institutions. The data for UK are given as 100% private. (Altbach et al. 2009) Appendix, Table 2, pp. 199–205.

The change from a funding model of block grants to an allocation by competition has been carried out almost everywhere, earlier or more radically in some countries than others. Nevertheless, the old-fashioned regime of high trust invested in the professoriate, whereby it could follow its own maxims and routines without any surveillance or accounting, has now been extinguished. The regime change regarding funding was, however, neither caused nor accompanied by new private spenders. The competitively allocated money comes overwhelmingly from the state or agencies administering public money. Higher education in Europe depends still overwhelmingly on the taxpayer.

Unfortunately, there are not many comparative studies on the funding of SSH in Europe, but it should not be controversial to say that the economic fundament of any scholarly activity needs close consideration.

Governance is of equal relevance for every discipline, but also the overall picture of science and scholarship. Studies on the governance of the HE sector discuss it primarily with the focus on academic autonomy. The legal status of universities usually attracts more attention and debates than the rules and regulations affecting lower levels in the universities, like disciplines. It is essential to determine whether the central government (or other levels of the state) takes direct care of the universities, or hands over the governance to independent boards operating in a similar way to (and very often modelled according to) the governing bodies of private US universities. In Germany, for instance, all educational affairs are the competence of the regions, the *Länder,* with the consequence that there is not one German ministry responsible for higher education, but 16.

Academic freedom and autonomy are held in high esteem by all members of the "republic of scholars." In reality, the level of autonomy varies heavily between countries. We can illustrate the differences by looking at the rules and practices of hiring the professoriate. The competence for the appointment of the higher ranks of the teaching staff, full professors and their counterparts in other systems, can fall to the appropriate university's level of management, but it also can be the prerogative of the government. In some countries recruitment is done exclusively inside a single university, in others via nationally organized competitive exams—*concours*—as in Argentina and for some disciplines (like Law, Medicine) in France or Italy. Whenever authority is to some degree heteronomous, political and other non-scholarly considerations might influence the outcome. Autonomy in the appointment of the personnel is not necessarily dependent on the level of centralization of the state. In highly centralized France the minister for higher education cannot influence the appointment of professors, whereas in federalized Germany the *Länder* ministers are in charge of such appointments and can even intervene in the selection process. In both countries, however, professors were, until recent changes in Germany, civil servants. Not even Bourdieu in his famous *Homo Academicus* (1988) pays much attention to the particularities of the selection process in universities. It is not surprising, therefore, that we have only a handful of serious studies about stakeholders in academic human resource management. The selection processes of the 'Mandarins' entering the European universities are still a kind of black box.[2] Unfortunately, our study does not offer deeper comparative insights but only glimpses towards possible future efforts. Nevertheless, we are sure that the trajectories of disciplines have depended upon configurations located at the disciplinary level of a university's organization (e.g. departments). In looking at the disciplines more closely, we again encounter an instance of the autonomy-heteronomy dimension.

Comparing Countries and Disciplines from Inside

From a comparative point of view on the development of disciplines, we discuss the following issues. What are the dynamics behind the growth of disciplines? When does a particular discipline expand and under which conditions does it get more funds? Is the number of students choosing a particular discipline restricted by any measure (employability, central planning, etc.)? What are the criteria for

[2]For the US see the classic study (Caplow and McGee [1958] 2001); for France and Germany: (Musselin 2010; Lutter and Schröder 2016) are not interested in the particularities of the hiring process but analyze the outcomes only.

extending the staff of universities? Which kind of indicators do those who make such decisions take into account? Are decisions about spending made at the individual university or by more centralized state agencies? Who has a say in the appointment of which academic status groups? Does anything like trans-national benchmarking play a role with regard to the development of disciplines? What, if any, are the mechanisms of reducing academic personnel, and at what organizational level are such decisions located? Are disciplines mentioned and regulated in laws and regulations at the national level or do disciplines and their offspring grow uncontrolled? It should be clear from this list that for significant answers one needs a high degree of familiarity with the formal and informal organization of academia. None of the questions can be answered via ready-made statistical data, and are rarely covered in the literature on Higher Education (Fielden 2008; Hénard and Mitterle n.d. [2009]; Paradeise et al. 2009).

Compared with other branches of science and scholarship, the fate of the disciplines studied in the chapters of this book is highly dependent on public funding. According to the OECD, in 2005 the disciplines belonging to the SSH obtained the following amounts of funding: in the Czech Republic only about 15% of the country's R&D expenditures; in Austria, Germany and the Netherlands around 22%; in Hungary 30%; in Italy 35%; and in Sweden 38%. These figures compare to only around 10% in the United States (no data are reported for Argentina, France and UK) (Santiago et al. 2008). None of the established typologies used to 'divide' Europe come close to the divides in the spread of funding for SSH. Post-communist Czech Republic and Hungary are very far away from each other, while "Catholic" Italy and "Protestant" Sweden subsidize such allegedly useless knowledge similarly. One should keep in mind, however, that the containerization of disciplines might hide severe differences between the disciplines put together.[3]

Some lacunas are the effect of the insufficient provision of data (see below), but could also be the consequence of our willful decision.

[3]It is worth remembering that distributing agencies seldom publish statistics about less aggregated entities than the container SSH, STEM, BioMed etc. In addition, it would be necessary to relate any of these figures also to the number of researchers and similar statistical populations.

We decided to study the historical development of seven disciplines by focusing on the social morphology of these entities over a period of about seven decades; decades experiencing the deepest imaginable transformation of academic markets. There is a danger that we are "preaching to the choir," but it is worth remembering the three major transformations implemented everywhere in Europe since the end of WWII in this field. First, an expansion of the student population followed by growth of the number of their professors. Second, the opening for women not only of extended study opportunities and but also academic jobs; in most countries, after the first two transformations, came a reorganization of the hierarchy of consecutive study programs, and academic grades. The final change is usually called the Bologna process, because of the place where the European ministers for higher education were assembled when they signed their agreement on the continent wide harmonization of academic curricula and systems of graduation. Some European countries managed these changes better than others. Sweden seems to present a case of smooth transition directed by a strong central force, where Germany embodies the ups and downs of shortages and (over)production of academics, resembling what is known in economics as the "pork cycle."

If we use as a cross-national measure, the percentage of people aged 25–64 holding a HE degree at the doctorate level is only around 1 out of 100 in all countries covered in this book. However, the enrollment in tertiary education differs markedly between these countries. Of those who are 20 years of age in Sweden only 24% is enrolled in an HE program, preceded by Germany with 27%, Italy 31%, Hungary 32%, Argentina 33%, United Kingdom 40%, The Netherlands 43%, and France 47% (OECD 2016). Of course, given the different dropout rates, not all of them end up with a tertiary degree.

A further difference between the countries emerges if we look at the academic fields in which the population of graduates earned their degrees. We find huge differences between countries, even if the classifications of disciplines leave a lot to be desired (Table 10.1).

The differences between the two branches of SSH are more pronounced than the gender ratios of the graduates. In all of our countries women are now (data for 2014) outperforming men, but the

Table 10.1 Percentage of tertiary-educated people, 25–64 year-old non-students, by fields of education

Country	Humanities, languages and arts	Social sciences, business and law
France	9	25
Germany	7	23
Italy	23	29
Netherlands	8	34
Sweden	7	25
England	16	30
OECD average	10	27

Note No data for Argentina and Hungary; UK: England only
(*Source* OECD (2016), Table A1.5., p. 45)

gender ratios[4] for the humanities are between 1.7 (Sweden, UK) and 2.6 (Italy), and for social sciences between 1.1 (UK) and 2.2 (Hungary). (OECD 2016, Table A3.3, p. 70).

This book contains reports of the development of a sample of SSH disciplines over a longer time period. By putting the data for eight countries side by side, the reader gets a fair picture of highly diverse trajectories and different growth lines per discipline.

Shaping Disciplines

Besides all of the rhetoric about the end of disciplines and the upswing of inter- and trans-disciplinarity, academic disciplines still play a crucial role for the training and promotion of the next generation of scholars and for the self-identification of academic staff. Both in everyday and professional communication, disciplines serve to form we-groups, to which someone belongs and that function as institutional structures at several levels. Students select study programs which, even if they are of interdisciplinary orientation, are built on the shoulders of disciplines. Trained professionals of such studies identify themselves as members of one or more disciplines or locate their

[4]Number of female graduates for one male graduate.

doings on the crossroads of distinct disciplines. Furthermore, the organizational structures of universities and research centers are often representations of the disciplinary organization of the academic world. In a word, scholarly disciplines are similar to occupations, licensed vocations, or can be even seen as tribes, thought collectives, etc. The very existence of such we-group identifications is further proof of the persistant significance of disciplines.[5]

Considering this stable structure of disciplines as institutions, one would expect to find data on disciplines in statistics on research, higher education etc. Our empirical findings do not confirm such hopes. Several EU member states do not have any data available about the size of scientific disciplines with regard to the number of professors teaching it, or being employed in universities or extramural research units. On the level of disciplines, information is available for the enrollment of students or graduates at different levels (B.A., M.A., and Ph.D.), but corresponding numbers for their instructors are seldom to be found. In most cases the reason is quite simple; employees are counted according to administrative units. For example, sociology and political science are very often put into one aggregated category (in other cases, sociology and anthropology form one group). Further, professors and other sectors of the staff are regularly counted according to rapidly changing organizational units, such as faculties, schools, etc.

A similarly confounding classification of scholars happens in the census. The International Standard Classification of Occupations (ISCO) does not pay attention to well established entities such as scholarly disciplines, therefore a researcher interested in a comparative study about the number of philosophers or sociologists in a particular population, for example, cannot tell with certainty what is being compared.

In ISCO 88 we find, for example "2442 SOCIOLOGISTS, ANTHROPOLOGISTS AND RELATED PROFESSIONALS" defined as "Sociologists, anthropologists and related professionals investigate

[5]Disciplines are broader than study programs. According to a recent report in DIE ZEIT in Germany there are 19,000 different study programs in tertiary education. This number alone would legitimize the usage of the language of disciplines. DIE ZEIT 19.10.2017, p. 69.

and describe the social structure of societies, the origin and evolution of humanity, and the interdependence between environmental conditions and human activities, and make the knowledge obtained available as a basis for policy decisions."

Also "2443 PHILOSOPHERS, HISTORIANS AND POLITICAL SCIENTISTS" are defined as "philosophers, historians and political scientists work, mostly by reasoning, in the field of epistemology, metaphysics or ethics. They conduct research and describe past events and activities, including the development of social and economic structures, or cultural and political institutions and movements, and make the knowledge obtained available as a basis for political, diplomatic and related policies."

Both groups are ambiguously combined, to say the least, but they remained the same in the next edition of ISCO 08, changing only the coding from 2442 to 2632, and 2443 to 2633.

In trying to establish time series data on these two groups, one finds the first group in ISCO-68 as "1-92 Sociologists, anthropologists and related scientists" but cannot identify the second group, which was not been classified at the time. In ISCO-58, neither group is represented but might be found under "0-61 University Teachers" or "0-Y Other Professional, Technical and Related Workers".

This short digression into historical statistics of occupation corroborates what Pierre Bourdieu stressed some years ago as the unwillingness of those who classify to be classified. On top of that, it makes the assessment of changing academic fields almost impossible.

We nevertheless made an attempt to portray and analyze the trajectory of seven disciplines over more than half a century. Figure 10.1 offers a comparative view on the long-term development of the disciplines in five countries for which reasonable comparable data were available. Because all disciplines in all countries saw a massive expansion in absolute numbers over the past seven decades, we plotted the numbers for each discipline as the relative size compared to each other. What can be seen in Fig. 10.1 are the relative changes in the size of disciplines within the larger subset of SSH disciplines studies in this volume.

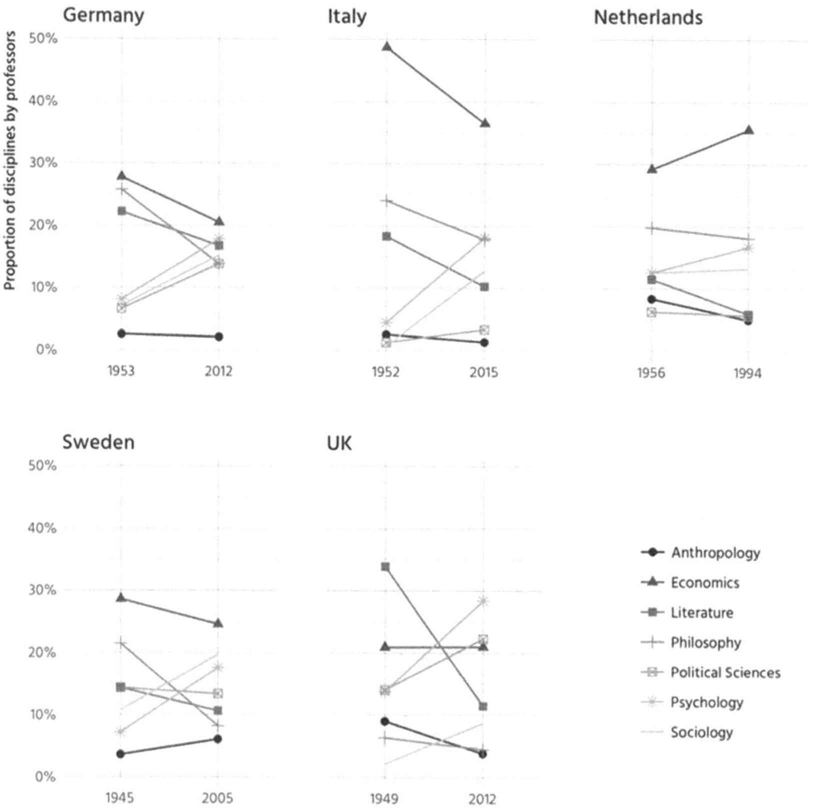

Fig. 10.1 A comparative view on the long-term development of the disciplines in five countries for which reasonable comparable data were available

In all but one case the units compared are the highest ranking faculty members, full professors and their equivalents elsewhere (in the case of United Kingdom alone we used the number of Ph.D. graduates deputizing their professors). There are indications that the growth rate of the highest rank in academia has been below those for the lower ranks everywhere but, for the shape of a discipline and the relative positioning of it *vis-à-vis* its neighbors and competitors, sticking with the "top-dogs" makes good sense. What one recognizes immediately is the diversity of slopes. In general the figure illustrates a point made earlier: Speaking about SSH as an aggregated entity is misleading with regard to the development of singular disciplines; they do not grow or decline in parallel. Since we excluded management studies from our comparison, the overall picture is relatively smooth. In Germany management grew over the same period by a factor of 30.9, compared with growth rates for the other disciplines ranging from 4.6 for anthropology to 19.9 for psychology; similar developments happened in other countries.

While the importance of some disciplines grew over time, others declined. Economics, which dominates the field of the SSH in most countries, declined most sharply in Italy, Germany, and Sweden, while it remained stable in the UK and increased its relative share in the Netherlands. Note again that business economics is not included, which saw the most rapid expansion over the same years. A similar point can be made for literature in Britain, Germany, Italy and the Netherlands, and philosophy in Sweden, Germany and Italy.

Figure 10.2 uses the same data but plots them as a comparison between disciplines. That is to say that the reader can again see the percentages of disciplines calculated at the country level and put into one graph per discipline only to illustrate the divergent slopes. The general slope depicts whether disciplines (across countries) experienced relative growth or decline and at what levels. While anthropology has consistently been, and has remained, a comparably marginal discipline, psychology and, to a lesser degree sociology and political science, have won most ground since the end of the Second World War. Economics shows the most heterogeneous patterns both in the size of its share and the dynamics of relative size change.

It is thus fairly safe to claim that such changes of relative prominence inside an arbitrary sample of disciplines serve only as a hint of real changes in the power relations between academic subfields. In all of the countries, the disciplines covered in this book, for the longest period of time under study, were assembled in one or other subunit of their university. As mentioned elsewhere, universities consist of departments (or their functional equivalents) at the disciplinary level and at a more aggregated level of *Fakultäten*, to use the unambiguous German term. Very seldom are departments the sole hierarchical level below the president, rector etc. At the intermediate level of the *Fakultät*, decisions

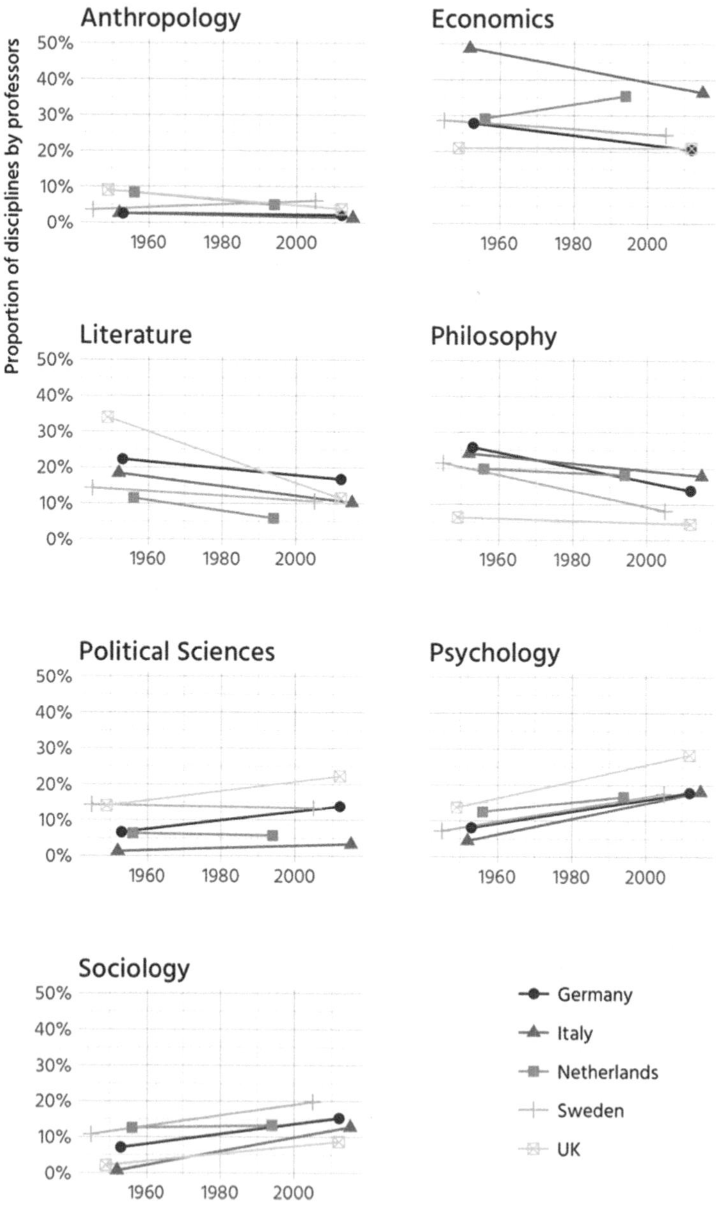

Fig. 10.2 A comparative view on the long-term development of the disciplines in five countries for which reasonable comparable data were available as a comparison between disciplines

are made that either favor or disadvantage individual disciplines. In the traditional Philosophical *Fakultät*, no single discipline dominated the others, whereas in intermediate units encompassing only the core of the social sciences, we see a quite different shift of domination. In the 1960s, such a school would have consisted of professors of economics, sociology, political science, statistics and economic history of roughly the same numbers. Today the same university subunit will be dominated by a discipline that was nearly inexistent half a century earlier: business economics. The reader might be reminded that we excluded business economics from our list of disciplines intentionally for this particular reason: cut-throat competition. One does not need to develop a misanthropic viewpoint to imagine that in a decision-making body, where one faction outnumbers all others, future decisions are liable to be made that benefit the dominant unit.

The picture is nevertheless incomplete, because we need an answer to the question of how and why it could have happened that the relative share of disciplines change over time. As an answer we would like to offer a model instead of a detailed narrative for the expansion of academic specialties. In such a model we need to specify the following parameters:

- Openness of admission to HE
- Rules for allocating additional (teaching) personnel
- Degree of centralization of the decision-making procedures
- Appeal of particular disciplines in the larger audience of post-adolescents.

We can distinguish between an A and B model. A would be an open admission university in which qualified entering students are completely free to choose a study program and B would be a model where the entry is restricted not only by the necessity of particular diplomas but also by numbers, examinations, or fees. In model A, the young cohorts decide the fate of disciplines whereas in model B either the governing board of each university or a centralized national body makes the decisions. However, in both models, the allocation of additional teaching personnel is, in one way or another, related to the number of students, since it is an established rule to expand the teaching staff if student numbers grow. In the reverse direction, such an automatism does not exist or works much slower. Dismissal of academic staff is possible, mostly on the lower levels of adjunct, part-time and non-tenured faculty members only. However, the increase of the number of students (growing demand in a particular field) triggers an immediate reaction the additional hiring of professors only in non-centralized governance systems of higher education. The more centralized the system, the less flexible are responses towards changing market conditions. The expansion of the student body is restricted only by infrastructure and quality requests. Elitist universities are forced not to accept too many applicants, but in some countries public universities cannot select students, thus becoming mass universities against their will. The lower the threshold, the higher the cross over. Model A functions like a free market only at a superficial level, where an increase in demand proportionately affects supply. Incoming students are not consumers and do not influence the increase in the professorate directly through their demand. In the world of the B model usually the ministry of education decides on additional funds for the enlargement of the professoriate. Since state bureaucracies are to some degree inert, lagging mostly behind recognized needs and cautious about new expenses, an increase in the number of professors follows the growth of students, usually with some time lag.

As long as the doors to the universities are open to all high school graduates, the fate of a discipline is in the hands of young people and their changing preferences. We do not know how these preferences develop or change over time, but it might be safe to name not less than six types of potential influences. First, the spectrum of "disciplines" offered in the high school curricula (which was largely replicated as a rule in nineteenth century Arts faculties). Even now, most European high school curricula contain some philosophy, a lot of literature, little to no sociology or political science, probably some psychology and rarely anthropology. Second, the composite bundle of what could be called youth culture and their labor market expectations. Third, the imitation of foreign models, under the aegis of 'modernity,' developed in dominant academic powers (like the US or Britain, after 1945).

Fourth, the impact of extra-mural scholarship in emerging disciplines promoted by learned societies or intellectual lobbies. Fifth, various intellectuals' demands for more or less applicable expertise coming from the state administration (such as for 'colonial studies' in colonial empires) or from other social agents (as illustrated by the unprecedented growth of management studies, responding to the needs of modern capitalist economies). Sixth, ideological options imposed by authoritarian states (like in Communist Eastern Europe) for the promotion of disciplines ('scientific socialism', Marxism–Leninism, etc.) in support of their political goals for education. We find the B model both in countries with centralized planning and in free market environment, Sweden and other Scandinavian countries on the one hand and the US and Britain on the other hand. The A model may apply to the situation in Germany and Italy.

The five countries plotted in Fig. 10.1 show some dissimilarity but we cannot say that the differences are related to any of the variables mentioned above. For sure, an open admission country like Italy shows a rapid decrease of economics during the six decades covered, but we encounter a similar slope in the UK for literature. However, any sharp change of academic personnel is also connected, to some extent, to the changing numbers of students enrolled. Therefore, it might not be an exaggeration to claim that the numerical shape of any discipline is often the consequence of highly unscholarly changes. This may be notably linked to the preferences of young people, unless it is the contrary: the curricular supply of new disciplines in the teaching agenda generates a growing demand from interested students, as could be observed in some socialist countries, mostly in the late 1960s, when sociology, demography and (later) political science were readmitted (first under Marxist disguises) among legitimate study options.

Whether one follows this explanatory sketch or not, we end by highlighting once again the advantages of a perspective that prefers institutional arrangements over alternative explanations. Looking at the mechanisms of the expansion of personnel demonstrates once more that factors only very loosely associated with the core of scholarship—cognitive assets and achievements, paradigms—seem to fulfil a much stronger role than is usually admitted.

References

Altbach, Philip G., Liz Reisberg, and Laura E. Rumbley. 2009. *Trends in global higher education: Tracking an academic revolution; a report prepared for the UNESCO 2009 World Conference on higher education.* Paris: UNESCO. http://unesdoc.unesco.org/images/0018/001832/183219e.pdf.

Bartelse, Jeroen, and Frans van Vught. 2007. Institional [sic!] profiles: Towards a typology of higher education institutions in Europe. *IAU Horizons* 13 (2–3): 9–11.

Boumans, Marcel. 2014. Haavelmo's epistemology for an inexact science. *History of Political Economy* 46 (2): 211–229.

Bourdieu, Pierre. 1988. *Homo academicus.* Stanford: Stanford University Press.

Caplow, Theodore, and Reece McGee. [1958] 2001. *The academic marketplace.* New Brunswick, NJ: Transaction Publishers.

Davies, Norman. 1996. *Europe: A history.* Oxford: Oxford University Press.

Esping-Andersen, Gøsta. 1990. *The three worlds of welfare capitalism.* Cambridge: Polity Press.

Fielden, John. 2008. Global trends in university governance. Education Working Paper Series 9.

Fleck, Christian, and Barbara Hönig. 2014. European sociology: Its size, shape, and 'excellence'. In *Routledge handbook of European sociology*, vol. 1, ed. Sokratis M. Koniordos and Alexandros-Andreas Kyrtsēs, 40–66. London: Routledge.

Gibbons, Michael (ed.). 1994. *The new production of knowledge: The dynamics of science and research in contemporary societies.* London: Sage.

Hall, Peter A., and David W. Soskice. 2001. *Varieties of capitalism: The institutional foundations of comparative advantage.* Oxford and New York: Oxford University Press.

Hallin, Daniel C., and Paolo Mancini. 2004. *Comparing media systems: Three models of media and politics. Communication, society and politics.* Cambridge: Cambridge University Press.

Hausman, Daniel M. 1992. *The inexact and separate science of economics.* Cambridge: Cambridge University Press.

Heilbron, Johan, Madeline Bedecarré, and Rob Timans. 2017. European journals in the social sciences and humanities. *Serendipities: Journal for the Sociology and History of the Social Sciences* 2 (1): 33–49. https://doi.org/10.25364/11.2:2017.1.3.

Helmer, Olaf, and Nicholas Rescher. 1959. On the epistemology of the inexact sciences. *Management Science: Journal of the Institute for Operations Research and the Management Sciences* 6: 25–52.

Hénard, Fabrice, and Alexander Mitterle. n.d. [2009]. Governance and quality guidelines in higher education: A review on governance arrangements and quality assurance guidelines. https://www.oecd.org/edu/imhe/46064461.pdf.

Kropp, Kristoffer. 2017. The cases of the European values study and the European social survey—European constellations of social science knowledge production. *Serendipities: Journal for the Sociology and History of the Social Sciences* 2 (1): 50–68. https://doi.org/10.25364/11.2:2017.1.4.

Lutter, Mark, and Martin Schröder. 2016. Who becomes a tenured professor, and why? Panel data evidence from German sociology, 1980–2013. *Research Policy* 45 (5): 999–1013. https://doi.org/10.1016/j.respol.2016.01.019.

Martins, Herminio. 1974. Time and theory in sociology. In *Approaches to sociology: An introduction to major trends in British sociology*, ed. John Rex, 246–294. International library of sociology. London: Routledge & Kegan Paul.

Musselin, Christine. 2010. *The market for academics*. Studies in higher education. New York: Routledge.

Nowotny, Helga, Peter Scott, and Michael Gibbons (eds.). 2001. *Re-thinking science: Knowledge and the public in an age of uncertainty*. Cambridge: Polity Press.

OECD. 2016. *Education at a Glance 2016*. Paris: OECD.

Paradeise, Catherine, Emanuela Reale, Ivar Bleiklie, and Ewan Ferlie. 2009. *University governance: Western European comparative perspectives*. Higher education dynamics. New York: Springer.

Santiago, Paulo, Karine Tremblay, Ester Basri, and Elena Arnal. 2008. *Special features: Equity, innovation, labour market, internationalisation. Tertiary education for the knowledge society*, vol. 2. Paris: OECD.

Schögler, Rafael, and Thomas König. 2017. Thematic research funding in the European union: What is expected from social scientific knowledge-making? *Serendipities. Journal for the Sociology and History of the Social Sciences* 2 (1): 107–130. https://doi.org/10.25364/11.2:2017.1.7.

Searle, John R. 2005. What is an institution? *Journal of Institutional Economics* 1 (1): 1–22.

Smith, Anthony D. 1983. Nationalism and classical social theory. *British Journal of Sociology* 34: 19–38.

van Vught, F.A. et al. 2010. U-Map: The European classification of higher education institutions. www.u-map.eu.

Wimmer, Andreas, and Nina G. Schiller. 2002. Methodological nationalism and beyond: Nation-state building, migration and the social sciences. *Global Networks* 2 (4): 301–334.

Ziegele, Frank. 2013. Classification of higher education institutions: The European case. *Pensamiento Educativo. Rivista de Investigación Educacional Latinoamericana* 50 (1): 76–95.

Index

C. Fleck et al. (eds.), *Shaping Human Science Disciplines*, Socio-Historical Studies of the Social and Human Sciences, https://doi.org/10.1007/978-3-319-92780-0

The manufacturer's authorised representative in the EU is Springer Nature Customer Service Centre GmbH, Europaplatz 3, 69115 Heidelberg, Germany. If you have any concerns regarding our products, please contact ProductSafety@springernature.com

Printed and bound by CPI Group (UK) Ltd, Croydon, CR0 4YY
30/09/2026
02225660-0013